Probiotic Dairy Products

# Probiotic Dairy Products

Edited by

Dr Adnan Tamime
Dairy Science and Technology Consultant
Ayr, UK

Blackwell
Publishing

SDT

Editorial Offices:
Blackwell Publishing Ltd, 9600 Garsington Road, Oxford OX4 2DQ, UK
Tel: +44 (0)1865 776868
Blackwell Publishing Professional, 2121 State Avenue, Ames, Iowa 50014–8300, USA
Tel: +1 515 292 0140
Blackwell Publishing Asia, 550 Swanston Street, Carlton, Victoria 3053, Australia
Tel: +61 (0)3 8359 1011



First published 2005 by Blackwell Publishing Ltd

2 2007

ISBN: 978-1-4051-2124-8



Library of Congress Cataloging-in-Publication Data

Probiotic dairy products / edited by Adnan Tamime.
p. ; cm.
Includes bibliographical references and index.
ISBN-13: 978-1-4051-2124-8 (alk. paper)
ISBN-10: 1-4051-2124-6 (alk. paper)
1. Dairy microbiology. 2. Dairy products in human nutrition.
[DNLM: 1. Dairy Products--microbiology. 2. Food Microbiology. 3. Food Technology.
4. Probiotics. QW85 P962 2005] I. Tamime, A. Y.
QR121.P76 2005
641.3'7--dc22

2005010456

A catalogue record for this title is available from the British Library

Set in 10/12 pt Times New Roman
by Sparks, Oxford – www.sparks.co.uk

The publisher's policy is to use permanent paper from mills that operate a sustainable forestry policy, and which has been manufactured from pulp processed using acid-free and elementary chlorine-free practices. Furthermore, the publisher ensures that the text paper and cover board used have met acceptable environmental accreditation standards.

For further information on Blackwell Publishing, visit our website:
www.blackwellpublishing.com

# Contents

# Preface to Technical Series

For more than 60 years, the Society of Dairy Technology (SDT) has sought to provide education and training in the dairy field, disseminating knowledge and fostering personal development through symposia, conferences, residential courses, publications, and its journal, the *International Journal of Dairy Technology* (previously known as *Journal of the Society of Dairy Technology*).

In recent years, there have been significant advances in our understanding of milk systems, probably the most complex natural food available to man. Improvements in process technology have been accompanied by massive changes in the scale of many milk/dairy processing operations, and the manufacture a wide range of dairy and other related products.

The Society has now embarked on a project with Blackwell Publishing to produce a Technical Series of dairy-related books to provide an invaluable source of information for practising dairy scientists and technologists, covering the range from traditional to modern large-scale operation. This, the first volume in the series, on *Probiotic Dairy Products*, under the editorship of Dr Adnan Tamime, complements the second volume on *Fermented Milks* in providing a wide-ranging review of this group of micro-organisms, which are increasingly recognised as playing a vital role in the maintenance of our health while also contributing to the microbiology of many fermented dairy products.

Andrew Wilbey
President, SDT
February 2005

# Preface

Fermented foods, including milk and dairy products, have played important roles in the diet of humans worldwide for thousands of years. Since the mid-1950s, there has been increasing knowledge of the benefits of certain micro-organisms [e.g. lactic acid bacteria (LAB) and probiotic gut flora] and their impact on human biological processes and, at the same time, of the identity of certain dairy or non-dairy components of fermented milks and their role in human health and body function. The purpose of this book, which is written by a team of international scientists, is to review the latest scientific developments in these fields with regard to the 'functional' aspects of fermented milk products and their ingredients.

Some scientific aspects reviewed in this publication are: (a) the latest knowledge regarding the gut microflora (e.g. identifying the beneficial gut microbiota in terms of probiotic and health aspects); (b) the use of a wide range of probiotic micro-organisms during the manufacture of different dairy products that have dominated the global markets for the past decades and are used as vehicles to increase the probiotic gut flora of humans; (c) the genomic sequences of certain strains of LAB; and (d) the use of prebiotic ingredients, such as galacto- and fructo-oligosaccharides, to enhance the viable count of probiotic microflora in humans.

Furthermore, numerous-related topics, for example, the current statutory regulations (national and international), analytical methods to enumerate these beneficial organisms, sensory profiling to improve the quality of the product and enhance consumer acceptability, bioactive components produced by the probiotic microflora, and the treatment of certain human diseases are also reviewed. It is of interest to note that the current research work on probiotic dairy products, which aims to understand the role of the intestinal microbiota, will underpin new strategies to improve the health status of consumers, and will contribute to a reduction in healthcare costs, particularly in ageing populations.

A.Y. Tamime
February 2005

# Contributors

**Dr A.Y. Tamime**
Dairy Science and Technology Consultant
24 Queens Terrace
Ayr KA7 1DX
Scotland
United Kingdom
Tel. +44 (0)1292 265498
Fax +44 (0)1292 265498
Mobile +44 (0)7980 278950
E-mail: adnan@tamime.fsnet.co.uk

**Prof. G.R. Gibson**
University of Reading
School of Food Biosciences
Whiteknights
P.O. Box 226
Reading RG6 6AP
United Kingdom
Tel. +44 (0)118 378 8715
Fax +44 (0)118 931 0080
E-mail: g.r.gibson@reading.ac.uk

**Miss B. O'Grady**
University of Reading
Food Microbial Sciences Unit
Science & Technology Centre
Whiteknights
P.O. Box 226
Reading RG6 6BZ
United Kingdom
Tel. +44 (0)118 378 4536
Fax +44 (0)118 931 0080
E-mail: b.ogrady@reading.ac.uk

**Prof. F. Dellaglio**
Università degli Studi di Verona
Dipartimento Scientifico e Tecnologico
Strada Le Grazie 15
37134 Verona
Italy
Tel. +39 045 8027917
Fax +39 045 8027051
E-mail franco.dellaglio@univr.it

**Dr G.E. Felis**
Università degli Studi di Verona
Dipartimento Scientifico e Tecnologico
Strada Le Grazie 15
37134 Verona
Italy
Tel. +39 045 8027930
Fax +39 045 8027051

**Prof. S. Torriani**
Università degli Studi di Verona
Dipartimento Scientifico e Tecnologico
Strada Le Grazie 15
37134 Verona
Italy
Tel. +39 045 8027917
Fax +39 045 8027928

**Dr E. Johansen**
Applied Biotechnology
Chr. Hansen A/S
10-12 Bøge Allé
DK2970 Hørsholm
Denmark
Tel. (+45) 45 74 84 64
Fax (+45) 45 74 89 94
E-mail: Eric.Johansen@dk.chr-hansen.com

**Dr K. Sørensen**
Genomics and Strain Development
Chr. Hansen A/S
10-12 Bøge Allé
DK2970 Hørsholm
Denmark
Tel. (+45) 45 74 83 54
Fax (+45) 45 74 89 94
E-mail: Kimlb.Soerensen@dk.chr-hansen.com

**Dr M. Saarela**
VTT Biotechnology and Food Research
Box 1500
FI-02044-VTT
Finland
Tel. +358 20 722 4466
Fax +358 20 722 7071
E-mail: Maria.Saarela@vtt.fi

**Miss A. Korslund Søndergaard**
Chr. Hansen A/S
10-12 Bøge Allé
DK2970 Hørsholm
Denmark
Tel. +45 45 74 85 60
Fax +45 45 74 88 16
E-mail: Anne.Soendergaard@dk.chr-hansen.com

**Prof. N.P. Shah**
Victoria University of Technology
School of Molecular Sciences
Werribee Campus
P.O. Box 14428
Melbourne City Mail Centre
Victoria 8001
Australia
Tel. +61 (0)3 9919 8289
Fax +61 (0)3 9919 8284
E-mail: Nagendra.shah@vu.edu.au

**Prof. V.V. Mistry**
Head Dairy Science Department
South Dakota State University
Brookings
SD 57007
USA
Tel. +1 (0)605 688 5731
Fax +1 (0)605 688 6276
E-mail: vikram_mistry@sdstate.edu

**Mr M. Hickey**
Derryreigh
Creggane
Charleville
Co. Cork
Ireland
Tel. +353 (0)63 89392
Mobile: +353 (0)87 2385653
E-mail: mfhickey@oceanfree.net

**Dr A. McCartney**
University of Reading
School of Food Biosciences
Food Microbial Sciences Unit
PO Box 226
Whiteknights
Reading RG6 6AP
United Kingdom
Tel. +44 (0)118 378 8593
Fax +44 (0)118 931 0080
E-mail: a.l.mccartney@reading.ac.uk

**Prof. C. Shortt**
Yakult UK Ltd.
12-16 Telford Way
Westway Estate, Acton
London W3 7XS
United Kingdom
Tel. +44 (0)20 8740 4111
Fax +44 (0)20 8740 4999
E-mail: cshortt@yakult.co.uk

**Ms S. Smart**
Yakult UK Ltd.
12-16 Telford Way
Westway Estate, Acton
London W3 7XS
United Kingdom
Tel. +44 (0)20 8740 2619
Fax +44 (0)20 8740 4999
E-mail: SSmart@yakult.co.uk

**Dr F. Angus**
Leatherhead Food International
Randalls Road
Surrey KT22 7RY
United Kingdom
Tel. +44 (0)1372 822217
Fax +44 (0)1372 822229
E-mail: fangus@leatherheadfood.com

**Dr G.A. O'May**
University of Dundee
Microbiology and Gut Biology Group
Ninewells Hospital Medical School
Dundee DD1 9SY
Scotland
United Kingdom
Tel. +44 (0)1382 496 341
Fax +44 (0)1382 633 952
E-mail: g.omay@dundee.ac.uk

**Prof. G.T. Macfarlane**
University of Dundee
Microbiology and Gut Biology Group
Ninewells Hospital Medical School
Dundee DD1 9SY
Scotland
United Kingdom
Tel. +44 (0)1382 496 250
Fax +44 (0)1382 633 952
E-mail: g.t.macfarlane@dundee.ac.uk

**Ms E. O'Connor**
Teagasc Moorepark
Dairy Products Research Centre
Fermoy
Co. Cork
Ireland
Tel. +353 (0)25 42229
Fax +353 (0)25 42340
E-mail: eoconnor@moorepark.teagasc.ie

**Mr E. Barrett**
Teagasc Moorepark
Dairy Products Research Centre
Fermoy
Co. Cork
Ireland
Tel. +353 (0)25 42229
Fax +353 (0)25 42340
E-mail: ebarrett@moorepark.teagasc.ie

**Prof. G. Fitzgerald**
University College Cork
Department of Microbiology
Cork
Ireland
E-mail: g.fitzgerald@ucc.ie

**Prof. C. Hill**
University College Cork
Department of Microbiology
Cork
Ireland
Tel. +353 (0)25 42229
Fax +353 (0)25 42340
E-mail: c.hill@ucc.ie

**Dr C. Stanton**
Teagasc Moorepark
Dairy Products Research Centre
Fermoy
Co. Cork
Ireland
Tel. +353 (0)25 42229
Fax +353 (0)25 42340
E-mail: cstanton@moorepark.teagasc.ie

**Prof. P. Ross**
Teagasc Moorepark
Dairy Products Research Centre
Fermoy
Co. Cork
Ireland
Tel. +353 (0)25 42229
Fax +353 (0)25 42340
E-mail: pross@moorepark.teagasc.ie

**Prof. T. Mattila-Sandholm**
Senior Vice President
Valio Ltd.,
Research & Development Department
P.O Box 30
FI-00039 VALIO
Finland
Tel. +358 50 3842931
E-mail: Tiina.Mattila-Sandholm@valio.fi

**Prof. W. M. de Vos**
Wageningen University
Department of Agrotechnology and Food Science
Laboratory of Microbiology
Hesselink van Suchtelenweg 4
Wageningen 6703 CT
The Netherlands
Tel: +31-317-483100
Fax: +31-317-483829
E-mail: willem.devos@wur.nl

# 1 Microbiota of the Human Gut

B. O'Grady and G.R. Gibson

## 1.1 Background

The human gastrointestinal (GI) tract is a highly specialised ecosystem that has evolved over time, both physiologically and microbiologically. At least in part, this is a consequence of the host and environmental pressures that it must counteract in order to maintain eubiosis. The GI tract is one of the most diverse and metabolically active organs in the human body. The human gut and its microbiota cannot be realistically considered as separate entities as they represent a dynamic biological system that has co-evolved from birth. The human GI tract is composed of highly adapted regions for mediation of its diverse functions, many of which impact markedly upon host health and welfare. Physiological considerations in each unique region influence the degree and type of colonisation and initial colonisers also modify the physiological conditions therein. This results in the development of distinct microhabitats along the length of the GI tract, which influence metabolism, protection and immune stimulation. Such effects are both local and systemic as the GI tract is connected to the vascular, lymphatic and nervous systems. The ability of the gut to sustain its beneficial microbiota, against harmful or opportunistic microbiota, in a desirable community structure, is critical for host health and reduction of disease risk. The focus of this chapter is to discuss how the complex interplays between the human GI tract and its indigenous microbiota affect host health and how certain beneficial microbial species, with their potential for manipulation, are crucial to this process.

## 1.2 The human gastrointestinal tract and its microbiota

The human gastrointestinal tract is sterile up until birth, when microbial colonisation begins during the delivery process. The inoculum may be largely derived either from the mother's vaginal or faecal flora (in a conventional birth) or from the environment (in a caesarean delivery). Hence, the microbiota that colonise the newborn tract are acquired post-natally. This is of extreme importance in the choice of delivery, as newborns delivered by caesarean section are exposed to a different microbiota than that of a vaginal delivery. Bacterial populations develop progressively during the first few days of life; facultative anaerobes predominate initially and create a reduced environment that allows for the growth of strict anaerobes (Stark & Lee, 1982). The choice of diet for the newborn is also of importance as the microbiota of breast-fed infants is predominated by bifidobacteria, whereas formula-fed infants have a more complex flora which resembles the adult gut in that bacteroides, clostridia, bifidobacteria, lactobacilli, Gram-positive cocci, coliforms and other groups are all represented in fairly equal proportions (Benno *et al.*, 1984). The occurrence of certain glycoproteins in human breast milk is thought to select for bifidobacteria (Gauhe *et al.*,

1954). During the weaning stage, the microbiota becomes more developed and the ecosystem is thought to be fairly stable at around 2 years of age. The prevalence of bifidobacteria in breast-fed infants is thought to confer protection by reducing the gut pH, which induces a concomitant reduction in other potentially harmful species. Moreover, bifidobacteria are able to exert directly antagonistic activities against gut pathogens. Newborns are susceptible to intestinal infections and atopic diseases as their immune system and GI tract develop. Mode of delivery and diet therefore have important implications, both at birth and later in life, as initial colonisation is involved in the development of the GI tract, its microbiota and in maturation of the immune system. During the first few years of life and upon weaning, the infant microbiota normalises. This composition will remain stable throughout most of adult life (Kimura *et al.*, 1997).

The GI tract begins with the oral cavity which is comprised of the mouth, nose and throat. In the oral cavity a particularly complex microbiota exists (Marsh, 1980). Bacteria can be found on the posterior and anterior tongue, sub- and supragingival plaque, buccal mucosa and vestibular mucosa (Willis *et al.*, 1999). These include members of the *Prevotella*, *Porphyromonas*, *Peptostreptococcus*, *Bacteroides*, *Fusobacterium*, *Eubacterium* and *Desulfovibrio* genera. Bacterial numbers fall dramatically to $<10^3$ colony forming units (cfu) $mL^{-1}$ of gastric contents as they encounter the stomach, which provides a highly effective barrier against invading micro-organisms, both pathogenic and benign. Few micro-organisms with the exception of acid-tolerant lactobacilli, yeasts – and notably *Helicobacter pylori* – can survive the harsh, strongly acidic and peristaltic nature of the stomach.

There is a high degree of variability between the stomach, small intestine and colon in terms of numbers and bacterial population types. This is due predominantly to different transit times, secretions and nutrient availability (Lambert & Hull, 1996; Guilliams, 1999). Micro-organisms are also determinants as they interact with and influence their surroundings to ensure their survival against competitors. This is achieved through innumerable mechanisms, such as increasing anaerobicity or through the production of deleterious compounds such as acid or antimicrobial substances. These compounds concurrently affect the host and can thereby have advantageous or devastating implications (Fooks & Gibson, 2002; see also Fuller & Perdigón, 2003).

The rapid transit time, low pH and presence of bile associated with the small intestine do not provide an environment that encourages growth of bacterial populations. The duodenum also has low microbial populations due to its short transit time and the secretion of intestinal fluids, which create a hostile environment (Sanford, 1992). However, there is a progressive increase in both numbers and species along the jejunum and ileum. The small intestine harbours enterococci, enterobacteria, lactobacilli, bacteroides and clostridia. These rapidly increase in numbers from $10^4$–$10^6$ cfu $mL^{-1}$ in the small intestine to $10^{11}$–$10^{12}$ cfu $mL^{-1}$ in the large intestine, as the flow of intestinal chyme slows upon entry into the colon (Salminen *et al.*, 1998).

The large gut is favourable for bacterial growth with a slow transit time, ready availability of nutrients and favourable pH. There are several hundred culturable species present, although a significant proportion are not cultivable by conventional methods (Suau *et al.*, 1999). The proximal colon represents a saccharolytic fermentation, due to its high substrate availability. Organic acids produced from fermentation result in a lower pH of 5.5–6.0 than the more neutral pH found in the distal colon. Transit in the distal colon is slower and nutrient availability is minimised, producing slower-growing populations that tend towards more proteolytic fermentations. The majority of bacteria are non-spore-forming anaerobes, of which the most numerically predominant are *Bacteroides* spp. and *Bifidobacterium* spp., *Eubacterium* spp., *Clostridium* spp., *Lactobacillus* spp., *Fusobacterium* spp. and various Gram-positive cocci (Salminen *et al.*, 1998). Bacteria present in lower numbers include *Enterococcus* spp., Enterobacteriaceae, methanogens and dissimilatory sulphate-reducing bacteria (Gibson & Roberfroid, 1995; Salminen *et al.*, 1998). Yeasts, including the opportunistic pathogen *Candida albicans*, are also present in the gut microbiota, although in healthy individuals counts do not exceed $10^4$ cfu $g^{-1}$ faeces (Bernhardt *et al.*, 1995; Bernhardt & Knoke, 1997). The vast majority (>90%) of the total cells in the body are present as bacteria in the colon. It is thought that over 60% of the faecal mass exists as prokaryotic cells. The intestinal microbiota has been estimated to consist of at least 400 different species although only 40 species predominate. Along the length the GI tract there are at least four different microhabitats (Freter, 1992):

- The surface of epithelium cells.
- The crypts of the ileum/caecum and colon.
- The mucus gel overlying the epithelium.
- The lumen of the intestine.

The classification of the microbiota as autochthonous and allochthonous complements the distinction between these different habitats of the GI tract (Savage *et al.*, 1968). Autochthonous micro-organisms colonise the GI tract, whereas allochthonous micro-organisms are transient and will predictably be found in the lumen. The slow transit time of the large intestine allows multiplication of the luminal microbiota and allochthonous micro-organisms exert equally important effects on the GI tract as their autochthonous counterparts.

## 1.3 Functions of the gastrointestinal microbiota

The GI tract, along with its microbiota, is one of the most metabolically active organs in the human body. The intestinal microbiota is involved in the fermentation of endogenous and exogenous microbial growth substrates. The metabolic end-products of carbohydrate fermentation are benign, or even advantageous to human health (Macfarlane & McBain, 1999). Major substrates available for the colonic fermentation are starches that for various reasons are resistant to the action of pancreatic amylases, can be degraded by bacterial enzymes, as well as dietary fibres such as pectins and xylans. Other carbohydrate sources available for fermentation in lower concentrations include oligosaccharides and a variety of sugars and non-absorbable sugar alcohols. Saccharolysis results in the production of short-chain fatty acids (SCFA) such as butyrate, acetate, propionate and lactate which contribute towards energy metabolism of the large intestinal mucosa and colonic cell growth; they can also be metabolised by host tissues such as the liver, muscle and brain. The production of SCFA concomitantly results in a lower pH that can protect against invading micro-organisms and also reduces the transformation of primary bile acids into secondary pro-carcinogenic bile acids (Cummings & Macfarlane, 1997). This is believed to be one of the mechanisms utilised by the beneficial gut microbiota that results in protection for the host.

Proteins and amino acids can be effective growth substrates for colonic bacteria, whilst bacterial secretions, lysis products, sloughed epithelial cells and mucins may also make a contribution. However, the diet provides, by far, the predominant source of nutrients, with around 70–100 g $d^{-1}$ of dietary residues being available to the colonic microbiota. These materials are degraded by a wide range of bacterial polysaccharidases, glycosidases, proteases and amino-peptidases to smaller oligomers and their component sugars and amino acids (Macfarlane & Gibson, 1994).

The gut profile of each adult represents a population of microbes that has evolved since birth that can best cope with the physiological and microbiological pressure encountered within this ecosystem. This stability provides resistance for the host, also known as the 'barrier effect', against invading micro-organisms, both pathogenic and benign. The indigenous gut microbiota is better adapted to compete for nutrients and attachment sites than the incoming micro-organism, which it may also inhibit through the production of compounds (Alderbeth *et al.*, 2000). The role of the intestinal microbiota in challenging invading micro-organisms and preventing disease through competitive exclusion is best demonstrated by the studies showing that germ-free animals are more susceptible to infection (Baba *et al.*, 1991). This demonstrates the individual role of beneficial micro-organisms in preventing infection through colonisation resistance.

Another important function of the gut microbiota is the production of vitamins B and K; this is best demonstrated again by studies where germ-free animals required a 30% increase in their diet to maintain their body weight, and supplementation with vitamins B and K as compared to animals with a microbiota (Hooper *et al.*, 2002).

However, the ability of the gut microbiota to utilise biologically available compounds can have negative outcomes. *H. pylori* can affect the absorption of vitamin C and important micronutrients for host health (Annibale *et al.*, 2002). Moreover, the fermentation of proteins and amino acids in the distal colon can lead to the production of toxic substances such as ammonia, phenols and amines that are undesirable for host health (Mykkanen *et al.*, 1998). This highlights the importance of ensuring a balance of beneficial bacteria to prevent the multiplication of opportunistic bacteria. Micro-organisms such as *Candida* spp. and *Clostridium* spp. play a role in gut homeostasis and can be tolerated in low numbers, but mediate harmful effects when a certain threshold level is exceeded.

The GI tract is in more contact with the external environment than our skin, which exposes ~2 $m^2$ compared to the GI tract, which exposes a surface area of ~200 $m^2$ (Guilliams, 1999). The microbiota of the GI tract is therefore heavily involved in gut maturation. As previously mentioned, the GI tract is sterile at birth, and exposure to the intestinal microbiota plays a critical role in stimulating local and systemic responses, allowing the maturation of the immune system. The intestinal microbiota also provides a source for non-inflammatory immune stimulation, throughout life, by stimulating the production of secretory IgA, which neutralises foreign bacteria and viruses (Moreau, 2000). The ability of the GI tract to perform its functions of nutrient uptake in conjunction with the exclusion of foreign antigens or micro-organisms is a complex and difficult process. This interplay between the host immune response and the gastrointestinal microbiota is critical to health; loss of this control may be clinically manifest through disorders such as inflammatory bowel syndrome (IBS).

It is apparent that the intestinal microbiota directly influences gastrointestinal health and systemically affects host health. In healthy individuals, the gastrointestinal microbiota exists in a state of eubiosis. However, this dynamic equilibrium can be 'disturbed' by the stresses of modern-day living, or antimicrobial intake, with serious repercussions for the host.

## 1.4 Influences on the GI tract and its microbiota

The intestinal microbiota of each individual originates from host genetics, environmental factors and microbiological influences. This culminates in a stable community of micro-organisms that is more unique than an individual's own fingerprint – even homozygotic twins develop distinct microbial profiles (Zoetendal *et al.*, 2001). Notwithstanding, the overall metabolism of a healthy gut ecosystem varies little from one individual to another, as evidenced by the ratios of major metabolic end-products. The human GI tract is more challenged than ever before with the high-stress lifestyles, processed foods and eating habits of modern-day living. Disturbances of microbiota can have serious implications, and this fragility merits careful consideration of the external influences on the GI tract and how they may disrupt host health. There are numerous factors that act upon the intestinal microbiota. These are briefly outlined in Table 1.1, and some of the more relevant influences are subsequently discussed.

The influence of diet on the neonatal intestinal microbiota has already been outlined; the GI tract of healthy humans remains relatively stable throughout life, apart from reports of a significant decrease of beneficial bifidobacteria in the elderly. The type of dietary intake has consequences in the colon as carbohydrate fermentations result in benign end-products. However, when carbohydrate levels become diminished, proteolytic fermentation in more distal regions produces toxic compounds that can predispose to diseases such as colorectal cancer or ulcerative colitis. The introduction of new protein-based diets such as the Atkins diet could potentially have serious long-term repercussions for gut health. High levels of processing and eating habits have simultaneously reduced levels of the dietary fibre which is of great importance in host health as it influences stool volume, colon motility, water absorption and faecal transit time. The impact of diet upon gut populations and their activity has been illustrated effectively by an experiment where a collection of bacterial species which suppressed the growth of *Escherichia coli* in mice fed a refined diet, failed to exert the same effect when an alternative diet was fed (Freter, 1988). Nonetheless, although diet can influence the gastrointestinal microbiota, there are other aspects that can disrupt gut homeostasis.

**Table 1.1** Influences on the composition of the gastrointestinal microbiota.

| |
|---|
| Type of feeding |
| Amount, chemical composition and availability of growth substrate |
| Availability of colonisation sites |
| Immunological interactions |
| Individual fermentation strategies by the bacteria |
| Intestinal transit time |
| Gut pH |
| Redox potential |
| Availability of inorganic electron acceptors |
| Production of bacterial metabolites |
| Presence of antimicrobial compounds |
| Xenobiotic compounds |
| Age of the host |
| Peristalsis |

Adapted from Fooks *et al.* (1999).

Chronic illness, immune suppression and the use of broad-spectrum antibiotics against infection can severely compromise the crucial balance between beneficial and harmful micro-organisms. The loss of beneficial genera sensitive to antibiotic therapy such as lactobacilli and bifidobacteria has implications for gastrointestinal health, as opportunistic pathogens can overgrow and the host has an increased risk for iatrogenic disease. A report of pseudomembranous colitis requiring surgical intervention has been documented following triple antibiotic therapy for *H. pylori* infection (Rai & Sundeepa, 2002).

The failure of antibiotics to treat infection as a consequence of increased microbial resistance, and the fear that new, more successful, antibiotics will not be developed has necessitated a new perspective on an age-old problem. These circumstances, in combination with the consumer demand for dietary supplements to maintain gastrointestinal health, have fuelled scientific research into alternative approaches.

Within this context, the potential for preventing dysbiosis or indeed fortifying the GI tract through modulation of the intestinal microbiota seems a tangible solution. As previous measures relied on the use of antibiotics 'against life', the use of probiotics 'for life' was essentially a return to the use of natural resources, found within our own bodies. The principle of using harmless bacteria to prevent disease has longevity, with the suggestion from Metchnikoff at the turn of the century that ingested bacteria could promote longevity and well-being (Metchnikoff, 1907). Micro-organisms associated with health benefits *in vivo* include members of the *Lactobacillus* and *Bifidobacterium* genera, although streptococci, enterococci, lactococci, bacilli and fungi such as *Saccharomyces* spp. and *Aspergillus* spp., have also been used. All are now a source for 'probiotics' and are the focus of intense research.

## 1.5 Beneficial microbiota: probiotics and health aspects

Probiotics have been defined as 'a preparation of or a product containing viable, defined micro-organisms in sufficient numbers, which alter the microbiota (by implantation or colonisation) in a compartment of the host and by that exert beneficial health effects in this host' (Havenaar & Huis In't Veld, 1992). The prebiotic concept was introduced by Gibson and Roberfroid (1995) as 'a non-digestible food ingredient that beneficially affects the host by selectively stimulating the growth and/or activity of one or a limited number of bacteria in the colon'. A synbiotic is a synergistic combination of a probiotic and prebiotic. The application of functional foods relies on the premise that the intestinal microbiota can be influenced by diet. The introduction of beneficial bacteria and/or prebiotics, which confer an advantage on the ingested beneficial bacteria and/or on beneficial bacteria already resident, allows for the modulation of the GI tract towards a more desired community. There are certain criteria which a strain must fulfil to be considered probiotic in nature. These represent a culmination of the physiological and manufacturing demands that a bacterial strain must survive, whilst still retaining its capability to exert a beneficial effect upon host health. However, the perception that intestinal colonisation is essential has been challenged in a review by Tannock (2003). The transit time of the intestinal tract is such that it allows for the multiplication of bacteria regardless of the loss of faecal material. This is important when considering the nature of allochthonous micro-organisms, which exert effects from their luminal habitat, and this indicates that colonisation is not a prerequisite for probiotic activity. Those strains that are considered probiotics and which are the focus of current

**Table 1.2** Micro-organisms that have been identified as probiotic microflora.

| Microflora | Species |
|---|---|
| Lactobacilli | *Lactobacillus acidophilus* |
| | *Lactobacillus rhamnosus* |
| | *Lactobacillus reuteri* |
| | *Lactobacillus casei* |
| | *Lactobacillus gasseri* |
| | *Lactobacillus plantarum* |
| | *Lactobacillus johnsonii* |
| Bifidobacteria | *Bifidobacterium bifidum* |
| | *Bifidobacterium longum* |
| | *Bifidobacterium breve* |
| | *Bifidobacterium adolescentis* |
| | *Bifidobacterium infantis* |
| Enterococci | *Enterococcus faecalis* |
| | *Enterococcus faecium* |
| Yeast | *Saccharomyces boulardii* |
| Lactococci | *Lactococcus lactis* subsp. *lactis* |

research are listed in Table 1.2. The ingestion of probiotic species can culminate in improved host health, as evidenced by human studies where the alleviation of symptoms of different disorders has been observed.

## 1.5.1 *Enzymatic activity and improved digestion*

Lactose maldigestion and intolerance are prevalent throughout the world. Lactose malabsorption is a consequence of the absence of, or decreased production of, β-galactosidase, which cleaves the lactose present in milk. Undigested lactose is clinically manifest by bloating, gas, cramps and diarrhoea. Several studies have demonstrated that the ingestion of yoghurts or milks containing live bacteria is better tolerated by lactose-intolerant individuals. This effect is largely due to digestion of the lactose by the probiotic lactase activity, effectively performing the functions of the defective human enzyme, and also as a result of a slower transit time of the product (Marteau *et al.*, 1990; de Vrese *et al.*, 2001). The ability of probiotic enzymes to aid digestion has also been studied in infants with sucrase deficiency. The ability to digest sucrose was augmented by the ingestion of *Saccharomyces cerevisiae*, a yeast that produces the sucrase enzyme (Harms *et al.*, 1987).

## 1.5.2 Clostridium difficile*-associated intestinal disease*

*Clostridium difficile* is a pathogenic bacterium that takes advantage of the compromised GI tract in ~20 to 40% of patients suffering from antibiotic-associated diarrhoea. It releases two toxins, A and B, which mediate diarrhoea and colitis. Eradication is difficult as prolonged antibiotic therapy further compromises the gastrointestinal flora, providing the pathogen with a competitive advantage. *C. difficile* also produces spores which can subsequently germinate inducing relapses, while the treatment of refractory patients proves onerous and the relapses can prove more debilitating than the original disease. Intervention with *Saccharomyces boulardii* in a placebo-controlled trial has shown that the use of this probiotic as an adjunct to the antibiotic therapy prevented relapse (McFarland *et al.*, 1994). Furthermore, a recent study has shown that probiotic intake reduced the incidence of *C. difficile* toxin-positive patients as compared to the placebo group (Plummer *et al.*, 2004).

### 1.5.3 *Antibiotic-associated diarrhoea*

The use of antibiotics to treat infection has devastating consequences on gastrointestinal balance (and particularly upon beneficial bacteria) that can result in diarrhoea in up to 20% of patients receiving antimicrobial therapy. Several well-designed, placebo-controlled randomised trials have demonstrated that *S. boulardii*, *Lactobacillus rhamnosus* GG and *Enterococcus faecium* SF68 were effective at reducing the incidence of diarrhoea in this patient population (Wunderlich *et al.*, 1989; McFarland *et al.*, 1995; Vanderhoof *et al.*, 1999). Evidently, antibiotic therapy disrupts the intestinal microbiota; however, it is unclear how these probiotics mediate their protection. It is conceivable that they may prevent the growth of opportunistic pathogens such as *C. difficile*, or fortify the epithelial barrier.

### 1.5.4 *Acute diarrhoea and gastroenteritis*

Acute diarrhoea can be caused by rotavirus infection, which is a serious contributor to infant morbidity and mortality, especially in developing countries. Several studies have been conducted with reports of decreases in the incidence and duration of diarrhoea accompanied by a decrease in the shedding of rotavirus, following probiotic ingestion of strains such as *Lb. rhamnosus* GG, *Bifidobacterium bifidum* and *Streptococcus thermophilus* (Saavedra *et al.*, 1994; Guandalini *et al.*, 2000). However, *Lb. rhamnosus* GG has shown the most promising results, with one study demonstrating that heat-inactivated *Lb. rhamnosus* GG exerted a beneficial effect comparable to living *Lb. rhamnosus* GG cells (Kaila *et al.*, 1995). Although the non-viable cells failed to elicit a similar immune response to their viable counterparts, the potential to exert beneficial effects on the host is relevant, especially in developing countries where product viability may be an issue. However, it is far more likely that probiotics need to be viable to exert inhibitory effects. This is because competitive mechanisms such as acid formation, antimicrobial excretion, substrate limitation and occupation of colonisation sites all require viability. Stimulation of the immune response does not, though this is only one facet of probiotic interaction.

Probiotics have also been applied in other gastrointestinal complaints such as *H. pylori* gastroenteritis, albeit less successfully. The *in vitro* anti-*Helicobacter* activity of lactobacilli, bifidobacteria and bacilli does not translate into *in vivo* efficacy, with suppression of *H. pylori* activity the only effect (Felley *et al.*, 2001). This field is in its infancy however, and the physiological restrictions imposed by the stomach require more careful consideration and research for a probiotic intervention than may be necessary for colonic targets.

### 1.5.5 *Extra-intestinal applications*

The indigenous microbiota of the human body is not restricted to the GI tract. Studies on urogenital tract infections, characterised by decreases in the protective lactobacilli genera, revealed that inserted capsules or oral dosages of a probiotic combination, containing *Lb. rhamnosus* GR-1 and *Lactobacillus fermentum* RC-14, restored the vaginal flora to a healthy composition in 90% of the women treated (Reid *et al.*, 2001). The nasopharyngeal microbiota of children suffering from recurrent otitis media has also been alleviated through the application of a nasal spray containing probiotic micro-organisms (Tano *et al.*, 2002).

### 1.5.6 *Other potential applications*

Probiotics are also being researched for a range of complex diseases and syndromes where there are strong indicators for the application of dietary interventions. A non-pathogenic strain of *E. coli* has been reported to prevent relapse in patients suffering with ulcerative colitis (Rembacken *et al.*, 1999). Evidence is accumulating for several studies in: (a) food allergy and atopic eczema (Kalliomaki *et al.*, 2001; Schmidt, 2004); (b) preventing traveller's diarrhoea (Hilton *et al.*, 1997); (c) Crohn's disease (Guslandi *et al.*, 2000); (d) colorectal cancer (Goldin *et al.*, 1996); (e) dental caries (Busscher *et al.*, 1999; Montalto *et al.*, 2004); and (f) relieving constipation (Attar *et al.*, 1999; Hamilton-Miller, 2004). These are a few, amongst other, exciting research avenues emerging for the uses of probiotics in benefiting host health.

### 1.5.7 *Product considerations*

Probiotics and prebiotics are marketed as health, or functional, foods whereby they are ingested for their purported positive advantages in the digestive tract and/or systemic areas such as the liver, vagina or bloodstream. Unlike new drugs or pharmaceuticals, which are screened intensively for safety and effectiveness, probiotics and prebiotics are less rigorously assessed. It is therefore relatively easy to launch a new product, and legislation against such products is loose. Nevertheless, consumers should be provided with an accurate assessment of physiological, microbial and safety aspects. Several criteria for the appropriate use of probiotics and prebiotics exist and may be summarised as follows. They should:

- Exert a proven beneficial effect on the consumer, preferably with a mechanistic explanation of how this occurred;
- Be non-pathogenic, non-toxic and free of adverse side effects;
- Maintain stability in the product;
- Contain a large number of viable cells (for probiotics);
- Survive well in the GI tract (the best products should be resistant to gastric acid, small gut secretions and have a good ability to influence bacteria already in the gut);
- Have good sensory and mouthfeel properties;
- Preferably be isolated from the same species as the intended use; and
- Have accurate product labelling and content.

Much effort has concentrated on identifying probiotic bacteria and characterising their beneficial credentials. It is generally considered that probiotic bacteria must possess certain properties. The probiotic must survive passage through the upper regions of the GI tract, and persist in the colon. There must be no adverse host response to the bacterium, its components or metabolic end-products. The probiotic should be antagonistic to mutagenic or pathogenic organisms in the gut, and must be genetically stable. For the successful introduction of the probiotic concept into the food market, chosen micro-organisms must be amenable to industrial processes and remain viable in the final food product (Ziemer & Gibson, 1998; Collins & Gibson, 1999). Advances in the genetics of probiotic strains [usually lactic acid bacteria (LAB) or yeasts] have enabled the determination of mechanisms involved in probiotic function, such as the production of antimicrobials, mucosal adhesins and organic acids. However, this also offers the possibility of modifying existing strains to increase survival and efficacy in the human GI tract.

Many of these traits also apply to prebiotics, although being non-viable dietary components the product survival characteristics are not applicable. The active ingredient should be defined, any product impurities quantified, and an operable dosage suggested.

One worrying aspect of the probiotic market is that labelling is not always accurate, in terms of microbial content and numbers. This should be addressed by the use of the molecular diagnostic procedures. It is imperative that labelling and actual contents should be identical. When growing micro-organisms in bulk culture, such as for use as probiotics, it is not unreasonable to assume that some genetic drift in their integrity occurs. Moreover, it will be the case that certain products become contaminated during processing and/or storage. This has been tested in various products available on the UK market. Briefly, the strategy was to carry out four isolation methods as follows: (a) recovery of strains present in high numbers ($>1 \times 10^6$ cfu $mL^{-1}$) on various selective agars; (b) targeted enrichment of strains present in low numbers only ($10^1$–$10^6$ cfu $mL^{-1}$) on various selective agars; (c) genetic probing of 16S rRNA by fluorescent *in situ* hybridisation (FISH) to recover culturable as well as non-culturable diversity; during FISH, fluorescently labelled probes hybridise rRNA of a particular bacterium in cells fixed on a microscope slide, and fluorescent cells are enumerated; and (d) enzymatic disruption of any 'protective' layers surrounding micro-organisms in products and subsequent recovery by both plating and FISH.

The colonies then recovered are then subjected to both phenotypic as well as genotypic methods of identification. The former involves the determination of morphological characteristics, microscopic traits and determination of selected biochemical aspects (analytical profile index). The problem with this approach is its unreliability. The level of phenotypic discrimination does not usually separate individual strains, or sometimes species, of the same genus. It also allows operator subjectivity. As such, the phenotypic approach is questionable and should not be relied upon. These tests were therefore supplemented by genetic studies involving DNA extraction from colonies, amplification by the polymerase chain reaction (PCR), 16S rRNA sequencing and comparing of diagnostic hypervariable regions with two reliable microbial diversity information databases.

This ongoing work has shown that something like 50% of current products do not match up with their labels in terms of microbial content and/or numbers of strains therein. Some products do not label their contents, and in certain cases some of these do not even contain microbes recognised as probiotics.

A variety of *in vitro* characteristics that could be applied to probiotic and prebiotic use prior to human studies is listed in Table 1.3. These would help to identify those strains or products that are robust and can exert the best possible chances of success. For probiotics, they are largely directed towards survival within the gastrointestinal milieu, while for prebiotics they outline composition effects upon the indigenous gut flora.

### 1.5.8 *Prebiotics*

Any dietary component that reaches the colon intact is a potential prebiotic, though much of the interest in the development of prebiotics is aimed at non-digestible oligosaccharides such as fructo-oligosaccharides (FOS), trans-galacto-oligosaccharides (TOS), isomalto-

**Table 1.3** *In vitro* tests (in addition to GRAS status) that may be applied to probiotics prior to human studies.

| Objective number | Objective description |
|---|---|
| 1 | **Identify products to be tested** – Probiotics should match up with their labels in terms of microbial content and/or number. |
| 2 | **Survival in the upper gut** – Products should be treated with digesta that resemble the gastric (e.g. HCl and pepsin) and small intestinal (e.g. bile salts, pancreatic enzymes) environments. This would determine those which have the capacity to survive transit through to the lower bowel. |
| 3 | **Survival in the colon** – A model of the human large intestine that reflects microbial events in the ascending, transverse and descending regions of the colon may be used to address probiotic persistence in the presence of a complex microbiota. One of the major problems with mixed culture work involving probiotics, is in differentiating the added strain from indigenous species or even genera. This applies to both the *in vitro* and *in vivo* situations. The advent of a range of molecular techniques, as applied to gut microbiology, has allowed this situation to change with the tools required for reliable identification studies now being readily available. |
| 4 | **Effects on gut microbial 'balance'** – Using model systems, the effects of test probiotics and prebiotics on the gut microflora composition and its dynamics may be addressed. Various molecular methodologies would be used to characterise whether the products stimulate an adverse or beneficial effect upon the microbiota composition. |
| 5 | **Attachment to gut cells** – The ability of probiotics to persist in the gastrointestinal tract is, at least partly, likely to be due to their comparative ability to adhere to the intestinal epithelium. *In vitro* cell lines could be used to compare attachment and predict whether residence is likely to be for a short- ($<1$ day), medium- (4 days) or long-term ($>4$ days) effect. |
| 6 | **Antibiotic resistance** – One problem with the addition of live bacteria to the human diet is the possibility that strains which carry genes for antibiotic resistance might transfer such resistance to the normal microbiota. This is especially problematic for probiotic strains that are genetically promiscuous. Antimicrobial transfer is likely to be carried out by conjugation rather than other routes such as transformation. A set of conjugation experiments whereby probiotics are grown in co-culture with selected members (predominant species) of gut genera should be carried out. These would include bacteroides, clostridia, enterococci, fusobacteria, eubacteria, *E. coli*, streptococci, as well as bifidobacteria and lactobacilli that are indigenous to the gut flora, but not used as commercial probiotics. |
| 7 | **Colonocyte death** – Bacteria able to kill colonocytes do so by invasive mechanisms and/or the elaboration of cellular toxins. The potential for killing, or inhibiting the growth of *in vitro* colonocytes may be assayed through use of an antimicrobial agent that kills the probiotic. Invasive species will be protected from the effects of this antimicrobial addition. |
| 8 | **Immune modulation** – Through the use of model systems this could be assayed by detecting common marker components such as cytokine expression. |
| 9 | **Haemolysis** – Many bacteria are $\alpha$- or $\beta$-haemolytic, and this should be avoided in probiotics, especially if there is a risk of intake being associated with blood, e.g. after dental surgery. This can easily be assayed using agar plates containing blood. |

oligosaccharides (IMO), xylo-oligosaccharides (XOS), soyoligosaccharides (SOS), gluco-oligosaccharides (GOS) and lactosucrose (see also Chapter 7). In Europe, FOS, GOS and lactulose have been shown to be prebiotics, through numerous volunteer trials, as evidenced by their ability to change the gut flora composition after a short feeding period (Gibson *et al.*, 2000). The Japanese market is more widespread. Non-digestible oligosaccharides (NDOs) are dietary substrates which meet the criteria of prebiotics (Gibson & Roberfroid, 1995). Oligosaccharides are sugars consisting of between approximately 2 and 20 saccharide units – that is, they are short-chain polysaccharides (Tomomatsu, 1994). Some occur naturally in several foods such as leek, asparagus, chicory, Jerusalem artichoke, garlic, artichoke, onion, wheat, banana and oats (Gibson *et al.*, 2000, 2003), as well as soybean. However, these foods contain only trace levels of prebiotics, so developments in functional foods have taken the approach of removing the active ingredients from such sources and adding them to more frequently consumed products in order to attain levels whereby a prebiotic effect may occur, for example in cereals, confectionery, biscuits, infant feeds, yoghurts, table spreads, bread, sauces and/or drinks.

One important development that is finding its way into functional foods is that of synbiotics. Here, a useful probiotic would be incorporated into an appropriate dietary vehicle with a suitable prebiotic. The premise is that the selective substrate would be metabolised by the live addition in the gut. This would enhance probiotic survival, as well as offer the advantages of both gut microbiota management techniques. A synbiotic has been defined as 'a mixture of probiotics and prebiotics that beneficially affects the host by improving the survival and implantation of live microbial dietary supplements in the gastrointestinal tract'. One example is a mixture of probiotic bifidobacteria with prebiotic fructo-oligosaccharides.

Prebiotics and synbiotics are much newer concepts than probiotics, and as such information on their health-promoting properties is more sparse. Nevertheless, the beneficial outcomes are likely to be similar to those of probiotics, with the chance that these are more enhanced, given the issues of reduced survivability.

## 1.6 Conclusions

The lack of adequate pharmaceutical development to counteract the rapid increase in antibiotic-resistant micro-organisms, the implications of diminished treatment for refractive patients, and the lack of availability of these antimicrobials in developing countries – where they are needed most – has demanded a new paradigm in the treatment of disease. This has been realised with the advent of probiotic, prebiotic and synbiotic strategies. These prophylactic approaches can be used in the prevention of disease, and they can also be applied as sole treatment or as adjuncts. This field has made astounding breakthroughs in recent years in several disease states, and continued research can only build on this. This only reinforces the knowledge that a beneficial microbial balance in the GI tract is crucial to host health as 'Disease usually results from inconclusive negotiations for symbiosis, an overstepping of the line by one side or the other, a biological misinterpretation of borders' (Thomas, 1974).

## References


Alderbeth, I., Cerquetti, M., Poilane, I., Wold, A.E. & Collignon, A. (2000) Mechanisms of colonisation and colonisation resistance of the digestive tract. *Microbial Ecology of Health and Disease*, **12**, 223–239.

Annibale, B., Capurso, G. & Delle Fave, G. (2002) Consequences of *Helicobacter pylori* infection on the absorption of micronutrients. *Digestive and Liver Disease*, **34**, S72–S77.

Attar, A., Lemann, M., Ferguson, A., Halphen, M., Boutron, M.C., Flourie, B., Alix, E., Salmeron, M., Guillemot, F., Chaussade, S., Menard, A.M., Moreau, J., Naudin, G. & Barthet, M. (1999) Comparison of a low dose polyethylene glycol electrolyte solution with lactulose for chronic constipation. *Gut*, **44**, 226–230.

Baba, E., Nagaishi, S., Fukata, T. & Arakawa, A. (1991) The role of intestinal microflora on the prevention of salmonella colonization in gnotobiotic chickens. *Poultry Science*, **70**, 1902–1907.

Benno, Y., Sawada, K. & Mitsuoka, T. (1984) The intestinal microflora of infants: composition of faecal flora in breast-fed and bottle-fed infants. *Microbiology and Immunology*, **28**, 975–986.

Bernhardt, H. & Knoke, M. (1997) Mycological aspects of gastrointestinal microflora. *Scandinavian Journal of Gastroenterology*, **32**, 102–106.
Bernhardt, H., Wellmer, A., Zimmerman, K. & Knoke, M. (1995) Growth of *Candida albicans* in normal and altered faecal flora in the model of continuous flow culture. *Mycoses*, **38**, 265–270.
Busscher, H.J., Mulder, A.F.J.M. & van der Mei, H.C. (1999) In vitro adhesion to enamel and in vivo colonization of tooth surfaces by lactobacilli from Bio-yoghurt. *Caries Research*, **33**, 403–404.
Collins, M.D. & Gibson, G.R. (1999) Probiotics, prebiotics and synbiotics: approaches for modulating the microbial ecology of the gut. *American Journal of Clinical Nutrition*, **69**, 1052–1057.
Cummings, J.H. & Macfarlane, G.T. (1997) Role of intestinal bacteria in nutrient metabolism. *Journal of Parenteral and Enteral Nutrition*, **21**, 357–365.
Felley, C.P., Corthesy-Theulaz, I., Rivero, J.L., Sipponen, P., Kaufman, M., Bauerfeind, P., Wiesel, P.H., Brassart, D., Pfeifer, A., Blum, A.L. & Michetti, P. (2001) Favourable effect of an acidified milk (LC-1) on *Helicobacter pylori* gastritis in man. *European Journal of Gastroenterology and Hepatology*, **13**, 25–29.
Fooks, L.J. & Gibson, G.R. (2002) Probiotics as modulators of the gut flora. *British Journal of Nutrition*, **88**, S39–S49.
Fooks, L.J., Fuller, R. & Gibson, G.R. (1999) Prebiotics, probiotics and human gut microbiology. *International Dairy Journal*, **9**, 53–61.
Freter, R. (1988) Mechanisms of bacterial colonization of the mucosal surfaces of the gut. *Virulence mechanisms of bacterial pathogens*, (ed. J.A. Roth), pp. 45–60, American Society for Microbiology, Washington DC.
Freter, R. (1992) Factors affecting the microecology of the gut. *Probiotics, the Scientific Basis*, (ed. R. Fuller), pp. 111–144, Chapman & Hall, London.
Fuller, R. & Perdigón, G. (eds.) (2003). *Gut Flora, Nutrition, Immunity and Health*, Blackwell Publishing, Oxford.
Gauhe, A., Gyorgy, P., Hoover, J.R., Kuhn, R., Rose, C.S., Ruelius, H.W. & Zilliken, F. (1954) Bifidus factor – preparation obtained from human milk. *Archives of Biochemistry*, **49**, 214–224.
Gibson, G.R. & Roberfroid, M.B. (1995) Dietary modulation of the human colonic microbiota – introducing the concept of prebiotics. *Journal of Nutrition*, **125**, 1401–1412.
Gibson, G.R., Berry Ottaway, P. & Rastall, R.A. (2000) *Prebiotics: New Developments in Functional Foods*, Chandos Publishing, Oxford.
Gibson, G.R., Rastal, R.A. & Fuller, R. (2003) The health benefits of probiotics and prebiotics. *Gut Flora, Nutrition, Immunity and Health*, (eds. R. Fuller and G. Perdigón), pp. 52–76, Blackwell Publishing, Oxford.
Goldin, B.R., Gualtieri, L.J. & Moore, R.P. (1996) The effect of *Lactobacillus* GG on the initiation and promotion of dimethylhydrazine-induced intestinal tumors in the rat. *Nutrition and Cancer*, **25**, 197–204.
Guandalini, S., Pensabene, L., Zikri, M.A., Dias, J.A., Casali, L.G., Hoekstra, H., Kolacek, S., Massar, K., Micetic-Turk, D., Papadopoulou, A., de Sousa, J.S., Sandhu, B., Szajewska, H. & Weizman, Z. (2000) *Lactobacillus* GG administered in oral rehydration solution to children with acute diarrhea: a multicentre European trial. *Journal of Pediatrics, Gastroenterology and Nutrition*, **30**, 54–60.
Guilliams, T.G. (1999) Healthy microbial organisms. *The Standard*, **2**, 1–8.
Guslandi, M., Mezzi, G., Sorghi, M. & Testoni, P.A. (2000) *Saccharomyces boulardii* in maintenance treatment of Crohn's disease. *Digestive Disease Science*, **45**, 1462–1464.
Hamilton-Miller, J.M. (2004) Probiotics and prebiotics in the elderly. *Postgraduate Medical Journal*, **80**, 447–451.
Harms, H.K., Bertele-Harms, R.M. & Bruer-Kleis, D. (1987) Enzyme substitution therapy with the yeast *Saccharomyces cervisiae* in congenital sucrase-isomaltase deficiency. *New England Journal of Medicine*, **316**, 1306–1309.
Havenaar, R. & Huis In't Veld, M.J.H. (1992) Probiotics: a general view. *Lactic acid Bacteria in Health and Disease*, Vol. 1, (ed. B.J.B. Wood), pp. 151–170, Elsevier Applied Science Publishers, Amsterdam.
Hilton, E., Kolakowski, P., Smith, M. & Singer, C. (1997) Efficacy of *Lactobacillus* GG as a diarrheal preventative in travellers. *Journal of Travel Medicine*, **4**, 41–43.
Hooper, L.V., Midtvedt, T. & Gordon, J.L. (2002) How host-microbial interactions shape the nutrient environment of the mammalian intestine. *Annual Review of Nutrition*, **22**, 283–307.
Kaila, M., Isolauri, E., Saxelin, M., Arvilommi, H. & Vesikari, T. (1995) Viable versus inactivated *Lactobacillus* strain GG in acute rotavirus diarrhoea. *Archives of Diseases in Children*, **72**, 51–53.
Kalliomaki, M., Salminen, S., Arvilommi, H., Kero, P., Koskinen, P. & Isolauri, E. (2001) Probiotic in primary prevention of atopic disease: a randomised placebo-controlled trial. *Lancet*, **357**, 1076–1079.

Kimura, K., McCartney, A.L., McConnell, M.A. & Tannock, G.W. (1997) Analysis of fecal populations of bifidobacteria and lactobacilli and investigation of the immunological responses of their human hosts to the predominant strains. *Applied and Environment Microbiology*, **63**, 3394–3398.

Lambert, J. & Hull, R. (1996) Upper gastrointestinal disease and probiotics. *Asia Pacific Journal of Clinical Nutrition*, **5**, 31–35.

Macfarlane, G.T. & Gibson, G.R. (1994) Metabolic activities of the normal colonic flora. *Human Health – The Contribution of Microorganisms*, (ed. S.A. Gibson), pp. 17–52, Springer-Verlag, London.

Macfarlane, G.T. & McBain, A.J. (1999) The human colonic microbiota. *Colonic Microbiota, Nutrition and Health*, (eds. G.R. Gibson and M.B. Roberfroid), pp. 1–25, Kluwer Academic Publishers, Dordrecht.

Marsh, P. (1980) The normal oral flora. *Oral Microbiology – Aspects of Microbiology Series*, (ed. P. Marsh), pp. 11–24, Nelson, Surrey.

Marteau, P.R., Flourie, B., Pochart, P., Chastang, C., Desjeux, J.F. & Rambaud, J.C. (1990) Effect of the microbial lactase (EC 3.2.1.23) activity in yoghurt on the intestinal absorption of lactose: an in vivo study in lactase deficient humans. *British Journal of Nutrition*, **64**, 71–79.

McFarland, L.V., Surawicz, C.M., Greenberg, R.N., Fekety, R., Elmer, G.W., Moyer, K.A., Melcher, S.A., Bowen, K.E., Cox, J.L. & Noorani, Z. (1994) A randomized placebo-controlled trial of *Saccharomyces boulardii* in combination with standard antibiotics for *Clostridium difficile* disease. *Journal of American Medical Association*, **271**, 1913–1918.

McFarland, L.V., Surawicz, C.M., Greenberg, R.N., Elmer, G.W., Moyer, K.A., Melcher, S.A., Bowen, K.E. & Cox, J.L. (1995) Prevention of β-lactam-associated diarrhea by *Saccharomyces boulardii* compared with placebo. *American Journal of Gastroenterology*, **90**, 439–448.

Metchnikoff, E. (1907) *The Prolongation of Human Life*. Heinemann, London.

Montalto, M., Vastola, M., Marigo, L., Covino, M., Graziosetto, R., Curigliano, V., Santoro, L., Cuoco, L., Manna, R. & Gasbarrini, G. (2004) Probiotic treatment increases salivary counts of lactobacilli: a double-blind, randomized, controlled study. *Digestion*, **69**, 53–56.

Moreau, M.C. (2000) Flore intestinale, prebiotique et effets sur la reponse immunitaire intestinale a IgA. *Archives de Pediatre*, **2000**, 247–248.

Mykkanen, H., Laiho, K. & Salminen, S. (1998) Variation in faecal bacterial enzyme activities and associations with bowel function and diet in elderly subjects. *Journal of Applied Microbiology*, **85**, 37–41.

Plummer, S., Weaver, M.A., Harris, J.C., Dee, P. & Hunter, J. (2004) *Clostridium difficile* pilot study: effects of probiotic supplementation on the incidence of *C. difficile* diarrhoea. *International Microbiology*, **7**, 59–62.

Rai, R. & Sundeepa, R. (2002) Pseudomembranous colitis requiring surgical intervention following triple therapy for *Helicobacter pylori* eradication. *ANZ Journal of Surgery*, **72**, 917–919.

Reid, G., Beuerman, D., Heinemann, C. & Bruce, A.W. (2001) Probiotic *Lactobacillus* dose required to restore and maintain a normal vaginal flora. *FEMS Immunology and Medical Microbiology*, **32**, 37–41.

Rembacken, B.J., Snelling, A.M., Hawkey, P.M., Chalmers, D.M. & Axon, A.T. (1999) Non-pathogenic *Escherichia coli* versus mesalizine for the treatment of ulcerative colitis: a randomised trial. *Lancet*, **354**, 635–639.

Saavedra, J.M., Bauman, N.A., Oung, I., Perman, J.A. & Yolken, R.H. (1994) Feeding of *Bifidobacterium bifidum* and *Streptococcus thermophilus* to infants in hospital for prevention of diarrhoea and shedding of rotavirus. *Lancet*, **344**, 1046–1049.

Salminen, S., Bouley, C., Boutron-Ruault, M.C., Cummings, J.H., Franck, A., Gibson, G.R., Isolauri, E., Moreau, M.C., Roberfroid, M.B. & Rowland, I.R. (1998) Functional food science and gastrointestinal physiology and function. *British Journal of Nutrition*, **80**, S147–S171.

Sanford, P.A. (1992) Digestive System Physiology, 2nd edn., published within a series called *Physiological Principles in Medicine*, (series eds. M. Hobsley, K.B. Saunders and J.T. Fitzsimons), Edward Arnold, London.

Savage, D.C., Dubos, R. & Schaedler, R.W. (1968) The gastrointestinal epithelium and its autochthonous bacterial flora. *Journal of Experimental Medicine*, **127**, 67–76.

Schmidt, W.-P. (2004) Model of the epidemic of childhood atopy. *Medical Science Monitor*, **10**, HY5–HY9.

Stark, P.L. & Lee, A. (1982) The microbial ecology of the large bowel of breast-fed and formula-fed infants during the first year of life. *Journal of Medical Microbiology*, **15**, 189–203.

Suau, A., Bonnet, R., Sutren, M., Godon, J.J., Gibson, G.R., Collins, M.D. & Dore, J. (1999) Direct analysis of genes encoding 16S rRNA from complex communities reveals many novel molecular species within the human gut. *Applied and Environment Microbiology*, **65**, 4799–4807.

Tannock, G.W. (2003) The intestinal microflora. *Gut Flora, Nutrition, Immunity and Health*, (eds. R. Fuller and G. Perdigón), pp. 1–23, Blackwell Publishing, Oxford.

Tano, K., Grahn Hakkansson, E., Holm, S.E. & Hellstrom, S. (2002) A nasal spray with alpha-haemolytic streptococci as long term prophylaxis against recurrent otitis media. *International Journal of Pediatrics Otorhinolaryngology*, **62**, 17–23.

Thomas, L. (1974) *The Lives of a Cell*, Penguin Group, Middlesex.

Tomomatsu, H. (1994) Health effects of oligosaccharides. *Food Technology*, **48**(10), 61–65.

Vanderhoof, J.A., Whitney, D.B., Antonson, D.L., Hanner, T.L., Lupo, J.V. & Young, R.J. (1999) *Lactobacillus* GG in the prevention of antibiotic-associated diarrhea in children. *Journal of Pediatrics*, **135**, 564–568.

de Vrese, M., Stegelmann, A., Richter, B., Fenselau, S., Laue, C. & Schrezenmeir, J. (2001) Probiotics-compensation for lactase insufficiency. *American Journal of Clinical Nutrition*, **73**, 421S–429S.

Willis, C.L., Gibson, G.R., Holt, J., Atherton, S. & Allison, C. (1999) Negative correlation between oral malodour and numbers and activities of sulphate-reducing bacteria in the human mouth. *Archives of Oral Biology*, **44**, 665–670.

Wunderlich, P.F., Braun, L., Fumagalli, I., D'Apuzzo, V., Heim, F., Karly, M., Lodi, R., Politta, G., Vonbank, F. & Zeltner, L. (1989) Double-blind report on the efficacy of lactic acid-producing *Enterococcus* SF68 in the prevention of antibiotic associated diarrhoea and in the treatment of acute diarrhoea. *Journal of Internal Medicine and Research*, **17**, 333–338.

Ziemer, C.J. & Gibson, G.R. (1998) Overview of probiotics, prebiotics and synbiotics in the functional food concept. *International Dairy Journal*, **8**, 473–479.

Zoetendal, E.G., Akkermans, A.D.L., Akkermans van Vliet, W.M., de Visser, R.A.G.M. & de Vos, W.M. (2001) The host genotype affects the bacterial community in the human gastrointestinal tract. *Microbial Ecology in Health and Disease*, **13**, 129–134.

# 2 Genomic Characterisation of Starter Cultures

F. Dellaglio, G.E. Felis, S. Torriani, K. Sørensen and E. Johansen

## 2.1 Introduction

Genomics is the determination and analysis of the total genome sequence of an organism. Ideally, it results in a complete list of all of the genes that the organism contains, and allows the construction of a good estimation of the metabolic pathways and potentials of the organism.

The first genome sequence of a lactic acid bacterium that was publicly available was that of *Lactococcus lactis* subsp. *lactis* (Bolotin *et al.*, 2001); this organism is commonly used as a starter culture during the manufacture of cheese and fermented milks. Typically, fermented dairy products are supplemented with probiotic strains to produce, for example, probiotic yoghurts or cheeses. Genome sequences of probiotic strains or species are of great interest to scientists, as they will assist in the elucidation and validation of claimed health benefits. While genome projects of 'classical' or traditional starter cultures are also ongoing and are expected to provide useful information on different technological aspects, this chapter will focus primarily on the genomics of probiotic micro-organisms. Many of the concepts will be applicable to both classical starter cultures as well as to probiotic starter cultures.

Genomics simplifies investigations with genetics by indicating which genes might be present; and it simplifies the investigations with physiology by indicating which metabolic pathways might be present. Genomics also simplifies strain development by indicating what is feasible, and suggesting the simplest approach to a desired objective. In addition, genomics permits experimentation that allows an understanding of the behaviour of the organism under various conditions through elucidation of the various regulatory mechanisms in use. As the number of complete genome sequences increases, comparative genomics will provide considerable insight into the evolution and genetic diversity of strains. Genomics based technologies may form the basis for new rapid quality assurance techniques and safety assessments, as well.

Prokaryote biology is being revolutionised by the advent of genome sequencing and analysis. Genomics is now regarded as a science encompassing the study of the structure, the content and the evolution of genomes. According to McAuliffe and Klaenhammer (2002), the beginning of the genome era is generally regarded to be July 1995, when The Institute for Genomic Research (TIGR) published the first complete bacterial genome sequence of *Haemophilus influenzae* (Fleischmann *et al.*, 1995). It should be noted, however, that the genome sequence of the ribonucleic acid (RNA) virus MS2 was completely determined in 1976 (Fiers *et al.*, 1976), and that the complete genome sequences of several other bacteriophages were known as well in advance of the release of the *H. influenzae* genome sequence.

Before the genomics era, molecular biology allowed elucidation of many biological phenomena, but its main limitation was the limited number of genes and gene products which could be analysed contemporarily. An organism, however, is characterised by the interaction of many genes and gene products, creating a complex network of inter-related components.

In order to understand the enormous impact of genomics on understanding the properties of lactic acid bacteria, one should examine the work of Germond *et al.* (2003), who characterised 130 *Lactobacillus delbrueckii* spp. isolates to obtain insight on the presence and stability of technological traits and to evaluate the evolutionary relationships among strains of this species. While the strains were carefully identified and four genes were characterised in detail in all strains, this enormous effort provided insight into only a small fraction of the genome. Currently, the genomic data available for *Lb. delbrueckii* spp. (http://genome.ornl.gov/microbial/lbul) contains more than 2000 candidate genes, while the complete genome sequence of the related species *Lactobacillus johnsonii* is predicted to contain more than 1800 (Pridmore *et al.*, 2004; see also Mills, 2004).

## 2.2 The 'Omic' approaches

### 2.2.1 *Background*

'Omic' approaches allow researchers to overcome the limitations of traditional approaches, obtaining a holistic, organism-wide picture. The 'omic' suffix indicates the study of several aspects, objects of different disciplines. Sequencing of a genome is far from providing an exhaustive knowledge of an organism. On the contrary, it opens the way to the integration of a great deal of information, via experimentation using the various 'omic' disciplines – that is, transcriptomics, proteomics, metabolomics, interactomics and phenomics.

*Transcriptomics* is the study of the transcriptome, the complete set of transcripts in a cell under a particular set of conditions. It is based on the use of micro-arrays, DNA chips or serial analysis of gene expression (SAGE) (Velculescu *et al.*, 1995), which allow monitoring of the accumulation of transcripts under different experimental conditions: differential gene expression in diverse conditions may help in determining the involvement of such genes in particular abilities of the organism.

*Proteomics* deals with the investigation of the complete set of proteins expressed by an organism – that is, the proteome. Typically, due to the sensitivity of the detection methods, only a subset of the proteins in a cell is analysed. Modifications of well-known techniques, such as the two-hybrid system, may prove useful in the evaluation of protein–protein and/or protein–other molecule interactions providing additional information on the biological networks, which characterise the micro-organisms. The complex of studies of those interactions is referred to as *interactomics*.

The observation of the phenotypic alterations after induced mutations constitutes *phenomics*; the systematic mutation and observation of all possible phenotypes is particularly fruitful in the identification of genes essential for particular functions or viability.

Analysis of the metabolite profile of an organism is referred to as *metabolomics*. This involves the identification and quantification of as many metabolic intermediates as possible, and provides insights into metabolic pathways evaluated in relation to genomic and biochemical data and organism physiology.

It has been suggested that the integration of all data collected and cross-validated with the indicated 'omics' approaches is the basis for the construction of a 'modular biology', instead of molecular biology. Organisms are composed of biological modules, and comprise complex systems of interactions between molecules (e.g. genes, proteins and other gene products) that produce different effects, such as biological functions and processes that explain the organism physiology (Ge *et al.*, 2003). The above-reported list of the approaches available for experimental evaluation and validation of genomic data should underline that a genome sequence is only the beginning of an amazing journey into the physiology of an organism.

When considering complex systems, such as mixed cultures or ecological niches, *metagenomics* attempts to elucidate and characterise the complex of genome content of all organisms, including those that cannot be cultured, contained in the environment under study.

Finally, *microbiome* is defined as the collective genome of the indigenous microbial population colonising a person (McAuliffe & Klaenhammer, 2002).

Genome scale analyses, comparisons and validations may produce important insights in both basic and applied research: from the understanding of strain physiology and evolution to engineering of metabolic pathways (Weinstock *et al.*, 2000; Papin *et al.*, 2003; Stephens & Laub, 2003; Zhang & Kim, 2003). It must also be emphasised that the analysis and interpretation of the large amount of data produced would be impossible without proper tools provided by bioinformatics (Kanehisa & Bork, 2003).

### **2.2.2** *Exploration of genomic sequences*

A genome sequence has little value on its own, but the value increases when the data are analysed and converted to biologically relevant information. This is done by computer analysis (data mining) to identify potential genes, potential regulatory regions and regulatory circuits, potential metabolic pathways, and potential functions of the cell. The real value is obtained when the biologically relevant data are converted to new knowledge and ultimately products via experimentation. This experimentation often takes the form of looking at gene expression under a number of conditions, and requires some sort of high-density micro-array technology, either on deoxyribonucleic acid (DNA) chips or microscope slides.

Genome sequencing must be followed by 'annotation' – that is, the assignment of useful biological information to raw sequence data. This includes the identification of RNAs (tRNA and rRNA), repetitive elements, and features such as promoters, terminators and operator regions and, most of all, identification of 'open reading frames' (ORFs). This is usually done with specific software, which allows *in silico* prediction of the genes, and it is available on the internet or as part of more extensive bioinformatics software packages. One serious problem with the *in silico* prediction of ORFs is that a number of the sequences identified are actually not protein-encoding regions; rather, these are ORFs, which are typically quite short and occur by chance. The overall result is that the number of genes predicted in a genome sequence is often overestimated (Skovgaard *et al.*, 2001), and a great deal of time must be invested in the manual elimination of these spurious ORFs from the annotated genome sequence.

After gene identification, biochemical and physiological functions, which are associated with predicted genes, are identified mainly through:

- sequence similarity of the encoded product with previously characterised proteins and structural RNAs available in public databases;
- the presence of motifs associated with particular functions; and
- the conservation of operon structures.

This *bioinformatic* approach to gene identification depends strongly on the availability of sequences in public databases; 30–60% of ORFs in newly sequenced genomes are putative proteins with 'unknown' function (McAuliffe & Klaenhammer, 2002). Moreover, whenever similar sequences are found, similarity hits are obviously not experimental validations of gene functions.

Sequence analysis can be performed using toolboxes available on various websites, such as EMBL/EBI (www.ebi.ac.uk), GenBank (www.ncbi.nlm.nih.gov) and TIGR (www.tigr.org). Many other valuable genomics tools are available, and have been recently reviewed by Siezen *et al.* (2004).

Once a genome sequence is obtained and annotated, it can be analysed to gain insight into the biology of the single organism and compared with other available genome sequences (i.e. comparative genomics), as will follow.

### **2.2.3** *Tools for converting genomic sequences to biologically relevant information*

*In silico* predictions alone cannot assess the role of every gene in the genome. They must be experimentally validated, with one or more genetic tools, in the step of functional genomics – that is, the discovery of the biological function of genes and the determination of the network of relationships of gene products (Stephens & Laub, 2003). The most powerful approaches undertaken to this aim are micro-array analysis and proteomics. Array technology allows the analysis of differential gene expression and global gene regulation (transcriptomics), while techniques such as 2D-gel electrophoresis or isoelectric focusing (IEF) may reveal real gene product characteristics, such as effective translation, post-translation modifications, isoelectric point, and molecular weight.

However, *in vivo* studies, which aim at the elucidation of the role of gene products, may be obtained with different techniques, depending on the genetic accessibility of the organism under study, such as: (a) targeted mutagenesis (Thomas & Capecchi, 1987); (b) gene knockout through homologous recombination (Thomas & Capecchi, 1987); and (c) RNA interference (Fjose *et al.*, 2001). These techniques are powerful tools to assign biological functions to predicted annotations (McAuliffe & Klaenhammer, 2002).

Regulation of the expression of genes allows an organism to respond to environmental conditions. This is especially relevant for probiotic strains, which are produced in one environment (e.g. in a stainless-steel tank in a factory) and expected to act in a very different environment (e.g. in the human gastrointestinal (GI) tract). Gene expression is regulated by specific DNA sequences called promoters, the genetic sequence that determines when and to what extent the associated genes are turned on. Promoters can be readily identified using genomics tools and subsequent biological experimentation. Ultimately, an understanding of gene regulation may allow the production of probiotic cells, which are optimally prepared to act in the GI tract.

The identification and confirmation of the activity of a promoter requires experimentation, and often involves the use of promoter probe vectors. Few promoter probe vectors for probiotic bacteria have been developed. One example is the plasmid pNZ7120 (Bron *et al.*, 2004a), which can be used to study gene expression in probiotic strains of the species *Lactobacillus plantarum*. This vector has been used in a screening experiment to isolate and characterise the bile salt-inducible promoters (Bron *et al.*, 2004b). Thirty unique loci, which respond to bile salts *in vitro*, were isolated using pNZ7120, and identified by comparison with the complete genome sequence of *Lb. plantarum* WCFS1 (Kleerebezem *et al.* 2003). These loci are involved in cell membrane function, cell wall function, redox reactions, gene regulation or had unknown functions. Two promoters, which respond to bile salts *in vitro*, were also shown to have a higher level of expression in the duodenum of the mouse, a location where the cells are expected to be exposed to bile salts. Thus, this system can be used to identify genes potentially involved in the survival of a probiotic strain in the mammalian GI tract.

Alternative approaches to identifying promoters and genes active in the GI tract could involve the use of techniques such as *In Vivo* Expression Technology (IVET; Slauch & Camilli, 2000), Selective Capture of Transcribed Sequences (SCOTS; Morrow *et al.*, 1999) or Signature Tagged Mutagenesis (STM; Holden & Hensel, 1998). All of these techniques involve the isolation of specific pieces of the DNA that are expressed in the GI tract, determination of the DNA sequence of the cloned DNA, and the use of the genomic DNA sequence to identify the actual genes, which have been isolated.

### 2.2.4 *What can genomics be used for?*

#### *Comparative genomics*

The availability of genome sequence information can be exploited for the understanding of the biology of the single organism, and be used in comparative studies with other available genome sequences. When the complete genome sequence of two or more members of the same genus or species is available, it is possible to look for similarities and differences, and to attempt to correlate these to the specific properties of the organisms in question. When only one genome sequence is available, micro-arrays can be designed to evaluate intraspecies genomic differentiation (Fitzgerald *et al.*, 2001; de Vos, 2001). This provides insights into the evolution and genetic divergence of strains, which may be useful for the evaluation of horizontal gene transfer, and can be used to furnish strain-specific targets for the design of strain-specific tests for patented starters and probiotic bacteria.

In addition, comparative genomics can also be very useful to gain insights into the taxonomy, phylogeny and evolution of starter cultures and probiotic species. It can be used to allow the identification of genes essential for life, or genes responsible for adaptation to particular niches or culture conditions and properties. Of course, these *in silico* predictions must be validated in biological tests, as described elsewhere.

#### *Tracking of strains*

In order for probiotic bacteria to exert their beneficial effects, it is necessary that they are present at the required location and in sufficient numbers. To confirm this, it is important that probiotic strains can be specifically identified and quantified in complex environments. These environments may contain other closely related bacteria, including members of the same species. Most approaches for identification of a particular strain are based on detection of DNA sequences specific to that strain. Having the complete genome sequence of a strain to be tracked will allow the rational design of oligonucleotide primers that can be tested for their specificity. DNA sequence-based techniques for tracking strains include fluorescence *in situ* hybridisation (FISH), *in situ* PCR and real-time PCR on purified isolates. FISH has been successfully used to follow the dynamics of the human gut microflora over an extended period of time (Harmsen & Welling, 2002).

Development of micro-arrays containing gene sequences from different bacteria, e.g. rRNA genes or other species-specific sequences, may be used to evaluate which bacterial species are present in different parts of the GI tract and, which are able to survive under different conditions (McAuliffe & Klaenhammer, 2002).

### *Strain characterisation*

The properties of an organism are encoded in the genetic material. A knowledge of the complete genome sequence of an organism is a valuable start to understanding the capabilities that a strain may have. Thus, a prediction can be made of all genes present in the strain. The function of many of these genes can be predicted based on their similarity to genes previously described in other organisms. Approximately one-third of the genes in any given genomic sequence will have an unknown function; however, elucidation of the function of genes of unknown function is an important element of genomics research.

The complete genomes of lactic acid bacteria often contain extrachromosomal DNA sequences called plasmids. While they are abundant in genus *Lactococcus*, both in terms of number and diversity, plasmids are much less common in intestinal *Lactobacillus* spp. and other organisms typically used as probiotics (Davidson *et al.*, 1996). The complete genome sequence of *Bifidobacterium longum* NCC2705 does, however, contain a single plasmid (Schell *et al.*, 2002), whilst *Lb. plantarum* WCFS1 contains three plasmids (Kleerebezem *et al.*, 2004). Since plasmids can be readily lost, it is important to know which genes are plasmid-borne so that the consequences of plasmid loss can be assessed. Occasionally, plasmids which code for undesirable traits, such as antibiotic-resistance, which should be identified. Fortunately, these can be eliminated by plasmid curing as illustrated by Danielsen (2002), who cured *Lb. plantarum* strain 5057 of a 10.9-kb plasmid containing the *tet*M gene.

Based on the list of genes, it is possible to reconstruct the metabolism of the strain. For example, the absence of the genes for a metabolic pathway can be taken as a good indication that the metabolic capability is lacking in the strain. Likewise, the presence of the genes for a metabolic pathway gives an indication that the strain can potentially use this metabolic pathway. It is, however, much more difficult to predict from a genome sequence the conditions under which a particular pathway will be active. Ultimately, predictions about metabolic activity must be confirmed by experimentation, for example, using transcriptomics, proteomics and metabolomics, as described elsewhere.

All bacteria have genes for the biosynthesis of the various structures of the cell including the polysaccharides, which are typically found on the cell surface. While it is possible to confirm from a genomic sequence that the strain has the capability to produce a cell surface polysaccharide, it is not possible to predict the structure of that polysaccharide from the genes that are present. This is a significant limitation to the use of genomics to understand probiotic cultures as these structures are believed to be very important in a number of probiotic properties.

A further complication of using genomics to understand the properties of a probiotic strain is that we often do not know which pathways are required for a particular probiotic characteristic. This would require information on the mode of action of a specific probiotic effect. On the other hand, genomics is a tool that is expected to provide a great deal of insight into the mode of action, as will be described later. Probiotic bacteria are subjected to a variety of stresses including: (a) during production of the culture; (b) during incorporation into milk products or dietary supplements; (c) during transport and storage; (d) during the actual consumption of the product containing the probiotic organisms; and (e) at the location where the probiotic organism is meant to exert its effect, for example in the GI tract. An analysis of gene expression (e.g. using micro-arrays) will illustrate which stresses are present, and may allow the design of countermeasures to alleviate these stresses. It may also suggest methods to precondition the cells so that they are better able to cope with these stresses. It is anticipated that probiotic strains will have an enhanced performance if they are stressed in many appropriate ways.

### *Strain improvement*

The knowledge acquired on the genetics and physiology of a probiotic starter strain gained from genomics can be used for strain improvement. For example, it might be possible to eliminate an undesirable property from a strain by simple mutagenesis and screening. The probability of success is greatly reduced if the strain has duplicate genes encoding this trait. Likewise, trying to select mutants with a high level of expression of a biochemical pathway would be futile if the relevant pathway did not exist. The presence of a gene on a plasmid increases the likelihood of loss as well as permits the intentional elimination of that gene via plasmid curing (Danielsen, 2002).

Strain improvement can be carried out either by traditional genetic methods (mutagenesis followed by screening) or by the use of recombinant DNA technology. A thorough understanding of the metabolism of a cell may allow the direct selection of mutants with a specific property, thereby eliminating the need for mutagenesis. Alternatively, high-throughput screening methods – again based on an understanding of the complete metabolism of the cell – may allow the isolation of a specific mutant without mutagenesis. Recombinant DNA technology allows the precise alteration of the DNA sequence of an organism. It is possible to specifically change a single nucleotide or a single codon in the chromosome of a microbe. For example, Sørensen *et al.* (2000) changed a serine codon to an amber codon in the *pyrF* gene of several industrial *Lac. lactis* subsp. *lactis* strains. These approaches are greatly facilitated by the possession of a complete genome sequence. The use of gene technology in dairy or food micro-organisms is controversial in many parts of the world. However, it is possible to use recombinant DNA technology to introduce changes such as the one described above while allowing a strain to maintain its 'generally recognised as safe' (GRAS) status in the USA. Such strains have been referred to as food-grade genetically modified organisms (GMOs) (Johansen, 1999), and acceptance of the use of these in dairy/food production, throughout the world, is conceivable.

### *Safety assessment*

The genome sequence of an organism provides a complete list of all of the genes that are present in that organism. Many of these genes will be unique to the organism, but others will have homologues in other organisms, and some of these will be fairly well characterised. Some genes have a strong association with undesirable properties (e.g. virulence factors, toxin production), and a simple *in silico* search can determine whether or not these genes are present. The presence of an undesirable gene is not necessarily a problem. It is possible that the gene is not expressed under conditions that the probiotic strain is likely to encounter. This can be assessed using transcriptomics and proteomics. Alternatively, the gene can be eliminated, either using traditional approaches or recombinant DNA technology as described above.

*Improving production conditions*

The industrial-scale production of probiotic strains and traditional dairy starter cultures takes place in large fermentation tanks where they are grown in milk and/or using particular growth substrates. They must be grown, harvested and stabilised to allow formulation into a stable and effective product. Transcriptomic analysis of the strains at different points in this process can provide important knowledge allowing for improved production conditions, such as optimisation of the starter culture media and improved formulation conditions. Furthermore, transcriptomics can also provide information on the behaviour of the strains in the final formulation, and may be useful for the development of preconditioning treatments, which can be applied in the factory to improve strain performance.

*Mode of action*

A number of beneficial effects are ascribed to probiotic micro-organisms. Many of these are based on animal studies or clinical trials with a small number of subjects. In most cases, a statistically significant effect is seen, but the actual mechanism behind this effect is unknown. Perhaps the most important contribution of genomics in the field of probiotics will be an elucidation of the mode of action of specific products with specific clinical effects. This area of research is still in its infancy, but will involve gene expression studies of the probiotic organism and tissues of the host. Ultimately, it is the interaction between these two genomes that determines the beneficial outcome.

## 2.3 State of the art

### 2.3.1 *Publicly available genome sequences*

Currently, there are more than 200 completed and published bacterial genome sequences, which are available on the internet [e.g. TIGR-Comprehensive Microbial Resource (CMR) (http://www.tigr.org/tigr-scripts/CMR2/CMRHomePage.spl), and the European Bioinformatics Institute genomes pages (http://www.ebi.ac.uk/genomes/bacteria.html)], in scientific journals, and another 500 or more are being determined and/or not yet published. The majority of those genome sequences concern pathogenic organisms and environmental isolates, but the number of beneficial food-related micro-organisms for which genome sequence data are available is increasing (Table 2.1), and additional genome statistics have been published by Klaenhammer *et al.* (2002).

It should be noted that for publication purposes, the norm or standard for the genome sequence should be 100% complete, and demonstrate the circular nature of a bacterial chromosome. For research purposes, however, this degree of completeness is not required. Most of the uses of a genomic sequence described in Section 2.2 can be achieved with a partial genome sequence (e.g. >95% complete and possibly consisting of a small number of contiguous sequences). This includes transcriptomics, proteomics and metabolomics, but obviously excludes analysis of the evolution of whole bacterial genomes. Even though a

**Table 2.1** Genome projects involving food-related organisms[1].

| Micro-organisms[2] | Strain | Genome size (Mb) | Reference/institution/contact person |
|---|---|---|---|
| **Sequencing completed** | | | |
| *B. longum* | NCC 2705 | 2.3 | Schell *et al.* (2002) |
| *Lb. johnsonii* | NCC 533 | 2.0 | Pridmore *et al.*, 2004 |
| *Lb. plantarum* | WCFS1 | 3.3 | Kleerebezem *et al.* (2003) |
| *Lb. casei* | DN-114 001 | NR[5] | Danone |
| *Lb. delbrueckii* subsp. *bulgaricus* | ATTC 11842[3] | 1.8 | INRA, Genoscope |
| *Lac. lactis* subsp. *lactis* | IL1403 | 2.3 | Bolotin *et al.* (2001) |
| *S. thermophilus* | CNRZ 1066 | 1.8 | Integrated Genomics Inc., INRA |
| *E. faecalis* | | NR | Genome Therapeutics |
| *E. faecalis* | V583 | 3.2 | Paulsen *et al.* (2003) |
| *E. faecium* | | NR | Genome Therapeutics |
| *E. faecium* | DO | NR | Joint Genome Institute, University of Texas |
| **Sequencing ongoing** | | | |
| *A. minutum* | ATCC 33267 | 1.9 | Integrated Genomics Inc. |
| *B. animalis* subsp. *lactis* | Bb-12 | 2.0 | Chr. Hansen A/S[6] |
| *B. animalis* subsp. *lactis* | DN-173 010 | NR | Danone[6] |
| *B. breve* | NCI 8807 | 2.4 | University College Cork |
| *B. longum* | DJO 10A | NR | Joint Genome Institute, University of Minnesota |
| *Bre. linens* | BL2/ATCC 9174 | 3.0 | Joint Genome Institute, Utah State University |
| *E. faecium* | ATCC 35667 | 2.1 | Integrated Genomics Inc. |
| *Lb. acidophilus* | ATCC 700396[4] | 2.0 | North Carolina State University and CalPoly Technical University, USA |

| | | | |
|---|---|---|---|
| *Lb. brevis* | ATCC 367 | 2.0 | Joint Genome Institute, University of California San Diego |
| *Lb. casei* | ATCC 334 | 2.5 | Joint Genome Institute, Utah State University |
| *Lb. delbrueckii* subsp. *bulgaricus* | ATCC11842[3] | 2.3 | INRA, Genoscope |
| *Lb. delbrueckii* subsp. *bulgaricus* | ATCC BAA-365 | 2.3 | Joint Genome Institute, University of Wisconsin |
| *Lb. delbrueckii* subsp. *bulgaricus* | DN-100 107 | NR | Danone |
| *Lb. gasseri* | ATCC 33323 | 1.8 | Joint Genome Institute, North Carolina State University |
| *Lb. helveticus* | CNRZ 32 | 2.3 | University of Wisconsin, Utah State University |
| *Lb. rhamnosus* | HN 001 | 2.4 | Fonterra Research, ViaLactica BioSciences |
| *Lb. sakei* | 23K | 1.9 | INRA |
| *Lac. lactis* subsp. *cremoris* | MG 1363 | 2.6 | University of Groningen |
| *Lac. lactis* subsp. *cremoris* | MG 1363 | 2.4 | INRA |
| *Lac. lactis* subsp. *cremoris* | SK11 | 2.3 | Joint Genome Institute, Utah State University, University of Minnesota |
| *Leu. citreum* | KM 20 | 2.0 | Korea Research Institute of Bioscience and Biotechnology (KRIBB) |
| *Leu. mesenteroides* subsp. *mesenteroides* | ATCC 8293 | 2.0 | Joint Genome Institute, North Carolina State University |
| *O. oeni* | IOEB 84.13 | 1.8 | Genome Express, INRA, Universities Dijon and Bordeaux |
| *O. oeni* | MCW PSU-1 ATCC BAA-331 | 1.8 | Joint Genome Institute, University of California, Davis |
| *P. pentosaceus* | ATCC 25745 | 2.0 | Joint Genome Institute, University of Wisconsin, Utah State University |
| *Pro. freudenreichii* | ATCC 6207 | 2.6 | DSM Food Specialties, Friesland Coberco Dairy Foods |
| *S. thermophilus* | LMD-9 | 1.8 | Joint Genome Institute, University of Nebraska |
| *S. thermophilus* | DN-001 147 | NR | Danone |

[1] Data compiled from GOLD™ Genomes OnLine Database (http://www.genomesonline.org/) and Klaenhammer *et al.* (2002).
[2] *B.* = *Bifidobacterium*; *Lb.* = *Lactobacillus*; *Lac.* = *Lactococcus*; *S.* = *Streptococcus*; *E.* = *Enterococcus*; *A.* = *Atopobium*; *Bre.* = *Brevibacterium*; *Leu.* = *Leuconostoc*; *O.* = *Oenococcus*; *P.* = *Pediococcus*; *Pro.* = *Propionibacterium*.
[3] Reported as complete in the GOLD™ database, but in progress in the Genoscope web site.
[4] Reported as completed by Klaenhammer *et al.* (2002), but incomplete in the GOLD™ database.
[5] NR = not reported.
[6] See also Masco *et al.* (2004).

large number of the sequences shown in Table 2.1 are listed as 'sequencing ongoing', it is certain that the analysis of the genomic data is already well underway.

## 2.3.2 *Evolutionary genomics of lactic acid bacteria*

As for other groups of micro-organisms, the availability of the first genome sequences of lactic acid bacteria allows a re-evaluation of their evolutionary relationships in the kingdom of prokaryotes, as well as at the family-genus level. Dairy or food micro-organisms belong to the lineage of Gram-positive bacteria, the *Firmicutes,* both with high guanine and cytosine (G+C) content (e.g. *Bifidobacterium* spp. and *Propionibacterium* spp.), and low G+C content (e.g. *Lactobacillus* spp., *Lactococcus* spp., *Enterococcus* spp., *Pediococcus* spp. and *Streptococcus* spp.). Genome trees constructed with five different approaches:

- Presence–absence of genomes in clusters of orthologous genes;
- Conservation of gene order among genomes;
- Distribution of percent identity between apparent orthologues;
- Sequence conservation in concatenated alignment of ribosomal proteins; and
- Comparative analysis of multiple trees inferred from representative protein families.

These suggest that the traditional association of high G+C Gram-positive bacteria (*Actinomycetes*) with low G+C Gram-positive bacteria (the *Bacillus-Clostridium* group) is not supported. This leads to the conclusion that the *Firmicutes* clade, as presently constituted, does not exist, and it has been proposed that a new clade should replace the existing one, which includes low G+C Gram-positive bacteria associated with the *Actinomycetes-Deinococcales-Cyanobacteria* group (Wolf *et al.*, 2001). A more recent report by the same group of researchers (Wolf *et al.*, 2002) takes into account the results of several other genome-tree analyses, which confirmed the relative unrelatedness of high G+C and low G+C Gram-positive bacteria; the latter is associated with mycoplasma, and forming an unresolved clade with *Actinobacteria*, *Cyanobacteria* and *Thermus-Deinococcus* group (Fig. 2.1).

At the genus level, the comparative analysis of dairy/food-related bacteria still suffers from the low number of genomes currently available. The trends emerging from the first available sequences are the minimisation of genomes due to the adaptation to a nutritionally complex environment via subsequent elimination of unnecessary systems, and their ability to undergo gene transfer through conjugation or transduction mediated by mobile genetic elements and phages (Klaenhammer *et al.*, 2002). Researchers investigating lactic acid bacteria taxonomy know that phylogenetic and metabolic properties of these organisms are often in contradiction with each other; however, the evaluation of differences in genome content and expression will, surely, clear some debated taxonomic points either through improved multiple alignment analysis or the inclusion of additional data, such as gene content and gene order conservation (Coenye & Vandamme, 2003; Siezen *et al.*, 2004). Preliminary analyses show that phylogenetically related lactobacilli, such as *Lactobacillus gasseri*, *Lb. johnsonii* and *Lactobacillus acidophilus*, have highly similar genome content, gene sequences and genome organisation, which are not maintained with another *Lactobacillus* species, i.e. *Lb. plantarum* (Siezen *et al.*, 2004). Since the natural diversity of lactic acid bacteria is expected to be enormous, the availability of genome sequences will make it possible to analyse the genetic diversity of strains on the basis of DNA-DNA hybridisation on micro-array slides.

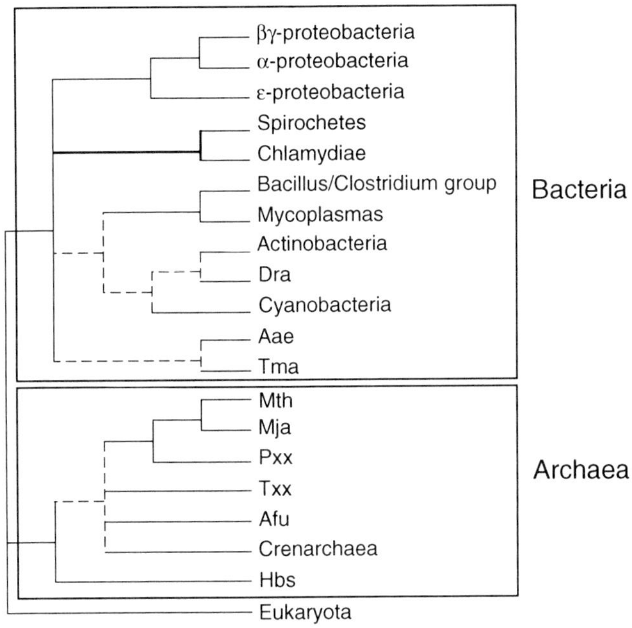


**Fig. 2.1** The tentative consensus of genome trees of prokaryotes. The black branches are conventional clades reproduced by genome-tree approaches; the heavy line branches are new clades that we consider to be firmly established by genome trees; broken line branches are tentative clades suggested by genome trees. Species/clades abbreviation are as follows:
(a) Bacteria: βγ-proteobacteria (*Haemophilus influenzae*, *Pasteurella multocida*, *Vibrio cholerae*, *Yersinia pestis*, *Salmonella typhimurium* LT2, *Escherichia coli* K12, *Escherichia coli* O157:H7 EDL933, *Escherichia coli* O157:H7, *Pseudomonas aeruginosa*, *Ralstonia solanacearum*, *Xylella fastidiosa* 9a5c, *Neisseria meningitidis* MC58, and *Neisseria meningitidis* Z2491), α-proteobacteria (*Rickettsia prowazekii* and *Rickettsia conorii*), ε-proteobacteria (*Helicobacter pylori* 26695 and *Helicobacter pylori* J99), Spirochetes (*Treponema pallidum* and *Borrelia burgdorferi*), Chlamydiae (*Chlamydia trachomatis* and *Chlamydophila pneumoniae*), Bacillus/Clostridium group (low G+C Gram-positive - *Clostridium acetobutylicum*, *Bacillus subtilis*, *Bacillus halodurans*, *Staphylococcus aureus* N315, *Lactococcus lactis*, *Streptococcus pyogenes* M1 GAS and *Streptococcus pneumoniae* TIGR4), Mycoplasmas (*Mycoplasma pulmonis*, *Mycoplasma pneumoniae*, *Mycoplasma genitalium*, *Ureaplasma urealyticum*), Actinobacteria (high G+C Gram-positive - *Mycobacterium tuberculosis* H37Rv, *Mycobacterium tuberculosis* CDC1551, *Mycobacterium leprae*), Dra (*Deinococcus radiodurans*), Cyanobacteria (*Synechocystis* spp. and *Nostoc* spp. PCC 7120), Aae (*Aquifex aeolicus*), and Tma (*Thermotoga maritime*).
(b) Archaea: Mth (*Methanothermobacter thermautotrophicus*), Mja (*Methanococcus jannaschii*), Pxx (*Pyrococcus horikoshii* and *Pyrococcus abyssi*), Txx (*Thermoplasmaacidophilum* and *Thermoplasma volcanium*), Afu (*Archaeoglobus fulgidus*), Crenarchaea (*Sulfolobus solfataricus* and *Aeropyrum pernix*), and Hbs (*Halobacterium* spp. NRC-1).
(c) Eukaryota: *Saccharomyces cerevisiae*.
After Wolf *et al.* (2002), data by permission from Elsevier.

### 2.3.3 *Complete genome sequences of potentially probiotic micro-organisms*

At present, the only genomes of potentially probiotic dairy/food-related bacteria, which are publicly available, are those of *B. longum* NCC2705, *Lb. johnsonii* NCC533 and *Lb.*

*plantarum* WCFS1. In addition, the complete genome sequence of *Lac. lactis* subsp. *lactis* strain IL1403 is also available; this species is commonly used as part of the microflora of probiotic dairy products. These findings will be briefly summarised, emphasising the useful information gained from these genome projects. In addition, an overview of what has been learned to date is shown in Table 2.2.

### Lactococcus lactis *subsp.* lactis *IL1403*

The first publicly available genome sequence of a lactic acid bacterium was that of *Lac. lactis* subsp. *lactis* strain IL1403 (Bolotin *et al.*, 2001); this was a plasmid-free laboratory strain derived from a cheese starter culture by plasmid curing (Chopin *et al.*, 1984). As the parental strain contains several plasmids, the genome sequence is actually incomplete. This is especially relevant because plasmids encode several technologically important traits, including the major cell wall protease responsible for the initial degradation of casein and the genes for lactose metabolism.

The genome sequence of *Lac. lactis* spp. clearly illustrates the effect of a change in the environment on genome structure. In nature, *Lac. lactis* spp. is found on plant or animal surfaces, while the strain that was sequenced is derived from a strain with a long history of use in dairy fermentations. As milk is a relatively rich environment, a number of biosynthetic capabilities are in the process of being lost. For example, even though genes for the biosynthesis of all 20 amino acids are present, dairy *Lactococcus* isolates are typically auxotrophic for, at least, six amino acids. Likewise, genes involved in the biosynthesis of riboflavin and folic acid are present, but the strains require these vitamins for growth in a chemically defined medium (Jensen & Hammer, 1993). Purine and pyrimidine biosyntheses are, however, conserved, and this is consistent with the fact that milk does not contain enough of these compounds to support the growth of *Lac. lactis* spp. auxotrophs (Dickely *et al.*, 1995).

Further evidence of the 'plastic' nature of the *Lactococcus* genome is the fact that strain IL1403 contains genes derived from six prophages and 43 insertion sequences (IS-elements) as well as genes that could potentially make the strain genetically competent – that is, able to take up DNA from the environment. The genome sequence of strain IL1403 shows evidence of incorporation of foreign DNA as well as the potential for rapid genome rearrangements mediated by multiple copies of similar IS elements. Indeed, Bolotin *et al.* (2001) postulate, based on the non-random distribution of IS elements, that a large portion of strain IL1403 genome was recently acquired from another *Lactococcus* strain.

The main industrial use of *Lactococcus* spp. involves the production of large amounts of lactic acid from lactose via homolactic fermentation. In addition, small amounts of flavour compounds can also be produced, for example, by heterolactic fermentation. As expected, the entire pathway from glucose to lactic acid is present. In addition, the genes required for the production of acetaldehyde, diacetyl (spontaneously forms from $\alpha$-acetolactate) and ethanol were also detected.

*Lactococcus lactis* spp. is not particularly adapted for life in the GI tract; rather, it is used to impart the desired organoleptic properties to the fermented dairy products, which are the most common carrier for the consumption of probiotic organisms. The survival of *Lactococcus* spp. and passage through the GI tract has, however, been demonstrated by Klijn *et al.* (1995).

**Table 2.2** Selected features inferred from the genomes of four species[1] relevant to probiotic dairy products.

| | *B. longum* NCC2705 | *Lb. johnsonii* NCC533 | *Lb. plantarum* WCFS1 | *Lac. lactis* subsp. *lactis* IL1403 |
|---|---|---|---|---|
| Known habitat(s) | Lower GI tract | Upper GI tract | Food, feed, GI tract | Dairy products |
| Genome size | 2 256 646 bp[2] | 1 992 676 bp | 3 308 274 bp | 2 365 589 bp |
| Predicted proteins | 1 730 | 1 821 | 3 052 | 2 310 |
| Biosynthesis of amino acids | All | None | All except branched chain amino acids | Has all the required genes but many are non-functional |
| Biosynthesis of vitamins | Folic acid, thiamine and nicotinate | None | Folic acid, pyridoxine, biotin, pantothenate, nicotinate, riboflavin | Genes for folic acid, menaquinone, riboflavin, and thioredoxin biosynthesis are present but may be inactive |
| Biosynthesis of nucleosides | Purines and pyrimidines | Pyrimidines only | Purines and pyrimidines | Purines and pyrimidines |
| Special features | Aerotolerance, breakdown of oligosaccharides, cell adhesion and a protease inhibitor | Many transport systems, no digestion of oligosaccharides, cell adhesion, bile tolerance and IgA protease | Versatile metabolism, breakdown of oligosaccharides, multiple stress response systems, and cell adhesion | Genome is evolving as strain adapts to the specific environment in a dairy fermentation |

[1] *B.* = *Bifidobacterium*; *Lb.* = *Lactobacillus*; *Lac.* = *Lactococcus*.
[2] bp = base pairs.

## Bifdobacterium longum *NCC2705*

The complete genome sequence of *B. longum* strain NCC2705 has revealed a number of features, which allow this strain to survive in the lower human GI tract (Schell *et al.*, 2002). While strain NCC2705 has not been reported to have probiotic properties, other members of genus *Bifidobacterium* have been attributed beneficial properties such as immunomodulation, prevention of diarrhoea, maintenance of GI tract balance, and amelioration of lactose intolerance. *B. longum* is a strict fermentative anaerobe and, as such, has no genes for aerobic or anaerobic respiratory components. It is moderately aerotolerant, and has homologues of genes, such as those encoding thiol peroxidase, alkyl hydroperoxide reductase and peptide methionine sulphoxide reductase, which could repair oxidative damage to proteins or lipids. These properties allow brief exposure to air, which would be encountered during the movement of this organism from one anaerobic environment to another.

While *B. longum* has the genes required for the fermentation of many monosaccharides, these are not expected to be abundant in the lower GI tract. Instead, this environment is expected to be rich in complex carbohydrates and, indeed, strain NCC2705 has a vast array of genes predicted to be involved in carbohydrate transport and oligosaccharide hydrolysis. In addition, a diversity of regulatory genes that control the expression of these genes was detected. In fact, more than 8% of the genome is associated with oligosaccharide metabolism. Fermentation of monosaccharides is necessary once the oligosaccharides are broken down to their monosaccharide subunits. *B. longum* NCC2705 has biosynthetic genes for all amino acids, pyrimidines, and purines, and homologues of most of the genes required for biosythesis of thiamine, folic acid and nicotinate. The biosynthetic pathways for riboflavin, biotin, cobalamin, pantothenate and pyridoxine are absent. Thus, this species would also be expected to grow outside the human GI tract, provided that suitable energy and vitamin sources were present, and that the oxygen tension was acceptably low. Additional features that could contribute to the ability of *Bifidobacterium* spp. to live in the GI tract include a potential fimbria structure and a teichoic acid-linked surface polysaccharide; a number of proteins, which would allow digestion of various natural polymers, and a potential serine protease inhibitor could be involved in the immunomodulatory activity of *Bifidobacterium* spp.

## Lactobacillus johnsonii *NCC533*

*Lb. johnsonii* resides in the upper regions of the human GI tract whereas *Bifidobacterium* spp. mainly colonise the large intestine. An analysis of the complete genome of strain NCC533 has revealed a number of properties, which reflect adaptation to the environment where this bacterium is typically found in the caecum at the ileal-colonic junction (Pridmore *et al.*, 2004). Strain NCC533 has been extensively studied, and has been shown to attach to epithelial cells, to modulate the host's immune system, and to inhibit pathogens. Based on the genome sequence, *Lb. johnsonii* is predicted to have a strict anaerobic metabolism, fermenting a variety of disaccharides and hexoses. Enzymes, such as xylanases and amylases for the digestion of larger polysaccharides, were not detected presumably because the upper GI tract contains sufficient mono- and disaccharides to fulfil the nutritional needs of *Lb. johnsonii*. In contrast to *B. longum*, *Lb. johnsonii* lacks the genes responsible for the *de novo* synthesis of most amino acids, although some interconversions are possible. Genes for the assimilation of ammonium and sulphur were not identified. Homologues for enzymes for the biosynthesis of vitamins, such as thiamine, nicotinate, riboflavin, biotin, cobalamin, pantothenate and pyridoxine, were not detected. The *de novo* synthesis of pyrimidines appears to be possible, while purine biosynthesis is not. Thus, *Lb. johnsonii* is restricted to environments such as the human upper GI tract, which are rich in vitamins, purines, amino acids and peptides. To facilitate the accumulation of the needed components from the environment, *Lb. johnsonii* has a wide variety of transporters, peptidases and proteases. *Lb. johnsonii* is highly specialised, and typically is not found outside the GI tract. A number of features of the strain NCC533 genome are expected to contribute to its probiotic properties. These include several genes, which encode proteins that are predicted to be attached to the outer surface of the cell. Among these are a putative lipoprotein, a mucus-binding protein, and a protein predicted to bind specifically to mucin and fimbriae, all of which are expected to enable the adhesion of the bacterial cell to cells in the GI tract. Three genes encoding bile salt hydrolases and genes for bile transport were also identified. These genes are expected to allow the cell to evade bile toxicity. An IgA protease may be involved in the interactions with the host immune system. Strain NCC533 has genes for the production of a bacteriocin (lactacin F), but one of these genes is inactivated by the insertion of an IS-element so it is unlikely that this strain actually produces lactacin F whereas other *Lb. johnsonii* strains might possess this capability.

### Lactobacillus plantarum *WCFS1*

*Lb. plantarum* is found in a wide variety of ecological niches including food, feed and the human GI tract. Some strains are sold as probiotics, and Kleerebezem *et al.* (2003) have determined the complete genome sequence of *Lb. plantarum* strain WCFS1 – that is, a human isolate which is not currently used as a probiotic. The versatility of *Lb. plantarum* is illustrated by the large genome size, which is nearly 50% greater than that of any other lactic acid bacterium, and has a very broad metabolic capacity. *Lb. plantarum* is also able to utilise a wide variety of carbon sources. This property results from a large number of genes involved in sugar transport and utilisation, and a versatile pyruvate metabolism, which can potentially produce both D- and L-lactate, formate, acetate, ethanol, acetoin and 2,3-butanediol. Strain WCFS1 has biosynthetic genes for most amino acids, but is unable to synthesise the branched-chain amino acids leucine, isoleucine and valine. In addition, two peptide uptake systems (multiple transporters for branched-chain amino acids and a broad array of peptidases) allow this strain to utilise exogenous amino acid sources. Genes involved in the biosynthesis of folic acid, pantothenate, biotin, pyridoxine, riboflavin and nicotinate were identified, again indicating the biosynthetic versatility of *Lb. plantarum*. The ability of this organism to live in a variety of niches is enhanced by a number of stress response systems, including a heat-shock response, a cold-shock response, an acid-stress system, alkaline-shock proteins, oxidative stress-related proteins, and systems for osmoprotection. In addition, a large number of regulatory genes ensure rapid adaptation to new environments. Furthermore, a number of extracellular proteins are predicted to be involved in adhesion due to homology to proteins involved in mucin binding, fibronectin binding and aggregation. Thus, *Lb. plantarum* has the ability to thrive both inside and outside the GI tract of the host.

### *Lactobacillus acidophilus* NCFM

The complete genome sequence of *Lb. acidophilus* NCFM has been determined, but the results obtained are not yet publicly available, though some elements of the analysis of these data have been published. These include: (a) genes involved in the digestion of fructo-oligosaccharides (Barrangou *et al.*, 2003); (b) genes that tolerate acidic conditions (Azcarate-Peril *et al.*, 2004); (c) and genes affecting the cell surface, cell division and adhesion to Caco-2 cells (Altermann *et al.*, 2004). Ultimately, these types of studies will help to elucidate the probiotic properties of *Lb. acidophilus*.

## 2.3.4 *Metagenomics*

Metagenomics has been defined as the culture-independent genomic analysis of microbial communities (Schloss & Handelsman, 2003). The name is derived from the statistical concept of meta-analysis (combining data from several independent studies to allow an overall analysis), and genomics as described herein. Metagenomics is a powerful technique for analysing micro-organisms in complex communities, and has the advantage that culturing of the organisms under laboratory conditions is not required. Typically, an environmental sample is collected, the entire microbial population is obtained, and total DNA is extracted and cloned to produce a metagenomic library. A large number of clones in this library are sequenced and the subsequent, *in silico* analysis provides a great deal of insight into the overall genetic and metabolic capabilities of the microbial population. Ultimately, by sequencing enough clones and using appropriately powerful software, it may become possible to assemble the complete genome sequence of the members of the consortium under study.

### *Metagenomic analyses*

*Bacteria inhabiting the human GI tract*
Clearly, a metagenomic analysis of the microbial flora of the human GI tract would provide a better understanding of this complex environment, and its influence on the health and well-being of the human host as well as provide a great deal of insight into the potential for probiotic foods to enhance these. Indeed, metagenomic projects have been initiated for the human oral cavity and the GI tract (Nelson, 2003), but results from these immense projects have, however, not yet become publicly available.

*Bacteriophages inhabiting the human GI tract*
In addition to the bacteria that are resident in the human GI tract, a variety of bacteriophages (i.e. viruses) are present. These can attack and kill the various bacteria that are present, and potentially change the environmental balance by favouring some bacteria over others. The characterisation of this viral community has been slow because growth of the bacteriophages in the laboratory requires growth of the host micro-organism. Since the majority of bacteria in the human GI tract cannot yet be cultured in synthetic media, characterisation of the viral community has been impossible. This inability to grow the bacteriophages *in vitro* can be overcome by extracting the bacteriophages from faecal matter, and characterising the total population by metagenomic analysis (Breitbart *et al.*, 2003). The results of such a study suggest that the viral community from a single individual contains more than 1000 bacteriophage genotypes. These are expected to have a profound effect on the population

dynamics of the bacteria in the GI tract, and could have important implications in maintaining the normal function and microbial balance in the GI tract. Understanding these effects could lead to the development of more effective probiotics. Indeed, bacteriophage-containing tablets and other preparations have been used for decades as prophylactic treatment to prevent GI tract disorders, especially in the former USSR and central Asia (Stone, 2002). It is likely that bacteriophages will also be incorporated into future probiotic products for use in the rest of the world.

## 2.4 Future perspectives

### 2.4.1 *Nutrigenomics*

A working draft of the human genome sequence was finished in 2000, and published in 2001. Currently, international efforts are under way to convert the draft sequence to a complete genome sequence; however, the main differences are the accuracy (less than one error in 10 000 base pairs) and degree of completeness (the only remaining gaps are in regions that cannot be reliably sequenced using today's technology). Whilst knowing a complete genome sequence will allow a greater understanding of human physiology and disease, impressive results have already been obtained using the draft genome sequence.

One specialised use of the human genome sequence will be to obtain an understanding of the interaction between the food we eat and the expression of genes in various types of cells and tissues. This line of research has been named nutrigenomics, and is a blend of the scientific disciplines of nutrition, physiology, molecular biology, genomics and bioinformatics. Nutrigenomics research is expected to show a very rapid growth because it fulfils the consumer's wish for safer food with better nutritional quality, and also fulfils the government's wish to keep healthcare budgets under control. Indeed, a number of institutions specialising in nutrigenomics research (with the associated web sites) have already been established in Europe (e.g. http://www.nutrigenomics.nl), North America (e.g. http://nutrigenomics.ucdavis.edu), and Asia/Pacific (e.g. http://www.nutrigenomics.org.nz). The old adage, "you are what you eat", accurately reflects the philosophy behind nutrigenomics.

Many potential applications for nutrigenomics research exist (Roberts *et al.*, 2001), and these will lead to a shift in the healthcare systems from the treatment of illnesses, especially those related to ageing and nutrition, to their prevention – primarily via informed and directed food choices. Determining the relationship between an individual's genetic make-up and their response to specific foods may allow the use of diet to optimise health in ways that previously were inconceivable. This could include the use of specific probiotic products to achieve a specific benefit. In the more distant future, probiotic products could be directed towards a specific subset of the population which, based on their genetic composition, will obtain a direct benefit from their consumption. A good example might be a fermented milk product containing a probiotic strain with anti-hypertensive properties; this would be directed towards people with high blood pressure or who had a genetic make-up that predisposed them to become hypertensive later in life.

### 2.4.2 *Mode of action of probiotics*

Many of the health benefits ascribed to probiotics are expected to be based on the modulation of human gene expression. These include immune modulation, anti-cancer effects, anti-hypertensive effects, lowering of blood cholesterol, the relief of symptoms of irritable bowel syndrome, and many others (Ouwehand *et al.*, 2002). The confirmation of which genes are affected – and to what extent – would go a long way towards determining the mode of action of various probiotics. Whilst performing the required experiments *in situ* would be overly invasive due to the need to collect cells that have been affected by the probiotic, valuable information may be derived by using transcriptomics on human cells grown in tissue culture in the presence or absence of a specific probiotic strain. It is also anticipated that the presence of human cells may affect gene expression of the probiotic strain, which could likewise be studied *in vitro* using transcriptomics. Ultimately, these types of studies could reveal the mode of action of a specific strain, and allow for the development of more effective probiotic products.

Another class of probiotic effect includes stabilisation of the gut microflora, and competitive exclusion of pathogens and other harmful microbes. This typically involves interactions between members of the microflora. Current studies are being focused on the effect of probiotic strains on the composition of the gut microflora, although it is anticipated that future studies will examine the effect of probiotic strains on the metabolic activity of the gut microflora.

The genomics of the gut microbiome, in combination with nutrigenomic methods, will ultimately provide a clear understanding of what exactly happens when an individual ingests a probiotic product. However, due to the immense amount of information that needs to be collected, and the knowledge that must be developed, this thorough understanding is still many years away.

### 2.4.3 *Development of new probiotics*

Once the mode of action of a specific probiotic strain is understood, the development of new strains will be simplified. Specific metabolic activities that are responsible for the probiotic effect will be identified, and screening for strains with the correct combination of activities will be possible. Alternatively, modern microbiological methods – including the use of recombinant DNA technology – will allow the production of strains with exactly the correct combination of properties. If a strain with a well-documented and well-understood probiotic effect is produced, the benefit(s) to the consumer would clearly outweigh any hypothetical risks in the use of gene technology, and consumer acceptance could rapidly occur.

An improved understanding of secretion processes, stress response and regulatory mechanisms will allow the rational design of new probiotic strains, especially when combined with the nutrigenomics concept. Additional benefits could include the use of probiotic strains to deliver bioactive molecules to precise locations in the GI tract. This understanding will also allow an improved industrial production of probiotic food ingredients, the pre-conditioning of strains to be used in industrial fermentation, and also development of food-grade genetically engineered micro-organisms.

## 2.5 Conclusions

Genomics is the study of an organism's genome, and an understanding of the gene content. Following a slow start during the 1970s and 1980s, genomics and the analyses of whole genomes accelerated during the 1990s as DNA sequencing became automated, and powerful computers enabled a rapid and more systematic manipulation of information. Today, it is possible to create a draft version of an entire bacterial genome in less than one week, although in practice it takes considerably longer. Initially, the sequencing of whole genomes was carried out by public institutions, but more recently private companies have shown a massive interest in the field of genomics. Indeed, a growing genomics industry exists today, which allows the outsourcing of all aspects of genome sequence determination and analysis. The list of completely sequenced genomes is increasing almost daily, and in particular the number of completed bacterial genomes sequenced is rapidly increasing.

A typical bacterial genome contains around three million base pairs, and codes for approximately three thousand genes. However, having this entire genome sequence available is not at all the same as having an understanding of the corresponding organism. Even in bacteria, there is still a whole new world to discover, and often more than one-third of the genes predicted from a genome analysis cannot be assigned a function. Fortunately, the genomics field has also grown to cover a large variety of disciplines, tools and methods, which ultimately will allow the assignment of functions to the remainder of these genes.

The original suffix 'ome' has been extended to a multitude of other contexts. The more common 'omics', such as transcriptomics, proteomics and metabolomics, are now fully recognised disciplines within genomics. Together with comparative genome analyses, meta-genomics and many other gene technologies, new bioinformatic tools will change our understanding of various organisms, their evolution and physiology. One limiting factor is the availability of whole genome sequences. This chapter has focused on the probiotic bacteria and their genomes. Due to the great commercial interest, the number of publicly available genomes of probiotic micro-organisms is limited to three at present (*B. longum* NCC2705, *Lb. johnsonii* NCC533, and *Lb. plantarum* WCFS1); however, it is anticipated that this list will grow slowly in the coming years.

In spite of the small number of sequences available, there is no doubt that the new genomic era will revolutionise the functional food market, even during the present decade. Genomics provides not only a unique identification for documentation of the bacterial strains, but also the right tools to uncover enhanced production methods and beneficial properties of the probiotic bacteria. Furthermore, analysis of the interaction with the human genome will allow elucidation of the modes of action of probiotic products as well as the development of future functional food products. This will also facilitate, in particular via the nutrigenomics approach, the development of new, improved and even safer probiotic products, which can be tailor-made to function towards one specific health condition, to be optimised for individual persons, to develop new food products, or to solve health concerns that are currently not considered to benefit from prebiotics.

## References

Altermann, E., Buck, L., Cano, R. & Klaenhammer, T. (2004) Identification and phenotypic characterization of the cell-division protein CdpA. *Gene*, **342**, 189–197.

Azcarate-Peril, A., Altermann, E., Hoover-Fitzula, R., Cano, R. & Klaenhammer, T. (2004) Identification and inactivation of genetic loci involved with *Lactobacillus acidophilus* acid tolerance. *Applied and Environmental Microbiology*, **70**, 5315–5322.

Barrangou, R., Altermann, E., Hutkins, R., Cano, R. & Klaenhammer, T. (2003) Functional and comparative genomic analyses of an operon involved in fructooligosaccharide utilization by *Lactobacillus acidophilus*. *Proceedings of the National Academy of Sciences USA*, **100**, 8957–8962.
Bolotin, A., Wincker, P., Mauger, S., Jaillon, O., Malarme, K., Weissenbach, J., Ehrlich, S.D. & Sorokin, A. (2001) The complete genome sequence of the lactic acid bacterium *Lactococcus lactis* ssp. *lactis* IL1403. *Genome Research*, **11**, 731–753.
Breitbart, M., Hewson, I., Felts, B., Mahaffy, J., Nulton, J., Salamon, P. & Rohwer, F. (2003) Metagenomic analysis of an uncultured viral community from human feces. *Journal of Bacteriology*, **185**, 6220–6223.
Bron, P., Hoffer, S., van Swam, I., de Vos, W. & Kleerebezem, M. (2004a) Selection and characterization of conditionally active promoters in *Lactobacillus plantarum*, using alanine racemase as a promoter probe. *Applied and Environmental Microbiology*, **70**, 310–317.
Bron, P., Marco, M., Hoffer, S., van Mullekom, E., de Vos, W. & Kleerebezem, M. (2004b) Genetic characterization of the bile salt response in *Lactobacillus plantarum* and analysis of responsive promoters *in vitro* and *in situ* in the gastrointestinal tract. *Journal of Bacteriology*, **186**, 7829–7835.
Chopin, A., Chopin, M.-C., Moillo-Batt, A. & Langella, P. (1984) Two plasmid-determined restriction and modification systems in *Streptococcus lactis*. *Plasmid*, **11**, 260–263.
Coenye, T. & Vandamme, P. (2003) Extracting phylogenetic information from whole genome sequencing projects: the lactic acid bacteria as a test case. *Microbiology*, **149**, 3507–3517.
Davidson, B., Kordias, N., Dobos, M. & Hillier, A. (1996) Genomic organization of lactic acid bacteria. *Antonie van Leeuwenhoek*, **70**, 161–183.
Danielsen, M. (2002) Characterization of the tetracycline resistance plasmid pMD5057 from *Lactobacillus plantarum* 5057 reveals a composite structure. *Plasmid*, **48**, 98–103.
Dickely, F., Nilsson, D., Hansen, E. & Johansen, E. (1995) Isolation of *Lactococcus lactis* nonsense suppressors and construction of a food-grade cloning vector. *Molecular Microbiology*, **15**, 839–847.
Fiers, W., Contreras, R., Duerinck, F., Haegeman, G., Iserentant, D., Merregaert, J., Jou, W., Molemans, F., Raeymaekers, A., van den Berghe, A., Volckaert, G. & Ysebaert, M. (1976) Complete nucleotide sequence of bacteriophage MS2 RNA: primary and secondary structure of the replicase gene. *Nature*, **260**, 500–507.
Fitzgerald, J., Sturdevant, D., Mackie, S., Gill, S. & Musser, J. (2001) Evolutionary genomics of *Staphylococcus aureus*: insights into the origin of methicillin-resistant strains and the toxic shock syndrome epidemic. *Proceedings of the National Academy of Sciences USA*, **98**, 8821–8826.
Fjose, A., Ellingsen, S., Wargelius, A. & Seo, H. (2001) RNA interference: mechanisms and applications. *Biotechnology Annual Review*, **7**, 31–57.
Fleischmann, R., Adams, M., White, O., Clayton, R., Kirkness, E., Kerlavage, A., Bult, C., Tomb, J., Dougherty, B., Merrick, J., McKenney, K., Sutton, G., FitzHugh, W., Fields, C., Gocayne, J., Scott, J., Shirley, R., Liu, L., Glodek, A., Kelley, J., Weidman, J., Philips, C., Spriggs, T., Hedblom, E., Cotton, M., Utterback, T., Hanna, M., Nguyen, D., Saudek, M., Brandon, R., Fine, L., Fritchman, J., Fuhrmann, J., Geohagen, N., Gnehm, C., McDonald, L., Small, K., Fraser, C., Hamilton, H. & Venter, C. (1995) Whole-genome random sequencing and assembly of *Haemophilus influenzae* Rd. *Science*, **269**, 469–512.
Ge, H., Walhout, A. & Vidal, M. (2003) Integrating 'omic' information: a bridge between genomics and systems biology. *Trends in Genetics*, **10**, 551–560.
Germond, J.-E., Lapierre, L., Delley, M., Mollet, B., Felis, G. E. & Dellaglio, F. (2003) Evolution of the bacterial species *Lactobacillus delbrueckii*: a partial genomic study with reflections on prokaryotic species concept. *Molecular Biology and Evolution*, **20**, 93–104.
Harmsen, H. & Welling, G. (2002) Fluorescence *in situ* hybridization as a tool in intestinal bacteriology. *Probiotics and Prebiotics: where are we going?*, (ed. G. Tannock), pp. 41–58, Caister Academic Press, Wymondham.
Holden, D. W. & Hensel. M. (1998) Signature tagged mutagenesis. *Methods in Microbiology*, **27**, 359–370.
Jensen, P. & Hammer, K. (1993) Minimal requirements for exponential growth of *Lactococcus lactis*. *Applied and Environmental Microbiology*, **59**, 4363–4366.
Johansen, E. (1999) Genetic engineering (b) Modification of bacteria. *Encyclopedia of Food Microbiology*, (eds. R. Robinson, C. Batt and P. Patel), pp. 917–921, Academic Press, London.
Kanehisa, M. & Bork, P. (2003) Bioinformatics in the post-sequence era. *Nature Genetics*, **33**, 305–310.
Klaenhammer, T., Altermann, E., Arigoni, F., Bolotin, A., Breidt, F., Broadbent, J., Cano, R., Chaillou, S., Deutscher, J., Gasson, M., van de Guchte, M., Guzzo, J., Hartke, A., Hawkins, T., Hols, P., Hutkins, R., Kleerebezem, M., Kok, J., Kuipers, O., Lubbers, M., Maguin, E., McKay, L., Mills, D., Nauta, A., Overbeek, R., Pel, H., Pridmore,

D., Saier, M., van Sinderen, D., Sorokin, A., Steele, J., O'Sullivan, D., de Vos, W., Weimer, B., Zagorec, M. & Siezen R. (2002) Discovering lactic acid bacteria by genomics. *Antonie van Leeuwenhoek*, **82**, 29–58.

Klijn, N., Weerkamp, A. & de Vos, W. (1995) Genetic marking of *Lactococcus lactis* shows its survival in the human gastrointestinal tract. *Applied and Environmental Microbiology*, **61**, 2771–2774.

Kleerebezem, M., Boekhorst, J., van Kranenburg, R., Molenaar, D., Kuipers, O., Leer, R., Tarchini, R., Peters, S., Sandbrink, H., Fiers, M., Stiekema, W., Lankhorst, R., Bron, P., Hoffer, S., Groot, M., Kerkhoven, R., de Vries, M., Ursing, B., de Vos, W. & Siezen, R. (2003) Complete genome sequence of *Lactobacillus plantarum* WCFS1. *Proceedings of the National Academy of Sciences USA*, **100**, 1990–1995.

Masco, L., Ventura, M., Zink, R., Huys, G. & Swings, J. (2004) Polyphasic taxonomic analysis of *Bifidobacterium animalis* and *Bifidobacterium lactis* reveals relatedness at the subspecies level: reclassification *Bifidobacterium animalis* as *Bifidobacterium animalis* subsp. *animalis* subsp. nov. and *Bifidobacterium lactis* as *Bifidobacterium animalis* subsp. *lactis* subsp. nov. *International Journal of Systematic and Evolutionary Microbiology*, **54**, 1137–1143.

McAuliffe, O. & Klaenhammer, T. (2002) Genomic perspectives on probiotics and the gastrointestinal microflora. *Probiotics and Prebiotics: where are we going?*, (ed. G. Tannock), pp. 263–309, Caister Academic Press, Wymondham.

Mills, D.A. (2004) Fermentation technology: the lactic acid bacteria genome project. *Journal of Food Science*, **69**, FMS28–FMS31.

Morrow, B., Graham, J. & Curtiss III, R. (1999) Genomic subtractive hybridization and selective capture of transcribed sequences identify a novel *Salmonella typhimurium* fimbrial operon and putative transcriptional regulator that are absent from the *Salmonella typhi* genome. *Infection and Immunity*, **67**, 5106–5116.

Nelson, K. (2003) The future of microbial genomics. *Environmental Microbiology*, **5**, 1223–1225.

Ouwehand, A., Salminen, S. & Isolauri, E. (2002) Probiotics: an overview of beneficial effects. *Antonie van Leeuwenhoek*, **82**, 279–289.

Papin, J., Price, N., Wiback, S., Fell, D. & Palsson, B. (2003) Metabolic pathways in the post-genome era. *Trends in Biochemical Sciences*, **28**, 250–258.

Pridmore, R., Berger, B., Desiere, F., Vilanova, D., Barretto, C., Pittet, A., Zwahlen, M., Rouvet, M., Altermann, E., Barrangou, R., Mollet, B., Mercenier, A., Klaenhammer, T., Arigoni, F. & Schell, M. (2004) The genome sequence of the probiotic intestinal bacterium *Lactobacillus johnsonii* NCC533. *Proceedings of the National Academy of Sciences USA*, **101**, 2512–2517.

Roberts, M., Mutch, D. & German, J. (2001) Genomics: food and nutrition. *Current Opinion in Biotechnology* **12**, 516–522.

Schell, M., Karmirantzou, M., Snel, B., Vilanova, D., Berger, B., Pessi, G., Zwahlen, M., Desiere, F., Bork, P., Delley, M., Pridmore, D. & Arigoni, F. (2002) The genome sequence of *Bifidobacterium longum* reflects its adaptation to the human gastrointestinal tract. *Proceedings of the National Academy of Sciences USA*, **99**, 14422–14427.

Schloss, P. & Handelsman, J. (2003) Biotechnological prospects from metagenomics. *Current Opinion in Biotechnology*, **12**, 303–310.

Siezen, R., van Enckevort, F., Kleerebezem, M. & Teusink, B. (2004) Genome data mining of lactic acid bacteria: the impact of bioinformatics. *Current Opinion in Biotechnology*, **15**, 105–115.

Skovgaard, M., Jensen, L., Brunak, S., Ussery, D. & Krogh, A. (2001) On the total number of genes and their length distribution in complete microbial genomes. *Trends in Genetics*, **17**, 425–428.

Slauch, J. & Camilli, A. (2000) IVET and RIVET: use of gene fusions to identify bacterial virulence factors specifically induced in host tissues. *Methods in Enzymology*, **326**, 73–96.

Sorensen, K., Larsen, R., Kibenich, A, Junge, M. & Johansen, E. (2000) A food-grade cloning system for industrial strains of *Lactococcus lactis*. *Applied and Environmental Microbiology*, **66**, 1253–1258.

Stephens C. & Laub, M. (2003) Microbial genomics: all that you can't leave behind. *Current Biology*, **13**, 571–573.

Stone, R. (2002) Bacteriophage therapy. Stalin's forgotten cure. *Science*, **298**, 728–731.

Thomas, K. R. & Capecchi, M. R. (1987) Site-directed mutagenesis by gene targeting in mouse embryo-derived stem cells. *Cell*, **51**, 503–512.

Velculescu, V., Zhang, L., Vogelstein, B. & Kinzler, K. (1995) Serial analysis of gene expression. *Science*, **270**, 484–487.

de Vos, W. (2001) Advances in genomics for microbial food fermentations and safety. *Current Opinion in Biotechnology*, **12**, 493–498.

Weinstock, G., Sajs, D., Hardham, J. & Norris, S. (2000) From microbial sequence to application. *Research in Microbiology*, **151**, 151–158.

Wolf, Y., Rogozin, I., Grishin, N., Tatusov, R. & Koonin, E. (2001) Genome trees constructed using five different approaches suggest new major bacterial clades. *BMC Evolutionary Biology*, **1**, 8.

Wolf, Y., Rogozin, I., Grishin, N. & Koonin, E. (2002) Genome trees and the tree of life. *Trends in Genetics*, **18**, 472–479.

Zhang, C. & Kim, S. (2003) Overview of structural genomics: from structure to function. *Current Opinion in Chemical Biology*, **7**, 28–32.

# 3 Production and Maintenance of Viability of Probiotic Micro-organisms in Dairy Products

A.Y. Tamime, M. Saarela, A. Korslund Søndergaard, V.V. Mistry and N.P. Shah

## 3.1 Introduction

Probiotic foods, including dairy products, have been classically defined as 'foods containing live micro-organisms believed to actively enhance health by improving the balance of micro-flora in the gut'. Currently, they are defined as 'microbial cells preparations or components of microbial cells that have a beneficial effect on health and well being of the host' (see the review by Gardiner *et al.*, 2002a). Several preparations employing *Lactobacillus* spp., *Bifidobacterium* spp. and *Enterococcus* spp. are well established in the market, and foods containing probiotic bacteria have been marketed in Japan since the 1920s. Today, there has been a tremendous increase in the number of microbial species included in probiotic dairy products (e.g. pasteurised milk, ice-cream, fermented milks, cheeses and baby feed milk powder). However, fermented foods remain the main vehicle of administration of probiotic organisms (Tamime *et al.*, 1995; Belem, 1999; Lourens-Hattingh & Viljoen, 2001a; Stanton *et al.*, 2002; Salama, 2002; Hawrelak, 2002; Shah, 2004). Among the fermented milk products, yoghurt is by far the most important vehicle for the delivery of probiotic organisms.

A number of health benefits derived by the consumption of foods containing *Lactobacillus acidophilus* and bifidobacteria are well documented (for further details, refer to Chapter 8), and more than 100 probiotic fermented milk products are available worldwide. To provide health benefits, it is essential that products sold with any health claims meet the criterion of a minimum concentration of probiotic bacteria of $10^6$ colony forming units (cfu) $mL^{-1}$ or $g^{-1}$ at the expiry date, as the minimum therapeutic dose per day is suggested to be $10^8$–$10^9$ cfu $mL^{-1}$ (Shah, 2000). However, studies have demonstrated different viabilities of probiotic organisms in fermented milks, especially in yoghurt (Anonymous, 1992; Iwana *et al.*, 1993; Shah *et al.*, 1995; Collins *et al.*, 1998; Vinderola *et al.*, 2000a; El-Rahman, 2000; Shah & Ravula, 2000a; Collins, 2001; Lourens-Hattingh & Viljoen, 2002; Varga *et al.*, 2003; La Torre *et al.*, 2003). Several factors have been claimed to affect the viability of probiotic cultures in fermented milks, including the final acidity of the product, availability of nutrients, dissolved oxygen and oxygen permeation through the package. The stability of probiotic cultures has been a major issue for yoghurt manufacturers and consumers, and in this chapter the technical and scientific aspects of probiotic dairy products will be reviewed.

## 3.2 Probiotic micro-organisms

'Traditional' lactic acid bacteria that are normally used during the manufacture of fermented milks and cheese belong to the genera *Lactococcus*, *Leuconostoc*, *Pediococcus*, *Streptococcus* and *Lactobacillus*; the former two genera are mesophilic, while the latter types are ther-

…lue and white moulds are used in cheesemaking, and
…iili production. In addition, a wide range of yeasts have
…he properties of these products will not be covered in this
…d in a separate book (*Fermented Milks*) to be published
…repared on behalf of the Society of Dairy Technology
…mime, 2005). However, *Saccharomyces boulardii* is the
…lentified (Lourens-Hattingh & Viljoen, 2001a; Surawicz,
…ss, *S. boulardii*, *Pediococcus* and propionic acid bacteria
…ns, though none has yet been shown, and further clinical
… health benefits to humans.

### 3 …

The probiotic micro-organisms, which have been used in fermented and unfermented milk products including cheese, are shown in Table 3.1, together with their main metabolic products. This provides some information on their possible role in flavour production, but the traditional lactic acid bacteria (i.e. starter cultures) are mainly responsible for much of the flavour and aroma (Tamime *et al.*, 2005).

The pediococci, lactobacilli, enterococci and bifidobacteria that are used as probiotic micro-organisms do not use the tricarboxylic acid cycle when fermenting the milk, although some of its enzymes may be present; nor is there a cytochrome system for harnessing energy from electrons of NADH. Energy is largely obtained via substrate-level phosphorylation and the ATPase of the cytoplasmic membrane. Carbohydrate is metabolised either through homofermentative or heterofermentative metabolic pathways. However, bifidobacteria metabolise the lactose in milk via the heterolactic fermentation (Marshall & Tamime, 1997), and details of all these metabolic pathways have been recently reviewed by Tamime *et al.* (2005).

### 3.2.2 *Examples of commercial starter cultures blends*

Yoghurt is normally manufactured using *Streptococcus thermophilus* and *Lactobacillus delbrueckii* subsp. *bulgaricus* as the starter culture. Probiotic micro-organisms grow slowly in milk; hence, the yoghurt starter culture is added to enhance the fermentation process and

**Table 3.1** Some selected characteristics of potentially probiotic micro-organisms used in dairy foods and their principal metabolic products.

| Starter organism | Metabolic product | Lactose fermentation |
|---|---|---|
| I. Lactic acid bacteria | | |
| *Pediococcus acidilactici* | DL lactate | Homofermentative |
| *Lactobacillus acidophilus*, *gasseri*, *helveticus*, and *johnsonii* | DL lactate | Homofermentative |
| *Lactobacillus casei*, *reuteri*, *plantarum*, *rhamnosus* and *fermentum* | DL lactate | Heterofermentative |
| *Bifidobacterium adolescentis*, *animalis* subsp. *animalis*, *bifidum*, *breve*, *infantis*, *animalis* subsp. *lactis*, and *longum* | L(+) lactate, acetate | Heterofermentative |
| *Enterococcus faecium*, and *faecalis* | L(+) lactate | Homofermentative |
| II. Yeasts | | |
| *Saccharomyces boulardii* | ? Ethanol, $CO_2$ | |

Data adapted from Masco *et al.* (2004) and Tamime *et al.* (2005).

*Lb. acidophilus*, *Bifidobacterium* spp. and *Lactobacillus casei* are incorporated as dietary adjuncts (Minelli *et al.*, 2004; Saito, 2004; Leroy & de Vuyst, 2004). Fermented milk with only *Lb. acidophilus* or *Lb. acidophilus* and *Bifidobacterium* spp. (known as AB cultures), *Lb. acidophilus*, *Bifidobacterium* spp. and *Lb. casei* (known as ABC cultures; see also Maiocchi, 2001) or *Lb. acidophilus*, *Bifidobacterium* spp. and *S. thermophilus* (known as ABT cultures; see Martín-Diana *et al.*, 2003) could be manufactured; however, a longer incubation period and product quality (i.e. milder flavour) are the two main factors that are sacrificed when fermenting milk with only AB, ABC or some blends of ABT cultures (e.g. ABT-1 and ABT-2). Thus, the normal practice is to make yoghurt with *S. thermophilus*, *Lb. delbrueckii* subsp. *bulgaricus* and probiotic micro-organisms (AB or ABC cultures).

Although the main choice of any probiotic microbial strain to be used as a starter culture or a blend with a starter culture is based on the health aspects beneficial to humans (Gardiner *et al.*, 2002a), the following aspects must be considered from a starter culture manufacturer's point of view: (a) the ability of the probiotic micro-organisms to grow in a medium to increase the cell counts; (b) the robustness of the organism to withstand the freezing and drying stages of preservation; and (c) the tolerance to acidity of the stomach and bile salts during their passage in the gastrointestinal (GI) tract. Nevertheless, blending strains into a starter culture for the manufacture of fermented milk products including the probiotic types is a critical procedure, and the measures taken to ensure the desired attributes achieved in the final product have been detailed by Tamime *et al.* (2005). The choice and the ratio of the strains in the starter culture are the key factors in this respect, which may include considerations of the fermentation time, texture, mildness, sugar tolerance, and post-acidification profiles (see Fig. 3.1). However, an important feature for probiotic products is the level and stability of the probiotic strains, and the current trend towards longer shelf-life products may be up to 52 days in some regions. For probiotic starter cultures used in these regions, it is particularly important that the strains are stable in order to be able to claim the health benefit(s). The probiotic/lactic acid bacteria blends are tested, for example, in commercial starter culture laboratories for the stability of probiotic microfloras during 28 days of shelf-life at 8°C, and the primary aim is to have a minimum count of $1 \times 10^6$ cfu $g^{-1}$ at the end of the storage period. If the probiotic counts in the blend(s) do not achieve such levels, the blend(s) is re-formulated (A.K. Søndergaard, unpublished data).

In addition, the interactions between probiotic strains and traditional starter cultures are another aspect that must be considered to achieve a high viable count at the end of the shelf-life of the product. For example, some strains of *S. thermophilus* and *Lb. delbrueckii* subsp. *bulgaricus* may inhibit certain strains of probiotic bacteria during the fermentation and storage of the product. The probiotic micro-organisms are, particularly, influenced by other bacteria during long fermentation times; however, during a short fermentation time, the growth of most probiotic bacteria has hardly started and they do not seem to be affected. A positive interaction between some probiotic strains is also known, for example between *Bifidobacterium* spp. and *Lb. acidophilus*.

Some probiotic micro-organisms may influence the flavour of the fermented product. For example, *Bifidobacterium* spp. will, when present in high numbers, produce a noticeable amount of acetic acid during a long fermentation time (Mahdi *et al.*, 1990; La Torre *et al.*, 2003), whilst *Lb. acidophilus* will produce acetaldehyde and lactic acid, and contribute to the characteristic 'bio' yoghurt flavour. Highly proteolytic probiotic strains may produce peptides, which confer a cheesy flavour/taste to the fermented milk product (Rašić & Kurmann, 1983).

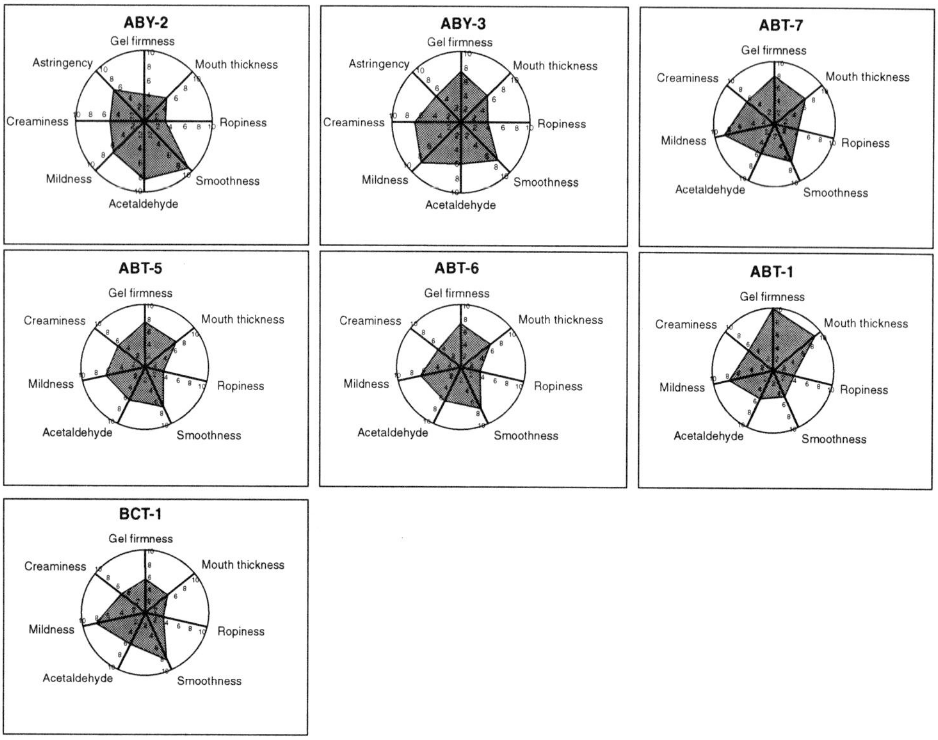


**Fig. 3.1** Sensory profiling of seven different blends of probiotic starter cultures. Note: It is evident that starter cultures ABY-2 and ABT-1 have the highest acetaldehyde content and firmness, respectively. The results are average of three replicates. ABY: A = *Lb. acidophilus* LA-5; B = *Bifidobacterium* spp. Bb-12; Y = yoghurt cultures or *S. thermophilus* and *Lb. delbrueckii* subsp. *bulgaricus*; BCT: B = *Bifidobacterium* spp. Bb-12; T = *S. Thermophilus*; C = *Lb. casei*. Data by permission of Chr. Hansen, Hørsholm, Denmark.

The main probiotic micro-organisms that are currently used belong to the genera *Lactobacillus* and *Bifidobacterium*. Other organisms that have been potentially identified as probiotic include: *Pediococcus acidilatici*, *Enterococcus* spp. and *S. boulardii*. Examples of these species that are currently in use worldwide are shown in Table 3.2, and the properties of *Enterococcus faecium* as a probiotic micro-organism in a fermented milk product called Gaio® have been reviewed by Bertolami and Farnworth (2003). In addition, *Lb. casei* Shirota is used as a single strain for the manufacture of Yakult (a Japanese fermented milk beverage), and the incubation period may last up to 5 days (Tamime & Marshall, 1997).

## 3.3 Economic value

For probiotic products, micro-organisms are selected for their health benefits. The incorporation of *Lb. acidophilus* and *Bifidobacterium* spp. into the yoghurt starter cultures produces a milk product of excellent 'therapeutic' value. Several preparations employing *Lb. acidophilus* and bifidobacteria are well established in the market. In France, products containing *Lb. acidophilus* and *Bifidobacterium* spp. have increased by approximately 300% to capture

**Table 3.2** The main species of micro-organisms that can potentially be used as probiotic cultures in dairy products.

| *Lactobacillus* spp. | Strain | *Bifidobacteriun* spp. | Strain | *Enterococcus* spp. | *Saccharomyces* spp. |
|---|---|---|---|---|---|
| *Lb. acidophilus* and/or *Lb. johnsonii* | LA-1/LA-5, La1, NCFM, DDS-1, SBT-2062 | *B. adolescentis* | | *E. faecalis* | *S. boulardii* |
| *Lb. delbrueckii* subsp. *bulgaricus* | Lb12 | *B. longum*[1] | BB-536, SBT-2928, UCC 35624, BB-46, BB-02 | *E. faecium* | |
| *Lb. delbrueckii* subsp. *lactis* | L1A | *B. breve* | Yakult | | |
| *Lb. casei* | Imunitass, Shirota | *B. bifidum* | Bb-11 | | |
| *Lb. plantarum* | 299v, Lp01 | *B. lactis*[2] | BB-12, Lafti™, B94, DR10, uR1, LMG 18314, DSM 10140, JCM 10602 | | |
| *Lb. rhamnosus* | GG,GR-1, LB21, 271 | *B. infantis* | Yakult, Danone, 01, 744 | | |
| *Lb. reuteri* | SD2112 or MM2 | *B. animalis*[3] | R101-8, LMG 10508, ATCC 25527, DSM 20104, JCM 1190 | | |
| *Lb. paracasei* | CRL 431 | | | | |
| *Lb. fermentum* | RC-14 | | | | |
| *Lb. helveticus* | B02 | | | | |
| *Lb. amylovorus* | | | | | |
| *Lb. crispatus* | | | | | |
| *Lb. gallinarum* | | | | | |
| *Lb. salivarius* | | | | | |

[1] The Chr. Hansen strain BB-02 is currently known as *Bifidobacterium longum* biotype *infantis*.
[2] Currently known as *Bifidobacterium animalis* subsp. *lactis* (Masco *et al.*, 2004).
[3] Currently known as *Bifidobacterium animalis* subsp. *lactis* (Masco *et al.*, 2004).
Adapted from Schillinger (1999), Lourens-Hattingh & Viljoen (2001b), Krishnakumar and Gordon (2001), Tamime (2002), Holm (2003) and Playne *et al.* (2003).

4% of total fresh milk sales (Hughes & Hoover, 1991). Presently, 11% of all yoghurt sold in France contains *Bifidobacterium* spp. In Europe, probiotic applications are mainly restricted to fermented milk products, and the economic value ($m) of such products in 1998 in selected countries were: 60 in Germany, 30 in The Netherlands, 28 in France, 24 in Spain, and 18 in the United Kingdom (Shortt, 1999). However, the economic values of probiotic yoghurt and probiotic drinking yoghurt sold in the United Kingdom in 2002 were £320.2 million and £68 million, respectively (Anonymous, 2003). Probiotic micro-organisms (i.e. included in sweet milk, whey drinks, cheeses and infant formulas) are manufactured in many countries, but no worldwide data are currently available regarding the economic value of these products (see also http://www.marketresearch.com/browse.asp?categoryid=510&SID=1498-1449-302165162-317076963 regarding the latest market value for probiotic products in Europe and the United States of America).

## 3.4 Types of probiotic dairy product

As mentioned elsewhere, a wide range of probiotic dairy products is available in different markets; typical examples include pasteurised milk, ice-cream, fermented milks, cheeses and baby feed milk powder (Tamime *et al.*, 1995; Playne *et al.*, 2003). The overall pattern of consumption of all types of fermented milks is steadily increasing in the majority of countries in the world, and this may be attributed the nutritional and health aspects associated with these products (IDF, 2002).

### 3.4.1 *Fermented milks and beverages*

A wide range of fermented milk products are made in different countries, but the classical example is yoghurt, which is manufactured as either set-, stirred- and/or drinking-types, with the products being flavoured by adding fruit preparations or fruit essences plus colouring matter (Tamime *et al.*, 1995; Tamime & Robinson, 1999; Bottazzi, 2002). Technical and scientific aspects regarding the production of yoghurt and other related products will be detailed by Tamime (2005). The manufacturing stages of probiotic yoghurt are very similar to 'classical' yoghurt, but natural/plain probiotic yoghurt is slightly sweeter in taste and the fermentation time is slightly longer compared with the 'classical' product and, as a consequence, the technical aspects of production will not be reviewed in this chapter (see Tamime, 2005).

Recently, traditional fermented milks, which have been developed to include probiotic micro-organisms, have been reviewed by Van de Water (2003), and some examples are as follows.

#### *Nordic cultured buttermilk (piimä, filmjölk) and kefir (drinking-type)*

Nordic cultured buttermilk is made by the microbial fermentation of pasteurised whole milk or skimmed milk. Typical starter cultures in fermented buttermilk are mesophilic lactic acid bacteria, such as *Lactococcus lactis* subsp. *cremoris*, *Lactococcus lactis* subsp. *lactis*, *Lactococcus lactis* subsp. *lactis* biovar. *diacetylactis* and *Leuconostoc mesenteroides* subsp. *cremoris*. The processed milk base is fermented at 20°C for ~20 h (final pH is 4.5–4.6), followed by stirring, cooling, flavouring (optional) and packaging (Mantere-Alhonen & Forsen, 1990; Kurmann *et al.*, 1992; Tamime & Marshall, 1997; Leporanta, 2001). The overall pattern of consumption of these products in different Nordic countries is shown in Table 3.3.

Originally, cultured buttermilk was consumed as a drink during meals and did not contain any flavouring. This is still the customary way of consumption in Finland; however, in Sweden, cultured buttermilk (filmjölk, fil) is often flavoured with fruits and berries, and consumed during breakfast (often with cereals) or as a snack in a manner similar to yoghurt.

Probiotic cultured buttermilk contains different micro-organisms beside the lactic starter cultures, such as *Lactobacillus rhamnosus* GG (piimä in Finland, hapupiim in Estonia, and syrnet melk in Norway), *Lb. rhamnosus* 271 (viilijuoma in Finland), *Lactobacillus reuteri*

**Table 3.3** Annual consumption (kg head$^{-1}$) of milk drinks and fermented dairy products in Nordic countries.

| | Country | | | | |
|---|---|---|---|---|---|
| Product | Denmark[1] | Finland[1] | Iceland[2] | Norway[2] | Sweden[2] |
| Yoghurt | 10.6 | 16.5 | 8.4 | 4.9 | 17.5[3] |
| Viili[4] | | 5.8 | | | |
| Skyr | | | 8.6 | | |
| Other products | 4.1 Buttermilk | 14.6 Piimä | 8.4 Súrmjólk | 11.5 | 13.1 Fil, Filmjölk |

[1] Consumption data for 2002.
[2] Consumption data for 2001.
[3] Consumption figures include kefir.
[4] Consumption data include also kermaviili, which is used in cooking preparations.
Data compiled from StatBank Denmark (http://www.statbank.dk/statbank5a/default.asp?w=1024), "Maito ja Terveys" (Milk and Health Journal) (http://www.maitojaterveys.fi/mvkulut.htm), Statistics Iceland: (http://www.hagstofa.is/template_lb_frameset_en.asp?PageID=797&ifrmsrc=/uploads/files/lh2002/L020506.xls&Redirect=False), Statistics Norway: (http://www.ssb.no/english/subjects/05/02/fbu_en/arkiv/tab-2002-12-16-06-en.html) and Livsmedels Sverige (http://www.livsmedelssverige.org/statistik/konsumtion.htm).

(piimä in Finland), *Lactobacillus casei* 431 (piimä in Finland), and with *Lb. acidophilus* alone or together with *Bifidobacterium* spp. (fil in Sweden, piimä in Finland and buttermilk in Egypt; El-Shafei, 2003). Some commercial probiotic fermented milks and related products that are sold in different European markets are shown in Table 3.4.

**Table 3.4** Examples of probiotic fermented milk products in the European market.[1]

| Type of product and trade name | Probiotic micro-organisms present in the products as stated by the manufacturer[2] |
|---|---|
| I. Non-drinkable fermented milks<br>Bifisoft, Bifidus, Bioghurt, Biofit, BiofardePlus, Biola, Biologic Bifidus, Cultura Dofilus, Dujat Bio Aktiv, Ekologisk Jordgubbs Youghurt, Fit & Aktiv, Fjällyoughurt, Fysiq, Gaio Dofilus, Gefilac, Gefilus, Lc1, Probiotisches Joghurt, ProViva, RELA, Verum, Vifit Vitamel, Vitality, Weight Watchers, Yogosan Milbona | *Lb. acidophilus*, *Lb. acidophilus* LA5, *Lb. rhamnosus* LGG, LB21 and 271, *Lb. casei* (also strain F19), *Lb. johnsonii*, *Lb. plantarum* 299v, *Lb. reuteri*<br>*Lac. lactis* subsp. *lactis* L1A<br>*B. bifidum*[3], *B. animalis* subsp. *lactis* BB-12, *B. animalis* subsp. *animalis* |
| II. Drinkable fermented milks (including cultured buttermilk, yoghurt drink, dairy drink)<br>A-fil, Actimel, Aktifit, AB-piimä, Bella Vita, Bifidus, Biofit, Biola, Casilus, Cultura, Emmifit, Everybody, Fit & Aktiv, Fundo, Gaio, Gefilac, Gefilus, Kaiku Actif, Lc1go, LGG+, Onaka, Öresundsfil, Philura, Probiotic drink, Proviva, Pro.x, Verum, ViktVäktarna, Vitality, Vive+, Yakult, Yoco acti-vit | *Lb. acidophilus*, *Lb. acidophilus* LA5, *Lb. casei* (F19, 431, Imunitass, Shirota), *Lb. rhamnosus* LGG (271 and LB21), *Lb. johnsonii*, *Lb. reuteri*, *Lb. plantarum* 299v, *Lactobacillus fortis*<br>*Lac. lactis* subsp. *lactis* L1A<br>*B. animalis* subsp. *lactis* BB-12, *B. bifidum*, *B. animalis* subsp. *animalis*, *B. longum* BB536 |
| III. Non-fermented dairy products (milk, ice-cream)<br>Gefilus, God Hälsa, RELA, Vivi Vivo | *Lb. rhamnosus* LGG, *Lb. plantarum* 299v, *Lb. reuteri* |

[1] Data compiled from the www web sites of the commercial dairy companies.
[2] Lactic starter cultures microflora are not listed.
[3] On some occasions these organisms have been referred to as 'bifidus' or lactobacilli, i.e. old nomenclature.

Kefir is a carbonated fermented milk product made by using a complex mixture of micro-organisms known as kefir grains. Typically, these include *Lac. lactis* subsp. *cremoris*, *Lac. lactis* subsp. *lactis*, *Lactobacillus kefir*, *Lactobacillus kefiranofaciens*, *Lactobacillus brevis*, *Lb. acidophilus*, *Leuconostoc* spp., *Acetobacter* spp., lactose-fermenting yeasts (*Kluyveromyces* spp.) and lactose-negative yeasts (*Saccharomyces* spp., *Candida* spp.). However, kefir grains can contain a wide variety of micro-organisms in addition to those mentioned above (Marshall, 1987; Tamime & Marshall, 1997; Garrotte *et al.*, 2001; Leporanta, 2001; Simova *et al.*, 2002; Santos *et al.*, 2003; Witthuhn *et al.*, 2004). The mixture of bacteria and yeasts in kefir grains appear as clusters of microbes held together with a matrix consisting mainly of the polysaccharide 'kefiran' produced by *Lb. kefiranofaciens* and protein. The activity of yeasts results in a product with a typical yeasty flavour, the formation of carbon dioxide and some ethanol (<2 mL 100 $mL^{-1}$). Nowadays, there are two main methods to produce kefir: (a) traditional, which uses kefir grains; and (b) modern, which uses direct-to-vat inoculation (DVI) starter cultures. In the traditional method, the processed milk is cooled to 18–25°C, inoculated with kefir grains (2–10 g 100 $mL^{-1}$) and incubated for 18–24 h followed by stirring and cooling. After incubation the grains are separated, washed and re-used (Kurmann *et al.*, 1992; Wszolek *et al.*, 2001; Schoevers & Britz, 2003). When DVI cultures are used, the milk is heat treated, cooled to inoculation temperature (32–35°C), inoculated with DVI cultures, and incubated for 10–15 h (final pH 4.4–4.5); the fermented product is stirred, cooled and packaged (K. Leporanta, Valio Ltd., personal communication). Commercial kefir starter cultures are available in different blends of thermophilic and mesophilic lactic acid bacteria and yeast cultures, which allows the flavour profiles of the product to be based on consumer acceptability. Recently, 'bio' kefir has been marketed in Poland containing *Bifidobacterium* spp. and/or *Lb. acidophilus* (Muir *et al.*, 1999; see also Farnworth & Mainville, 2003).

In recent years, kefir has gained increasing interest in the U.S. and elsewhere as a health product with probiotic claims. Traditionally, kefir is consumed as a non-flavoured fermented milk drink but, currently in the U.S., fruit and berry-flavoured varieties are produced. Kefir with a probiotic strain, such as *Lb. rhamnosus* GG, is produced in Estonia and Latvia.

### *Non-drinking fermented milk products*

Viili and långfil are variations of viscous fermented milk products made with mesophilic starter cultures. Both products were commonly produced on farms to preserve milk during summer time when milk production was at its highest. Långfil is produced with the same mesophilic lactic acid bacteria species as cultured buttermilk, but these strains produce large amounts of exopolysaccharides, which makes the product much more viscous than cultured buttermilk. During the production of långfil, the milk is heated to a high temperature, cooled to 18–20°C, mixed with starter cultures in the tank, packaged and fermented for 18–20 h (Mantere-Alhonen & Forsen, 1990; Oberman & Libudzisk 1998; Leporanta, 2003). Långfil is mainly consumed in northern parts of Sweden and, similar to filmjölk and fil, it is eaten during breakfast (usually with berries or cereal) or as a snack. To our knowledge there are no probiotic långfil products on the market.

Viili is a viscous fermented milk product, which is manufactured in Finland. The industrial production of viili began during the late 1950s. Viili is produced from milk by fermenting it with mesophilic starter cultures (*Lac. lactis* subsp. *cremoris*, *Lac. lactis* subsp. *lactis* biovar. *diacetylactis* and *Leu. mesenteroides* subsp. *cremoris*) together with a mould (*G. candidum*) in the retail container. The fermentation time is ~20 h at 20°C (final pH is ~4.3). Traditionally, viili was made from non-homogenised milk, which resulted in the formation of a cream layer on the surface of the milk. *G. candidum* grew on this layer and formed a velvety growth similar to Camembert and Brie. There is also a non-mouldy viili-type product, and the high-fat variety is used only in cooking (Mantere-Alhonen & Forsen, 1990; Kurmann *et al.*, 1992; Leporanta, 2003).

A wide variety of viili-type products are available in Finland including low-fat, low-lactose and berry-flavoured variants. Viili is consumed mainly at breakfast and as a snack. A product containing *Lb. rhamnosus* GG is currently the only probiotic viili product available on the market.

### *Skyr, ymer and strained yoghurt (concentrated fermented milks)*

Skyr is a traditional and concentrated fermented milk product that is made and consumed in Iceland. Skyr is produced from skimmed milk or skimmed milk powder by fermentation with yoghurt starter cultures and lactose-fermenting yeasts. The milk is heated to a high temperature, cooled to 40°C (traditionally, cheese rennet is added), incubated for ~5 h and then cooled to 20°C; fermentation is then continued at 20°C for 18 h, allowing the yeasts to grow (or until the pH falls to 4.2). Whey is removed using mechanical separators or by ultrafiltration (UF); the denatured protein mass is recovered by UF and added to the skyr mass to increase the yield of the product, and the final pH ranges between 3.8 and 4.0. Skyr is consumed during breakfast or as a snack, and it is often flavoured with berries and fruits (Gudmundsson, 1987; Wolpert, 1988; Kurmann *et al.*, 1992).

Ymer is a concentrated fermented milk product, which was developed in Denmark in the 1930s. Today, it is produced from heat-treated milk that has been homogenised and ultrafiltered (Mogensen, 1980). When using UF milk, the product is more concentrated (i.e. the protein content is 6 g 100 $g^{-1}$ and the solids-non-fat (SNF) is 11 g 100 $g^{-1}$). The UF milk is re-heat-treated, homogenised and fermented with *Lac. lactis* subsp. *cremoris* and *Lac. lactis* subsp. *lactis* biovar. *diacetylactis* at 18–20°C for 18–20 h (final pH 4.4–4.6). After fermentation, the product is stirred, cooled, left to stand at 5°C for 1 day and then re-stirred prior to packaging (Delaney, 1977; Ulrich, 1980; Kurmann *et al.*, 1992; Oberman & Libudzisk, 1998). To our knowledge, there are no probiotic-containing ymer or skyr products on the market.

Concentrated, strained or Greek-style yoghurt is a fermented milk product that is manufactured in many countries in the Middle East (Tamime & Robinson, 1999). The different methods available to manufacture this product are: cloth bag method, mechanical separator, UF or product formulation. In the ultrafiltered Greek-style product, the viable cell count of bifidobacteria ranges between $2 \times 10^5$ and $4 \times 10^7$ cfu $g^{-1}$, depending on the type of milk used (Mahdi *et al.*, 1990). The effect of levels of milk solids, fat, fat-substitute and vegetable oils in the milk base on the quality of concentrated yoghurt made with ABT culture or enterococci species was reported by Amer *et al.* (1997), Taha *et al.* (1997) and El-Samragy (1997).

### *Quality appraisal of probiotic fermented milks*

Probiotic micro-organisms can be incorporated into fermented milk using different methods.

(1) The most popular way is to add the probiotic micro-organisms together with the starter cultures (i.e. as a DVI culture). Since the fermentation rarely occurs in conditions optimal for probiotic species, such organisms do not usually grow markedly during a mixed fermentation with the 'traditional' starter cultures.
(2) The probiotic organisms may be grown in one batch of milk in order to achieve a high viable count, whilst another batch of milk is fermented with the 'traditional' starter cultures. After the fermentation stages, the two batches are mixed together to produce a probiotic fermented milk.
(3) A probiotic micro-organism(s) may be used as a starter culture; however, the fermentation time may be up to several days. A typical example of the use of a probiotic culture alone is during the manufacture of Yakult, which is fermented with *Lb. casei* strain Shirota (K. Leporanta, Valio Ltd., personal communication). The health properties of Yakult have been recently reviewed by Matsuzaki (2003), whilst the growth activity of *Lb. acidophilus* in different mammalian milks has been reported by Drakoularakou *et al.* (2003).

During the production of probiotic fermented milks, several aspects must be considered, including the following: (a) many probiotic strains grow slowly in non-supplemented milk; (b) the production conditions (especially the traditional fermentation temperatures) are often unsuitable for their growth; and (c) the metabolites of probiotic micro-organisms may be undesirable due to the formation of off-flavours (e.g. bifidobacteria produce acetic acid, which gives a vinegar-like taste) (Gomes & Malcata, 1999; Saxelin *et al.*, 1999; Saarela *et al.*, 2000; Ostlie *et al.*, 2003). However, if the food matrix supports the growth of probiotic micro-organisms and no off-flavour formation occurs, growth during the production of fermented milks can lower the processing costs and increase the adaptation of probiotic organisms, leading to enhanced viability. In the cases where probiotic and traditional starter micro-organisms are both present during the fermentation stage, it is important to use compatible and suitable blends of probiotic/starter cultures (Saxelin *et al.*, 1999; Ouwehand *et al.*, 2000). In the extreme case, the starter cultures may produce inhibitory compound(s) (e.g. hydrogen peroxide, high amounts of lactic acid) that are harmful to the probiotic culture(s) and lead to a decrease in viable count in the product (Katla *et al.*, 2001; Vinderola *et al.*, 2002). Nevertheless, some starter cultures may enhance the growth and survival of probiotic micro-organisms by producing growth-promoting substrates or by reducing the oxygen content in the milk (Kailasapathy & Rybka, 1997; Dave & Shah, 1997a, 1997b; Saarela *et al.*, 2000; Vinderola *et al.*, 2002). Another important factor, which should not be overlooked, is the growth temperature in mixed fermentations. Some of the traditional products described above are fermented at 20 or 30°C, which are sub-optimal temperatures for the growth of probiotic micro-organisms and, in particular, the strains that originate from the human GI tract (optimum growth temperature of 37°C). Increasing the fermentation temperature to favour the growth of probiotic micro-organisms is not recommended because it can lead to an unacceptable flavour profile in the products (Mantere-Alhonen & Forsen, 1990). Therefore, mixed fermentation with probiotic micro-organisms has the best chance of success when the probiotic strain is combined with a thermophilic starter (e.g. a blend of *Lb. acidophilus* and/or bifidobacteria and yoghurt starter cultures) (Saxelin *et al.*, 1999; Gardini *et al.*, 1999; Saarela *et al.*, 2000). Alternatively, the probiotic micro-organisms may be added at high numbers to a 'traditional' starter culture to produce a fermented milk product, irrespective of sub-optimal growth temperature for the probiotic species (see Baron *et al.*, 2000).

### 3.4.2 *Cheeses*

Success in the use of probiotic bacteria with fermented liquid milk products has inspired the development of other dairy products with probiotics (Heller, 2001; Ross *et al.*, 2002; Fondén *et al.*, 2003; Heller *et al.*, 2003; Boylston *et al.*, 2004). Examples of probiotic bacteria that are used in dairy products are listed in Table 3.2.

The production of cheeses, especially the ripened types, with probiotic bacteria presents unique challenges because of the need for co-survival of these bacteria with the conventional lactic acid bacteria, mould or yeasts that are used for cheesemaking. These micro-organisms may be antagonistic, competitive or symbiotic towards each other. Some key characteristics of cheeses/cheesemaking that are relevant to the inclusion of probiotics are as follows:

- Relative low moisture content;
- Presence of salt (i.e. salt in moisture ratio);
- Starter culture organisms (e.g. acid production, flavour production during the maturation stage and competition for nutrients);
- Extended storage over 3 months, which can influence the biochemical activities, alteration of redox potential and re-organisation of the cheese structure.

Probiotic micro-organisms must survive the entire shelf-life of the cheese to be of any therapeutic value to the consumer, they must not produce metabolites during cheesemaking or maturation that are detrimental to the quality of the cheese, and they should not interfere with the normal activity of other essential micro-organisms in the cheese. In addition, probiotic micro-organisms should survive the cheesemaking process and the entire life of the cheese, must not produce anti-microbial compounds, and/or should be able to grow in starter culture media, such as whey-based or phage inhibitory media.

Various cheese varieties have been successfully used as carriers of probiotic micro-organisms. Some cheeses may be particularly suitable for the delivery of probiotic bacteria relative to fermented milks such as yoghurt, because of lower acidity and the existence of a complex cheese matrix of protein and fat that will provide protection to probiotic micro-organisms during their passage through the gastrointestinal tract (Stanton *et al.*, 1998; see also Donnelly, 2003). Studies have demonstrated that Turkish white brined, Feta-type, Cheddar, Philippine white soft, Edam, Emmental, Domiati, Ras, soft, Herrgård cheeses, Quarg, and cheese-based dips compare favourably with yoghurt as delivery systems for viable probiotic micro-organisms (Ghoddusi & Robinson, 1996; Murad *et al.*, 1998; Gardiner *et al.*, 1999a; Barraquio *et al.*, 2001; Osman & Abbas, 2001; Psomas *et al.*, 2001; Shehata *et al.*, 2001, 2004a, 2004b, 2004c; Ahola *et al.*, 2002; Antonsson *et al.*, 2002; Mehanna *et al.*, 2002; Abdou *et al.*, 2003; El-Kholy *et al.*, 2003; de Vrese, 2003; Kasimoğlu *et al.*, 2004; Medici *et al.*, 2004; Milanovic *et al.*, 2004; Tharmaraj & Shah, 2004; Tungjaroenchai *et al.*, 2004; Weinrichter *et al.*, 2004a, 2004b; Yilmaztekin *et al.*, 2004).

#### *Methods of introduction in cheese*

Cheesemaking involves many steps, each of which serves a unique function in the development of flavour and texture, and Fig. 3.2 is used as an example to show the manufacturing stages of Cheddar cheese. Many of these cheesemaking variables interact to produce a final product with the desired qualities. The inclusion of probiotic micro-organisms, therefore,

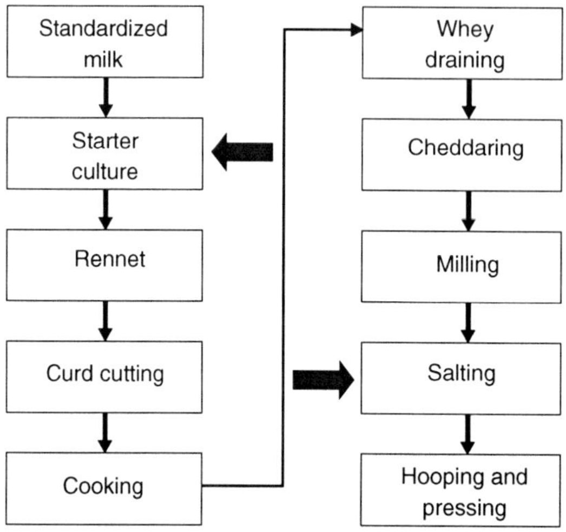


**Fig. 3.2** The manufacturing stages of Cheddar cheese. Note: The bold arrows show steps where probiotic organisms may be introduced.

must be balanced with impacts on various other cheesemaking factors and cheese characteristics. This complexity introduces challenges to optimise the application of probiotic micro-organisms in cheese. Thus, the time and method of addition of the probiotic culture during cheesemaking must be selected carefully and may be dependent on strain or cheese variety.

Probiotic bacteria may be introduced into cheese as adjunct cultures along with the lactic starter cultures (Stanton *et al*., 1998; McBrearty *et al*., 2001; Perko *et al*., 2002). In a variation of this method, probiotic Cheddar cheese was manufactured by standardising the cheese milk with cream that had been fermented with *Bifidobacterium infantis* (Daigle *et al*., 1999). The introduction of probiotic organisms along with lactic starter cultures directly into milk prior to cheesemaking raises the risk of losing large numbers of the probiotic cells to whey or domination by lactic starter culture. This can particularly be a problem with slow-growing probiotic bacteria, but with proper strain selection the method has been used successfully. A spray-dried probiotic milk powder containing *Lactobacillus paracasei* NFBC 338 (Rifir) was produced with a probiotic survival rate of 84.5%, and this was used in the manufacture of Cheddar cheese as an adjunct culture (Gardiner *et al*., 2002b). The initial *Lb. paracasei* count of $2 \times 10^7$ cfu $g^{-1}$ increased to $3.3 \times 10^7$ cfu $g^{-1}$ in 3 months. *Lactobacillus gasseri* grows slowly in milk, so it is difficult to use it independently for cheesemaking; however, when used in combination with *S. thermophilus*, it maintained viability in a semi-hard cheese (Gardiner *et al*., 2002b). When introduced into cheese milk for the manufacture of a semi-hard cheese, *Lb. acidophilus* LF221 maintained a level of approximately $6.8 \times 10^6$ cfu $g^{-1}$ for 6 weeks (Rogelj *et al*., 2002). High viability ($10^7$ cfu $g^{-1}$) of *Bifidobacterium bifidum*, *Lb. casei* and others has also been reported in Queso Fresco cheese (Viderola *et al*., 2000b; Suarez-Solis *et al*., 2002). Microencapsulation may be used to protect probiotic organisms and improve viability (Abou-Dawood, 2002; Kailasapathy, 2002; Godward & Kailasapathy, 2003a; Picot & Lacroix, 2003), and encapsulated *B. bifidum*, *B. infantis* and *Bifidobacterium longum* have been used in the manufacture of Crescenza cheese. While there were no differences in the sensory quality of the cheeses, the product made with bifidobacteria only had traces of lactose and high amounts of lactic and acetic acids (Gobetti *et al.*, 1998; see also the review by Boylston *et al.*, 2004). Coliform counts decreased in cheeses with non-capsulated bifidobacteria (Abou-Dawood, 2002).

Another method of introduction, particularly into semi-hard and hard cheeses, is through the addition of a dried culture during salting of curd, as is done with some accelerated ripening enzyme preparations. This method minimises the losses of bacterial cells to whey and eliminates the effects of competition with lactic acid bacteria during milk ripening (Dinakar & Mistry, 1994). In one study (Dinakar & Mistry, 1994), two preparations of bifidobacteria were used; one was a *B. bifidum* preparation in which live bacteria were immobilised by forming gelled beads in carrageenan; these beads were then freeze-dried. The second preparation was a commercial powder preparation containing *B. bifidum*. These preparations were added separately to the cheese curd at milling, such that the viable counts of bifidobacteria in the cheese were $10^6$ cfu $g^{-1}$. Cheese composition, proteolysis, lactic acid production and sensory characteristics during maturation were not affected by the added bifidobacteria. Acetic acid and ethanol, which normally are metabolites of bifidobacteria, were not detected. It is important to note, however, that the bifidobacteria remained viable and increased in numbers during 24 weeks of ripening. After 24 weeks, both preparations produced counts of ~$10^7$ cfu $g^{-1}$ of cheese. However, an alternative technique is to enhance the growth of probiotic bacteria in milk hydrolysate (i.e. increase of the biomass of the cells) before using them in cheesemaking (Gomes *et al.*, 1998).

Some cheese varieties, such as Cottage cheese, offer another method for the incorporation of probiotic organisms. A unique step in Cottage cheese manufacture is the addition of a fermented cream dressing for flavour and texture development. Probiotic organisms, such as *Lb. rhamnosus* GG, have been used to ferment the cream dressing with no adverse impact on flavour (Tratnik *et al.*, 2000). During storage, the viable counts of the probiotic organism doubled. *B. infantis* has also been evaluated in a similar application and showed high metabolic activity during the storage of Cottage cheese (Blanchette *et al.*, 1996; Daigle *et al.*, 1998).

While some studies have demonstrated a limited impact of added bifidobacteria on cheese quality (Dinakar & Mistry, 1994; Daigle *et al.*, 1999), others have suggested an impact on sensory qualities due to the formation of acetic acid in cows' (Gomes *et al.*, 1995) as well as goats' (Gomes & Malcata, 1998) milk cheeses.

### *Strain selection*

The importance of proper strain selection for probiotic applications in fermented milk manufacture is well recognised, and also applies to cheeses (Gilliland, 2001). The strains should be of human origin, non-pathogenic, and must survive the low pH conditions in the GI tract (Abou-Dawood, 2002). For cheese applications, the probiotic strains must also be compatible with the cheese starter cultures, they should not produce flavour and texture defects, and must also survive during the entire storage period of cheese. For example, in one study that used two strains for Cheddar cheesemaking it was observed that, with *Bifidobacterium animalis* subsp. *lactis* BB-12 (see Table 3.2; Masco *et al.*, 2004), the moisture content of cheese was higher, and there was more proteolysis and better flavour than with *B. longum* BB536 (McBrearty *et al.*, 2001). The production of acetic acid in cheese was also observed. Such variability in impact on cheese has also been observed with other varieties. Lactose was completely hydrolysed when biotype *infantis* and *B. longum* BB-02 and BB-46 (see Table 3.2) were used for manufacturing Canestrato Pugliese cheese from ewe's milk (Corbo *et al.*, 2001). A high activity of β-galactosidase was also observed, similar to another study, with *B. infantis* in Cheddar cheese (Daigle *et al.*, 1999).

Strain selection for cheesemaking should take into account the applicability for health benefits as well as optimum growth in the cheese matrix. All strains of probiotic micro-organisms do not have the capacity to perform all known beneficial health-related functions (Gilliland, 2001) because the probiotic properties are strain-specific. For use in manufacturing Edam cheese, it was shown that *B. bifidum* and Edam starter cultures grew optimally as a mixed culture (Sabikhi & Mathur, 2000, 2002).

The definition of probiotic bacteria now extends beyond the traditionally recognised bifidobacteria and a few lactobacilli. Within the lactobacilli group, probiotic cheeses with human-derived *Lb. paracasei* have been manufactured with no impact on cheese composition (Gardiner *et al.*, 1998; Stanton *et al.*, 1998). In one study (Stanton *et al.*, 1998), *Lb. paracasei* NFBC 338 and NFBC 364 grew to $2.9 \times 10^8$ cfu $g^{-1}$ in matured cheese in 3 months, and maintained the numbers for up to 200 days (Fig. 3.3). One of the prerequisites of probiotics is that they be of human origin. For this reason, other organisms from the human intestinal system have also been evaluated for probiotic uses. An example of this is *E. faecium*, a thermophilic organism which has been used in cheesemaking. One strain, PR88, is also believed to be useful in the treatment of irritable bowel syndrome. When used as an adjunct starter for cheese, this bacterium survived 9 months of maturation, and improved the cheese flavour via increased proteolysis (Gardiner *et al.*, 1999b).

Propionibacteria have also been suggested to have probiotic properties (Mantere-Alhonene, 1995; Jan *et al.*, 2002); some strains are used in the manufacture of Swiss cheese and, thus, offer the possibility of adding to the value of the cheese. *Propionobacterium freudenreichii* subsp. *shermanii* SI41 is able to survive acid and bile salts and, thus, function as a probiotic (Jan *et al.*, 2002). What is also interesting and useful is that *P. freudenreichii*

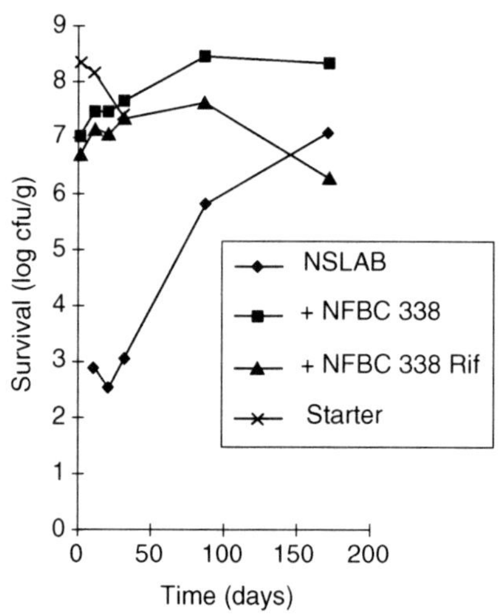


**Fig. 3.3** The incorporation of *Lactobacillus paracasei* NFBC in Cheddar cheese. After Stanton *et al.* (1998); data by permission of *International Dairy Journal*.

subsp. *shermanii* has also been shown to produce a bifidogenic factor that serves as a growth factor for bifidobacteria (Kaneko, 1999). Yeast from infant faeces and Feta cheese has been isolated, and has potential for use in industrial fermentation (Psomas *et al.*, 2001).

It is recognised that probiotic bacteria have certain health-promoting properties (Gilliland, 2001), but they may also serve a useful function by producing compounds of 'health' value. For example, in Finland *Lb. acidophilus* and *Bifidobacterium* spp. were used to produce a low-fat cheese (Ryhänen *et al.*, 2001). These organisms apparently produced bioactive peptides with anti-hypertensive properties during maturation of the cheese, thus adding to the nutritional and probiotic values of the product. The inclusion of probiotic cultures, such as *Lb. rhamnosus* GG, in cheese may also have an impact on dental caries (Abou-Dawood, 2002). Some strains of bifidobacteria produce anti-microbial compounds, which reduce the levels of pseudomonas in Cottage cheese (O'Riordan & Fitzgerald, 1998). There also are commercial probiotic cultures of *Lb. rhamnosus* and *P. freudenreichii* subsp. *shermanii* with anti-clostridial effects and activity against contaminating yeasts and moulds (Hansen, 1997).

### **3.4.3** *Ice-cream and frozen desserts*

Ice-cream and frozen desserts have potential as carriers of probiotic organisms, but freeze stress must be considered with respect to viability during manufacture and extended storage. Frozen yoghurt technology may be adopted for the inclusion of probiotic cultures into ice-cream and frozen desserts. Various combinations of lactobacilli and bifidobacteria have been used in these products (see Tamime *et al.*, 1995). Addition may be direct (i.e. blending of ice-cream mix and probiotic cells immediately prior to freezing), or it may involve fermentation of the milk for proliferation of probiotic cells prior to blending with the ice-cream mix (Christiansen *et al.*, 1996; Ravula & Shah, 1998; Haynes & Playne, 2002). In either case, the protection of cells against freeze damage is of importance. Encapsulation and freeze-drying, and co-encapsulation of different micro-organisms (e.g. *Lb. acidophilus* 2401, 2404, 2409, and 2415, *B. infantis* 1912, *B. animalis* subsp. *lactis* 1941, 920 and Bb-12, *B. longum* 5581) have been evaluated (Godward & Kailasapathy, 2000, 2003b; Talwalker & Kailasapathy, 2003a; Kailasapathy & Sultana, 2003; see also Masco *et al.*, 2004) with the result that free cells and freshly encapsulated cells without freeze-drying demonstrated the best survival rates. Properly selected strains, such as *Lactobacillus johnsonii* La1, survive the relatively high sugar content of ice-cream as well as the sub-lethal injuries caused by freezing. Counts of $10^7$ cfu $g^{-1}$ were maintained for 10 weeks (Kebary *et al.*, 1998; see also Kebary *et al.*, 2004; El Shazly *et al.*, 2004; El Tahra *et al.*, 2004a, 2004b; Hamed *et al.*, 2004; Rao & Prakash, 2004) or 8 months of storage (Alamprese *et al.*, 2002). Some strains do not resist the freezing and churning that occurs during ice-cream manufacture (Hagen & Narvhus, 1999; Haynes & Playne, 2002), but others such as *B. longum* and *B. infantis* are able to survive these processes and storage up to 11 weeks (Davidson *et al.*, 2000) or 52 weeks (Haynes & Playne, 2002), and are not influenced by the fat content of the product. A study involving 13 strains of *Lb. acidophilus* and 11 of bifidobacteria demonstrated that these effects were strain-dependent (Ravula & Shah, 1998). Because ice-cream is a non-fermented product, the impact of the probiotic micro-organisms on flavour should be a consideration. Some organisms, such as *Lb. reuteri* (Hagen & Narvhus, 1999) or *B. bifidum* (Ma, 1995) for example, produce a slightly acid flavour due to fermentation. Manufacturing conditions that limited fermentation may be adopted to minimise such flavours (see also Ordonez *et al.*, 2000a, 2000b).

### 3.4.4 *Miscellaneous dairy products*

#### *Probiotic 'sweet' milk*

In the 1970s, sweet acidophilus milk, which has similar organoleptic properties to pasteurised milk, was introduced in the U. S. market, but it accounted for only 1% of the total volume of liquid milk sales (Shortt, 1999). This product does not entail fermentation of the milk, and concentrated *Lb. acidophilus* preparation is added to cold pasteurised milk before packaging; the anticipated viable count being ~$5 \times 10^6$ cfu $mL^{-1}$ (Salji, 1992). However, Young and Nelson (1978) reported that the level of *Lb. acidophilus* declined by a factor of 10 during a 2-week storage period, and this was attributed to the level of inoculum and the strain used.

Milk containing *B. longum* ATCC 15708 at a dose of $5 \times 10^8$ cfu $mL^{-1}$ improved lactose digestion *in vivo*, and the milk was better tolerated by lactose maldigesters (Jiang *et al.*, 1996). Another product similar to acidophilus milk is known as BRA sweet milk; this contains *B. infantis*, *Lb. reuteri* and *Lb. acidophilus*, the micro-organisms being added to pasteurised cold milk before packaging (Rothschild, 1995).

#### *Fat spread*

A prototype reduced-fat (60 g 100 $g^{-1}$) edible table biospread was made from milk fat and soy oil containing mixed cultures of *Lb. casei* ACA-DC 212.3 and *B. infantis* ATCC 25962 (Charteris *et al.*, 2002). The process was modified to enhance the viability of the probiotic micro-organisms (i.e. both cultures showed 1 $\log_{10}$ decline after processing). The rate of decline in the viability of bifidobacteria during shelf-life was greater than that of the lactobacilli, and it was concluded that more development work is required.

#### *Dried products*

There is a significant market for dried probiotic pharmaceutical products, and a variety of products are available in the form of dietary supplements. This discussion will not focus on these products (Tamime & Marshall, 1997; Kaur *et al.*, 2002), but rather on other dried products that are subsequently used in the manufacture of dairy and other products.

Dried products are manufactured either with freeze-drying methods or spray-drying. While the cost of production is important, survival during the drying process and subsequent storage is equally important. Unless appropriate drying conditions are selected (Lian *et al.*, 2002), cell damage and loss of viability will occur during spray-drying, and viability during storage is inversely related to the storage temperatures (Gardiner *et al.*, 2000). Various factors should be considered with respect to the viability of probiotic micro-organisms in dried products (Prajapati *et al.*, 1986; Gilliland *et al.*, 2001), including:

- Drying method
- Type and size of packaging
- Temperature and humidity of storage
- Powder quality

- Rehydration procedure
- Handling of rehydrated product

Protective compounds, such as gum acacia, were used to protect *Lb. paracasei* NFBC 338 during spray-drying and storage (Desmond *et al.*, 2001, 2002). This method improved survival during drying and storage, as well as resistance to bile. On the other hand, the same organism was spray-dried without any protection with survival rates of 84.5% for use in Cheddar cheesemaking as described earlier (Gardiner *et al.*, 2002b). Compression coating of *Lb. acidophilus*-containing powders in combination with sodium alginate and hydroxy-propyl cellulose was used to increase storage stability 10-fold after 30 days of storage at 25°C compared to free cells (Chan & Zhang, 2002).

Dried preparations of probiotics are of particular interest for the manufacture of infant formulae. An important objective of the infant formula industry is to manufacture products that are functionally similar to human milk. Whilst the processing technology is available to manufacture an infant formula with a gross composition similar to that of human milk, methods continue to be refined for other factors (Lonnerdal, 2003). For example, the gut flora of infants fed human milk is different from that of formulae-fed infants (Edwards *et al.*, 2002). Various methods have been proposed for the introduction of probiotic organisms into the infant gut. These methods include the incorporation of probiotics in dried preparations (see Saavedra *et al.*, 2004), which will enhance the proliferation of probiotic organisms in the gut. In one example, blending freeze-dried preparations of *Lb. reuteri*, *Lb. acidophilus* and *B. bifidum* led to the development of an infant formula to prevent diarrhoea. This type of formulation was effective in preventing diarrhoea in infants when consumption of the three organisms was $10^8$ to $10^{10}$ cfu $day^{-1}$ (Halpin-Dohnalek *et al.*, 1999). An example of a commercialised milk powder (i.e. Neslac) containing *B. animalis* subsp. *lactis* BB-12 for older infants has been reported by Playne *et al.* (2003) (see also Chouraqui *et al.*, 2004; Masco *et al.*, 2004). Another approach is to include dried prebiotics such as oligosaccharides (Goni-Cambrodon & Gudiel-Urbano, 2001; Kunz & Rudolff, 2002), and lactulose (Strohmaier, 1997) into infant formulae.

### *Long shelf-life fermented milk drinks or beverages*

As mentioned earlier (see Section 3.4.1), the fresh probiotic fermented drinks/beverages market is booming in the United Kingdom and worldwide, but long shelf-life dairy products do not contain either the lactic starter cultures or the probiotic micro-organisms. However, this problem can be overcome by using a specially designed straw (known as LifeTop™ Straw; or alternatively referred to as a probiotic straw), which was developed and patented by BioGaia in Sweden (Thorball *et al.*, 2001; Anonymous, 2001; see also www.biogaia.com or www.reuteri.com). A freeze-dried *Lb. reuteri* culture is suspended in oil droplets and attached to the inside of a two-jacketed straw. The straw is packed in an outer packaging container that is made of laminated aluminium foil that is impermeable to moisture and oxygen. According to the manufacturer, the straw contains $10^8$ cfu, has a shelf-life of 12 months at 25°C and, when the consumer drinks 100 mL of the beverage, 99% of the probiotic bacteria are released. The same concept could be applied to other types of probiotic bacteria, and may be used, especially by children, when drinking pasteurised or flavoured milk drinks.

*Milk- and water-based cereal puddings*

*Lb. rhamnosus* GG, *Lb. acidophilus* LA-5 and 1748, and *B. animalis* subsp. *lactis* BB-12 were successfully used for the production of milk- and water-based puddings with and without prebiotics (e.g. polydextrose and Litesse™) (Helland *et al.*, 2004). All strains showed good growth and survival in milk-based puddings (e.g. viable counts ranged between 8 and 9.1 $\log_{10}$ cfu $g^{-1}$), significantly ($P$ <0.05) higher concentrations of lactic acid were produced after storage, the pH level was reduced to <4.4, and the highest concentration of diacetyl (18 mg $kg^{-1}$) was detected in puddings inoculated with *Lb. rhamnosus* GG. In addition, puddings prepared with or without the addition of Litesse™ were not significantly different.

## 3.5 Viability of probiotic micro-organisms

In order to obtain the desired health effects, probiotic bacteria must be able to grow in milk (some organisms may not be able to grow in milk, e.g. *Lb. rhamnosus* GG), and survive in sufficient numbers. It has been suggested that probiotic organisms should be present in a food to a minimum concentration of $10^6$ cfu $g^{-1}$, or the daily intake should be about $10^9$ cfu $g^{-1}$. Such high numbers have been suggested to compensate for possible losses in the numbers of the probiotic organisms during passage through the stomach and intestine. In Japan, the Fermented Milks and Lactic Acid Bacteria Beverages Association have developed a standard, which requires a minimum of $10^7$ viable cfu $mL^{-1}$ to be present in dairy products. Studies have demonstrated that several probiotic micro-organisms grow poorly in milk, and the viability of these organisms is often low in yoghurt. A number of brands of commercial yoghurt have been analysed in Australia and in Europe for the presence of *Lb. acidophilus* and bifidobacteria. Most of the products contained very low numbers of these organisms, especially bifidobacteria (see Tamime, 2002). Viability and activity of the bacteria are important considerations, because these bacteria must survive in the food during its shelf life, during transit through the acidic conditions of the stomach, and resist degradation by hydrolytic enzymes and bile salts in the small intestine.

The viability of probiotic bacteria in yoghurt depends on the strains used, the interaction between species present, the production of hydrogen peroxide due to bacterial metabolism, and the final acidity of the product. Viability also depends on the availability of nutrients, growth promoters and inhibitors, the concentration of sugars, dissolved oxygen and oxygen permeation through package (especially for *Bifidobacterium* spp.), inoculation level, and fermentation time (see also Oliveira & Damin, 2003). Bifidobacteria are anaerobic in nature, and therefore a high oxygen content may affect their growth and viability. *Lb. acidophilus* is reported to have a high cytoplasmic buffering capacity (pH 3.72 to 7.74), which allows it to resist changes in cytoplasmic pH and to gain stability under acidic conditions. *Lb. acidophilus* is more tolerant of acidic conditions than *Bifidobacterium* spp., and growth of the latter is significantly retarded below pH 5.0. The tolerance of *Bifidobacterium* spp. to acidic conditions is strain-specific. *B. longum* has shown better survival in acidic conditions and bile concentrations compared to *B. infantis*, *B. adolsecentis* and *B. bifidum*. Furthermore, *B. longum* is also easier to grow in milk, whilst *B. animalis* subsp. *animalis* has good survival properties in fermented milks. However, this last species is not of human origin (Lankaputhra & Shah, 1996). Recently, Mättö *et al.* (2004) confirmed that the tolerance of *Bifidobacterium* spp. to acidic conditions and bile is strain-specific, but *B. animalis* subsp. *animalis* has shown better survival than the other bifidobacterial species.

*Lb. delbrueckii* subsp. *bulgaricus* affects the survival of *Lb. acidophilus* and bifidobacteria due to the acid and hydrogen peroxide produced during the fermentation stage. Due to its proteolytic nature, *Lb. delbrueckii* subsp. *bulgaricus* grows rapidly and produces acid quickly; whilst appearing to liberate the essential amino acids such as valine, glycine and histidine that are required to support the growth of bifidobacteria. *S. thermophilus* does not inhibit the growth of probiotic organisms, and may stimulate the growth of probiotic organisms due to the consumption of oxygen. However, some trials conducted at Chr. Hansen Laboratory show that some *S. thermophilus* strains inhibited the growth of bifidobacteria.

### 3.5.1 *Composition of the fermentation medium*

Probiotic bacteria are used for the fermentation of milk to a limited extent because of their slow growth in milk. Although *Lb. acidophilus* and *Bifidobacterium* spp. show some level of β-galactosidase activity, the reason for such poor growth is related to the low concentration of free amino acids and small peptides in milk, which are insufficient to support the growth of these organisms. Therefore, adding casein or whey protein hydrolysates, yeast extract, glucose and vitamins can enhance the growth of *Lb. acidophilus* and *Bifidobacterium* spp. in milk (see also Lucas *et al*., 2004; Desai *et al*., 2004). The addition of milk protein increases the buffering capacity of fermented milks and allows better survival of probiotic organisms. Bifidobacteria are capable of utilising lactulose and oligosaccharides. As other intestinal bacteria are unable to utilise these complex carbohydrates, these compounds are referred to as prebiotics or 'bifidus factors'. Prebiotics are included in most probiotic products in order to promote the growth of bifidobacteria in the intestine (for further details, refer to Chapter 7).

In general, probiotic bacteria grow better in rich synthetic media, viz. tryptose peptone yeast (TPY) and de Man, Rogosa and Sharpe (MRS) broths, than in milk (Shah, 2000). However, these media are complex, costly for the large-scale propagation of probiotic bacteria, and may also impart off-flavour(s) unless extensively washed before incorporation. To manufacture a quality product – both in terms of texture and viability of probiotic bacteria – a milk-based medium is usually required because of the presence of casein.

The slow growth of probiotic micro-organisms in milk leads to the risk of overgrowth of undesirable micro-organisms, and strains that do not grow well tend to produce unpleasant flavours. Normally, it takes 4 h to complete the fermentation process with *Lb. delbrueckii* subsp. *bulgaricus* and *S. thermophilus*, compared to 20 or 24 h with probiotic cultures alone. For this reason, fermented milk products containing *L. acidophilus* and bifidobacteria are often produced in conjunction with other cultures such as *Lb. delbrueckii* subsp. *bulgaricus* and *S. thermophilus* in the case of yoghurt manufacture. Both groups of cultures are either added together, or fermentation takes place in two steps (see Section 3.6.4).

### 3.5.2 *Viability as affected by oxygen*

To maintain confidence in probiotic products, it is important to demonstrate good survival of the bacteria in products throughout the shelf-life of the product. Since bifidobacteria are anaerobic micro-organisms, oxygen toxicity is an important and critical problem. During yoghurt production, oxygen can easily invade and dissolve in the milk. To exclude oxygen during the production of probiotic milk products, special equipment is required to provide an anaerobic environment. Oxygen can also enter the product through packaging materials during storage. A satisfactory growth of a number of *Bifidobacterium* spp. (i.e. without the use of anaerobic conditions) has been reported in a whey-based medium containing L-cysteine (0.05 g 100 $mL^{-1}$) and yeast extract (0.3 g 100 $mL^{-1}$) (Dave & Shah, 1997d, 1998). L-cysteine is added to reduce the redox-potential in order to allow the growth of bifidobacteria.

Oxygen affects probiotic cultures in two ways. First, it is directly toxic to the cells; certain probiotic cultures are sensitive to oxygen and they die in its presence. Second, in the presence of oxygen, certain cultures, particularly *Lb. delbrueckii* subsp. *bulgaricus*, produce peroxide. A synergistic inhibition of probiotic cultures due to acid and hydrogen peroxide has been reported (Lankaputhra & Shah, 1996); for this reason, removal of *Lb. delbrueckii* subsp. *bulgaricus* from some starter cultures (i.e. ABT starter cultures) has achieved some success in improving the survival of probiotic organisms. Several studies have focused on the prevention of detrimental effects of oxygen on probiotic cultures, including the use of anti-oxidants or oxygen scavengers (Dave & Shah, 1997c; see also Talwalkar & Kailasapathy, 2003b; Talwalker *et al.*, 2004).

## 3.6 Methods to improve the viability of probiotic micro-organisms in the product

### 3.6.1 *Selection of bacterial strain(s)*

The parameters for screening micro-organisms for potentially valuable probiotic strains should include the need for the strain to be viable and metabolically active within the GI tract. In addition, it is important that the viability of the strain and stability of its desirable characteristics be maintained during commercial production as well as in the final product (Godward *et al.*, 2000; see also Talwalker & Kailasapathy, 2004). High viable counts and survival rates during passage through the stomach are necessary to allow live probiotics from the fermented milk products to play a biological role in the human intestine. Survival against the acid conditions of the stomach and bile salts are, therefore, of prime importance.

The tolerance of micro-organisms to acid and bile is strain-specific. Many strains of *Lb. acidophilus* and *Bifidobacterium* spp. intrinsically lack the ability to survive harsh conditions in the gut, and may not be suitable for use as dietary adjuncts in fermented milks. Some strains of *Lb. acidophilus* are reported to survive best under acidic conditions, and at modest bile concentrations (Clark *et al.*, 1993; Clark & Martin, 1994). *B. longum* and *Bifidobacterium pseudolongum* have shown the best tolerance to acid and bile salts (Lankaputhra & Shah, 1995). Thus, the selection of appropriate strains on the basis of their acid and bile tolerance would help to improve the viability of these probiotic bacterial strains (Takahashi *et al.*, 2004).

### 3.6.2 *Type of packaging container*

Bifidobacteria are anaerobic, while *Lb. acidophilus* is microaerophilic. Since bifidobacteria are anaerobic, oxygen toxicity is an important consideration, as oxygen can dissolve readily in milk. Dave & Shah (1997b) studied the survival of *Lb. delbrueckii* subsp. *bulgaricus*, *S. thermophilus* and probiotic organisms in yoghurt stored in glass bottles and plastic containers. The increase in numbers and survival of *Lb. acidophilus* during storage was directly related to the dissolved oxygen content, which was higher in yoghurts stored in plastic containers than glass. The counts remained higher in products stored in glass bottles than in those stored in plastic cups, and the bifidobacteria also multiplied better in glass than in plastic. Survival of the bifidobacteria in yoghurt prepared in glass bottles was substantially higher than that prepared in plastic containers. Better survival and viability of bifidobacteria in de-aerated milk has also been observed (Klaver *et al.*, 1993). Thus, it may be important to store the products in glass containers or to increase the thickness of the packaging materials – that is, to reduce the permeability rate of oxygen – used for AB, ABC or ABT products.

### **3.6.3** *Rate of inoculation*

Since probiotic organisms grow poorly in milk, a large inoculum size (5–10 mL 100 $mL^{-1}$) is required, compared to 1 mL 100 $mL^{-1}$ for the yoghurt starter cultures. Similarly, probiotic organisms do not grow well in the presence of other organisms, including *Lb. delbrueckii* subsp. *bulgaricus* and *S. thermophilus*. The size of the inoculum of the primary starter culture may also influence the population of probiotic bacteria. A small inoculum of the yoghurt organisms can result in over-acidification of the product, and this results in poor survival of the probiotic bacteria.

Dave and Shah (1997a) studied the effect of the concentration of starter addition on the viability of *Lb. delbrueckii* subsp. *bulgaricus*, *S. thermophilus* and probiotic bacteria in yoghurt made from four commercial starter cultures. Two starter cultures contained *S. thermophilus* and *Lb. delbrueckii* subsp. *bulgaricus*, *Lb. acidophilus*, and bifidobacteria as the constitutive microbiota, whereas the other two cultures were ABT types. The starter cultures were in the freeze-dried direct-to-vat set (DVS) form and were used at a rate of 0.5, 1.0, 1.5 or 2.0 g 10 $L^{-1}$ in separate containers. The incubation conditions were as recommended by the starter culture manufacturers. Although *S. thermophilus* multiplied better with a lower level of inoculum, the final counts of the organism remained slightly higher at the higher level of inoculum, though primarily this may be due to the higher initial numbers.

The final pH at the end of fermentation is the most crucial factor for the survival of probiotic organisms, as a pH below 4.4 at this point causes a substantial decrease in probiotic bacteria numbers. Hence, the inoculum level must be carefully adjusted and monitored.

### **3.6.4** *Two-stage fermentation*

Inhibitory substances such as acid and hydrogen peroxide produced by yoghurt starter bacteria are mainly responsible for the poor survival of probiotic cultures. Although yoghurt starter cultures produce inhibitory substances against probiotic cultures, the former are essential in yoghurt manufacture to speed-up the fermentation process, and to provide the typical yoghurt flavour. Generally, yoghurt starter bacteria grow faster than probiotic bacteria during the fermentation period and produce acids, which could reduce the viability of probiotic bacteria.

One method of improving the viability of probiotic organisms is to add probiotic organisms after fermentation of the milk. This allows the use of strains of probiotic bacteria that cannot grow in the presence of other organisms. However, the survival of probiotic organisms may be lower if the bacteria are added after fermentation.

Another method would be to carry out the initial fermentation with probiotic cultures, followed by completion of fermentation with, for example, *Lb. delbrueckii* subsp. *bulgaricus* and *S. thermophilus* (Lankaputhra & Shah, 1997). Using this approach, the fermentation time could be slightly longer than the traditional fermentation process. An initial fermentation could be carried out with probiotic cultures for 2 h, followed by fermentation with the yoghurt starter cultures for 4 h. This would allow the probiotic organisms to be in the final stage of their lag phase or the early stage of their log phase, and result in higher counts of probiotic organisms at the end of the 6-h fermentation. The counts of probiotic bacteria have been found to increase substantially in products prepared using a two-step fermentation process.

### 3.6.5 *Microencapsulation technique*

The numbers of probiotic bacteria in frozen fermented dairy desserts or frozen yoghurt are reduced significantly by the presence of acid, freeze-injury, a high sugar concentration of the product and oxygen toxicity (Tamime *et al.*, 1995). About 16 g sugar 100 $g^{-1}$ product are added to frozen fermented dairy desserts, and this has been found to affect the growth of probiotic bacteria. Microencapsulation is a process whereby the cells are retained within the encapsulating membrane in order to reduce cell injury or cell loss (Fig. 3.4). This may have applications in several products, such as cheese (Godward & Kailasapathy, 2003a), yoghurt (ChienJung, 2000; Adhikari *et al.*, 2000; Sultana *et al.*, 2000; WenRong & Griffiths, 2000; Hansen *et al.*, 2002; Godward & Kailasapathy, 2003c; Picot & Lacroix, 2004), and simulated

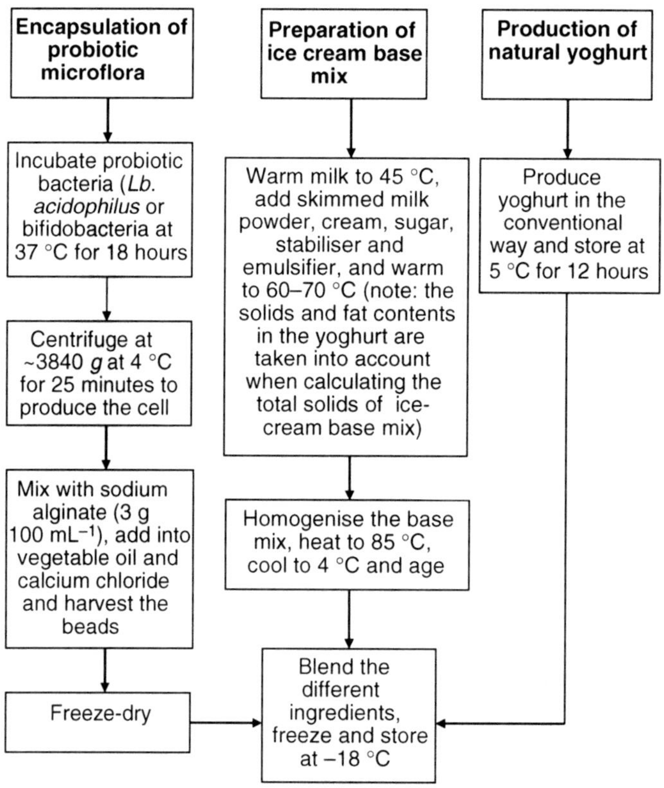


**Fig. 3.4** Preparation of probiotic fermented frozen dessert. Note: mix the yoghurt, ice-cream base and syrup (i.e. sucrose at 65% brix) in the following proportions: 45, 45 and 10, respectively. Adapted from Shah & Ravla (2004).

gastric juice and bile solutions (WenChian *et al.*, 2003; see also Kailasapathy & Sultana, 2003; Sridar *et al.*, 2003; Krasaekoopt *et al.*, 2003, 2004; Chandramouli *et al.*, 2004; Hsiao *et al.*, 2004; Anjani & Kailasapathy, 2004; HungChi *et al.*, 2004; see section 3.4.2).

Gelatin or vegetable gums have been used for the microencapsulation of bacteria, and have been reported to provide protection to acid-sensitive probiotic organisms. The entrapment of living microbial cells in calcium alginate is simple and inexpensive; moreover, alginate is non-toxic so that it may be safely used in foods. Alginate gels can be solubilised by sequestering calcium ions such that the entrapped cells are released (Rao *et al.*, 1989; Sheu & Marshall, 1993).

Encapsulated probiotic organisms, when incorporated into fermented frozen dairy desserts, showed an improved viability of $>10^5$ cfu $g^{-1}$ in the product compared to counts of $<10^3$ cfu $g^{-1}$ when non-encapsulated organisms were used (Shah & Ravula, 2000b, 2004).

### 3.6.6 Supplementation of the milk with nutrients

During yoghurt making, *S. thermophilus* dominates the early stage of fermentation. As the redox potential of milk is reduced and the pH is lowered from 6.5 to 5.5, growth of *Lb. delbrueckii* subsp. *bulgaricus* is stimulated. *S. thermophilus* dominates the early stage of fermentation and, then below pH 5.0, *Lb. delbrueckii* subsp. *bulgaricus* dominates to produce acetaldehyde and lactic acid, yielding the characteristic green apple flavour. Continued acid production lowers the pH of yoghurt to almost 4.6, which is the iso-electric point of casein; this induces gelation and the fermentation is terminated at pH 4.5.

*Lb. delbrueckii* subsp. *bulgaricus* produces essential amino acids owing to its proteolytic nature (Shihata & Shah, 2000, 2002), and its associative growth relationship with *S. thermophilus* is well established. The streptococci also produce growth factors for the former organism. However, *Lb. delbrueckii* subsp. *bulgaricus* also produces lactic acid during refrigerated storage. This process is known in the industry as 'post-acidification' and, if it occurs during refrigerated storage, it causes a loss in viability of the probiotic bacteria.

To overcome losses in the viability of probiotic bacteria due to acid produced during fermentation and storage (post-acidification), the current trend is to use starter cultures that are devoid of *Lb. delbrueckii* subsp. *bulgaricus*, such as ABT. *S. thermophilus*, which is less proteolytic than *Lb. delbrueckii* subsp. *bulgaricus*, is the main organism responsible for fermentation in some ABT cultures (e.g. ABT-1 and ABT-2). This blend of starter cultures increases the fermentation time significantly (up to 10 h), as there is no associative growth without *Lb. delbrueckii* subsp. *bulgaricus*. These ABT starter cultures necessitate the incorporation of micronutrients (peptides and amino acids) via casein hydrolysates in order to reduce the fermentation time and improve the viability of probiotic organisms.

Dave and Shah (1998) studied the effects of some dairy and non-dairy ingredients (whey powder, WP; whey protein concentrates, WPC; and acid casein hydrolysates, ACH) on the viability of *Lb. acidophilus* and bifidobacteria in yoghurt made from four commercial starter cultures. The added ingredients considerably affected the incubation time, as some provided peptides and amino acids for bacterial growth. The addition of WP, WPC and ACH improved the viability of bifidobacteria, as did nitrogen sources in the form of peptides and amino acids, and was responsible for an improved viability of the bifidobacteria. However, the addition of milk protein (casein and whey) hydrolysates enhanced the acidification rate of *S. thermophilus* and reduced the growth rate of the probiotic micro-organisms (*Lb. acidophilus* LA-5 and *Lb. rhamnosus* LR-35) in fermented milks during the manufacturing stages, although the survival count of the latter bacteria was improved after storage (Lucas *et al.*, 2004).

### 3.6.7 *The use of oxygen scavengers*

Oxygen content and redox potential have been shown to be important factors for the viability of bifidobacteria during storage. Ascorbic acid (vitamin C) acts as an oxygen scavenger, and is permitted in fruit juices and other products as a food additive. Furthermore, milk and milk products supply only 10–15% of the daily requirements of vitamin C (Räsic & Kurmann, 1978). Thus, the fortification of yoghurt with ascorbic acid would increase its nutritive value.

The oxygen content and redox potential are reported gradually to increase during the storage of yoghurt in plastic cups, but the redox potential remains lower in the presence of ascorbic acid (Dave & Shah, 1997c). *S. thermophilus* is aerobic, and counts would be expected to remain lower in the presence of ascorbic acid. The viability of *Lb. delbrueckii* subsp. *bulgaricus* (being micro-aerophilic to anaerobic) is expected to improve with increasing concentrations of ascorbic acid.

Although the addition of ascorbic acid helps to improve the survival of *Lb. acidophilus*, the oxygen-scavenging effect may not be sufficient to improve the viability of anaerobic bifidobacteria (Dave & Shah, 1997c).

### 3.6.8 *The addition of cysteine*

Media used for the enumeration of bifidobacteria often contain L-cysteine (0.5–0.1 g 100 $mL^{-1}$) in order to improve recovery. Cysteine, a sulphur-containing amino acid, provides amino nitrogen as a growth factor while reducing the redox potential, both of which favour the growth of anaerobic bifidobacteria.

Cysteine at 250 mg $L^{-1}$ appears to improve the survival of *Lb. acidophilus* and *Bifidobacterium* spp. It should be noted that a low level of cysteine (50 mg $L^{-1}$) would promote the growth of *S. thermophilus* and reduce incubation time, particularly in ABT starter cultures (Dave & Shah, 1997d). A slight decrease in redox potential is beneficial for the survival of *S. thermophilus*; however, when the cysteine concentration is raised above 50 mg $L^{-1}$, the depression in redox potential affects bacterial growth. The growth of *Lb. delbrueckii* subsp. *bulgaricus* would be improved with low levels of cysteine, but suppressed at higher levels.

## 3.7 Future developments

It is clear that, over the past three decades, there has been growing interest in the incorporation of probiotic micro-organisms in dairy products, mainly with regard to health-associated benefits to the consumer (see Chapter 8). This has been reflected in the increase in the numbers of these products in worldwide markets, and also by increases in the consumption of probiotic dairy products such as fermented milks. Indeed, much knowledge has been acquired with regard to the growth, survival and new isolates of probiotic bacteria, in addition to the development of new technologies for the manufacture of different dairy products.

According to Mattila-Sandholm *et al.* (2002), future technological prospects that should be considered or addressed in view of the 'functionality' of probiotic micro-organisms include the following:

- Fermentation and drying technologies
- Microencapsulation
- Strain characterisation including daily dosage, stability, viability, and non-viability
- Food matrix formulation
- Targets prebiotics

Furthermore, the probiotic dairy industry should work closely not only with the regulatory authorities in any one particular country, but also with the medical profession in order to substantiate the health claims associated with these beneficial micro-organisms. In addition, there is a need to determine how best to communicate with the consumer. Clearly, overcoming these difficulties will help to ensure the acceptability of probiotic dairy products by the consumer and, hence, safeguard the future of the industry.

## References

Abdou, S.M., El-Alfy, M.B., Hefny, A.A. & Nasr, W.I.A. (2003) Probiotic Ras cheese made from concentrated milk by ultrafiltration or recombination techniques. *Egyptian Journal of Dairy Science*, **31**, 319–333.

Abou-Dawood, S.A.I. (2002) Survival of nonencapsulated and encapsulated *Bifidobacterium bifidum* in probiotic Karish cheese. *Egyptian Journal of Dairy Science*, **30**, 43–52.

Adhikari, K., Mustapha, A., Grün, I.U. & Fernando, L. (2000) Viability of microencapsulated bifidobacteria in set yogurt during refrigerated storage. *Journal of Dairy Science*, **83**, 1946–1951.

Ahola, A.J., Yli-Knuuttila, H., Suomalainen, T., Poussa, T., Ahlstrom, A., Meurman, J.H. & Korpela, R. (2002) Short-term consumption of probiotic-containing cheese and its effect on dental caries. *Archives of Oral Biology*, **47**, 799–804.

Alamprese, C., Foschino, R., Rossi, M., Pompei, C. & Savani, L. (2002) Survival of *Lactobacillus johnsonii* La1 and influence of its addition in retail-manufactured ice-cream produced with different sugar and fat concentrations. *International Dairy Journal*, **12**, 201–208.

Amer, E.S.N., Girgis, S.S., Taha, H. & Abed-El-Moeety, S.H. (1997) Effect of milk total solids and type of starter on the quality of Labneh. *Egyptian Journal of Dairy Science*, **25**, 179–192.

Anjani, K. & Kailasapathy, K. (2004) Survival of co-encapsulated probiotics and prebiotics in yoghurt. *Milchwissenschaft*, **59**, 396–399.

Anonymous (1992) Yoghurt and probiotics. *Choice*, **11**, 32–35.

Anonymous (2001) Choose life top. *Dairy Industries International*, **66**(11), 45.

Anonymous (2003) *Short Life Dairy Products*, Müller Market Report 2003, Müller Dairy (UK) Ltd., Market Drayton.

Antonsson, M., Ardö, Y., Nilsson, B.F. & Molin, G. (2002) Screening and selection of *Lactobacillus* strains for use as adjunct cultures in production of semi-hard cheese. *Journal of Dairy Research*, **69**, 457–472.

Baron, M., Roy, D. & Vuillemard, J.-C. (2000) Biochemical characteristics of fermented milk produced by mixed-cultures of lactic starters and bifidobacteria. *Lait*, **80**, 465–478.

Barraquio, V.L., Militate, S.T., Gonzaga, J.O. & Emata, O.C. (2001) Fermented milk drink as starter adjunct in the manufacture of probiotic white soft cheese. *The Philippine Agricultural Scientist*, **84**, 382–387.

Blanchette, L., Roy, D., Belanger, G. & Gauthier, S.F. (1996) Production of Cottage cheese using dressing fermented by bifidobacteria. *Journal of Dairy Science*, **79**, 8–15.

Belem, M.A.F. (1999) Application of biotechnology in the product development of nutraceuticals in Canada. *Trends in Food Science & Technology*, **10**, 101–106.

Bertolami, M.C. & Farnworth, E.R. (2003) The properties of *Enterococcus faecium* and the fermented milk product Gaio®. *Handbook of Fermented Functional Foods*, (ed. E.R. Farnworth), pp. 59–75, CRC Press LLC, Boca Raton.

Bottazzi, V. (2002) [Siamo e molto bravi in biochimica: produciamo lo yoghurt]. *Industria del Latte*, **XXXVIII**(1-2), 5–64.

Boylston, T.D., Vinderola, C.G., Ghoddusi, H.B. & Reinheimer, J.A. (2004) Incorporation of bifidobacteria into cheeses: challenges and rewards. *International Dairy Journal*, **14**, 375–387.

Chan, E.S. & Zhang, Z. (2002) Encapsulation of probiotic bacterial *Lactobacillus acidophilus* by direct compression. *Trans IChemE*, **80**(C), 78–82.

Chandramouli, V., Kailasapathy, K., Peiris, P. & Jones, M. (2004) An improved method of microencapsulation and its evaluation to protect *Lactobacillus* spp. in simulated gastric conditions. *Journal of Microbiological Methods*, **56**, 27–35.

Charteris, W.P., Kelly, P.M., Morelli, L. & Collins, J.K. (2002) Edible table (bio)spread containing potentially probiotic *Lactobacillus* and *Bifidobacterium* species. *International Journal of Dairy Technology*, **55**, 44–56.

ChienJung, H. (2000) [Studies on the manufacture of yoghurt by microentrapment of Bifidobacteria]. *Journal of the Chinese Society of Animal Science*, **29**, 125–131.

Chouraqui, J.-P., van Egroo, L.D. & Ficot, M.C. (2004) Acidified milk formula supplemented with *Bifidobacterium lactis*: impact on infant diarrhea in intestinal care settings. *Journal of Pediatric Gastroenterology and Nutrition*, **38**, 288–292.

Christiansen, P.S., Edelsten, D., Kristiansen, J.R. & Nielsen, E.W. (1996) Some properties of ice-cream containing *Bifidobacterium bifidum* and *Lactobacillus acidophilus*. *Milchwissenschaft*, **51**, 502–504.

Clark, P.A. & Martin, J.H. (1994) Selection of bifidobacteria for use as dietary adjuncts in cultured dairy foods: III - tolerance to simulated bile concentrations of human small intestines. *Cultured Dairy Products Journal*, **29**(3), 18–21.

Clark, P.A., Cotton, L.N. & Martin, J.H. (1993) Selection of bifidobacteria for use as dietary adjuncts in cultured dairy foods: II - tolerance to simulated pH of human stomachs. *Cultured Dairy Products Journal*, **28**(4), 11–14.

Collins, J.K. (2001) Demonstration of functional properties of probiotic lactic acid bacteria. *Industria del Latte*, **XXXVII**(1-2), 33–60.

Collins, J.K., Thornton, G. & Sullivan, G.O. (1998) Selection of probiotic strains for human applications. *International Dairy Journal*, **8**, 487–490.

Corbo, M.R., Albenzio, M., de-Angelis, M., Sevi, A., Gobbetti, M. & de-Angelis, M. (2001) Microbiological and biochemical properties of Canestrato Pugliese hard cheese supplemented with bifidobacteria. *Journal of Dairy Science*, **84**, 551–561.

Daigle, A., Roy, D. & Vuillemard, J.C. (1998) Growth response and metabolic activity of *Bifidobacterium infantis* ATCC 27920G in creams enriched with native phosphocaseinate retentate. *Milchwissenschaft*, **53**, 625–629.

Daigle, A., Roy, D., Belanger, G. & Vuillemard, J.C. (1999) Production of probiotic cheese (Cheddar-like cheese) using enriched cream fermented by *Bifidobacterium infantis*. *Journal of Dairy Science*, **82**, 1081–1901.

Dave, R.I. & Shah, N.P. (1997a) Effect of level of starter culture on viability of yoghurt and probiotic bacteria in yoghurts. *Food Australia*, **49**, 164–168.

Dave, R.I. & Shah, N.P. (1997b) Viability of yoghurt and probiotic bacteria in yoghurts made from commercial starter cultures. *International Dairy* Journal, **7**, 31–41.

Dave, R.I. & Shah, N.P. (1997c) Effectiveness of ascorbic acid as an oxygen scavenger in improving viability of probiotic bacteria in yoghurts made with commercial starter cultures. *International Dairy Journal*, 7, 435–443.

Dave, R.I. & Shah, N.P. (1997d) Effectiveness of cysteine as redox potential reducing agent in improving viability of probiotic bacteria in yoghurts made with commercial starter cultures. *International Dairy Journal*, **7**, 537–545.

Dave, R.I. & Shah, N.P. (1998) Ingredient supplementation effects on viability of probiotic bacteria in yoghurt. *Journal of Dairy Science*, **81**, 2804–2816.

Davidson, R.H., Duncan, S.E., Hackney, C.R., Eigel, W.N. & Boling, J.W. (2000) Probiotic cultures survival and implications in fermented frozen yogurt characteristics. *Journal of Dairy Science*, **83**, 666–673.

Delaney, A.-G. O. (1977) Factory-scale experiments on the production of ymer by ultrafiltration. *Milchwissenschaft*, **32**, 651–653.

Desai, A.R., Powell, I.B. & Shah, N.P. (2004) Survival and activity of probiotic lactobacilli in skim milk containing prebiotics. *Journal of Food Science*, **69**, FMS57–FMS60.

Desmond, C., Stanton, C., Fitzgerald, G.F., Collins, K. & Ross, R.P. (2001) Environmental adaptation of probiotic lactobacilli towards improvement of performance during spray drying. *International Dairy Journal*, **11**, 801–808.

Desmond, C., Ross, R.P., O'Callaghan, E., Fitzgerald, G. & Stanton, C. (2002) Improved survival of *Lactobacillus paracasei* NFBC 338 in spray-dried powders containing gum acacia. *Journal of Applied Microbiology*, **93**, 1003–1011.

Dinakar, P. & Mistry, V.V. (1994) Growth and viability of *Bifidobacterium bifidum* in Cheddar cheese. *Journal of Dairy Science*, **77**, 2854–2864.

Donnelly, L. (2003) Gut feeling. *Dairy Industries International*, **68**(5), 19–20.

Drakoularakou, A.P., Kehagias, C., Karakanas, P.N., Koulouris, S. & Timbis, D. (2003) A study of the growth of *Lactobacillus acidophilus* in bovine, ovine and caprine milk. *International Journal of Dairy Technology*, **56**, 59–61.

Edwards, C.A., Parrett, A.M., Stephen, A.M., Henry, C. J. Marks, J. & Shortt, C. (2002) Intestinal flora during the first months of life: new perspectives. *British Journal of Nutrition*, **88**(Suppl. 1), S11–S18.

El-Kholy, W.I., El-Shafei, K. & Sadek, Z.I. (2003) The use of immobilized *Lactobacillus* metabolites as a preservative in soft cheese. *Arab Universities Journal of Agricultural Sciences*, **11**, 607–622.

El-Rahman, A.E.R.M.A. (2000) Survival of probiotic bacteria in yoghurt and their antimicrobial effect. *Alexandria Journal of Agricultural Research*, **45**, 63–80.

El-Samragy, Y.A. (1997) Labneh or yoghurt-cheese: a review. *Egyptian Journal of Dairy Science*, **25**, 165–178.

El-Shafei, K. (2003) Production of cultured buttermilk using mixed cultures containing polysaccharide producing *Leuconostoc mesenteroides*. *Egyptian Journal of Dairy Science*, **31**, 43–60.

El-Shazly, A., El Tahra, M.A.A. & Abo-Sera, M.M. (2004) Effect of different methods for the manufacture of frozen yogurt on its properties. *9th Egyptian Conference for Dairy Science and Technology, Cairo 9–11 October 2004, Research Papers I*, 183–194.

El Tahra, M.A.A., El Shazley, A. & Abo-Sera, M.M. (2004a) Effect of fat content on the properties of fresh and stored frozen yoghurt. *9th Egyptian Conference for Dairy Science and Technology, Cairo 9–11 October 2004, Research Papers I*, 195–207.

El Tahra, M.A.A., El Shazley, A. & Abo-Sera, M.M. (2004b) Effect of fat content and probiotic cultures on the properties of frozen yoghurt. *9th Egyptian Conference for Dairy Science and Technology, Cairo 9–11 October 2004, Research Papers I*, 209–225.

Farnworth, E.R. & Mainville, I. (2003) Kefir: A fermented milk product. *Handbook of Fermented Functional Foods*, (ed. E.R. Farnworth), pp. 77–111, CRC Press LLC, Boca Raton.

Fondén, R., Saarela, M., Mätto, J. & Mattila-Sandholm, T. (2003) Lactic acid bacteria (LAB) in functional dairy products. *Functional Dairy Products*, (eds. T. Mattila-Sandholm & M. Saarela), pp. 244–262, Woodhead Publishing Ltd., Cambridge.

Gardiner, G.E., Ross, R.P., Collins, J.K., Fitzgerald, G. & Stanton, C. (1998) Development of a probiotic Cheddar cheese containing human-derived *Lactobacillus paracasei* strains. *Applied and Environmental Microbiology*, **64**, 2192–2199.

Gardiner, G.E., Stanton, C., Lynch, P.B., Collins, J.K., Fitzgerald, G. & Ross, R.P. (1999a) Evaluation of Cheddar cheese as a food carrier for delivery of a probiotic strain to the gastrointestinal tract. *Journal of Dairy Science*, **82**, 1379–1387.

Gardiner, G.E., Ross, R.P., Wallace, J.M., Scanlan, F.P., Jagers, P.P.J.M., Fitzgerald, G.F., Collins, J.K. & Stanton, C. (1999b) Influence of a probiotic adjunct culture of *Enterococcus faecium* on the quality of Cheddar cheese. *Journal Agriculture and Food Chemistry*, **47**, 4907–4916.

Gardiner, G.E., O'Sullivan, E., Kelly, J., Auty, M.A.E., Fitzgerald, G.F., Collins, J.K., Ross, R.P. & Stanton, C. (2000) Comparative survival rates of human-derived probiotic *Lactobacillus paracasei* and *Lactobacillus salivarius* strains during heat treatment and spray drying. *Applied and Environmental Microbiology*, **66**, 2605–2616.

Gardiner, G.E., Ross, R.P., Kelly, P.M. & Stanton, C. (2002a) Microbiology of therapeutic milks. *Dairy Microbiology Handbook*, (ed. R.K. Robinson), 3rd edn. pp. 431–478, John Wiley & Sons Inc., New York.

Gardiner, G.E., Bouchier, P., O'Sullivan, E., Kelly, J., Collins, J.K., Fitzgerald, G., Ross, R.P. & Stanton, C. (2002b) A spray-dried culture for probiotic cheese manufacture. *International Dairy Journal*, **12**, 749–756.

Gardini, F., Lanciotti, R.M., Guerzoni, M.E. & Torriani, S. (1999) Evaluation of aroma production and survival of *Streptococcus thermophilus*, *Lactobacillus delbrueckii* subsp. *bulgaricus* and *Lactobacillus acidophilus* in fermented milks. *International Dairy Journal* , **9**, 125–143.

Garrotte, G.L., Abraham, A.G. & De Antoni G.L. (2001) Chemical and microbiological characterisation of kefir grains. *Journal of Dairy Research*, **68**, 639–652.

Ghoddusi, H.B. & Robinson, R.K. (1996) The test of time. *Dairy Industries International*, **63**(7), 25–28.

Gilliland, S.E. (2001) Probiotics and prebiotics. *Applied Dairy Microbiology* (eds. E.H. Marth & J.L. Steele), 2nd edn., pp. 327–344, Marcel Dekker, New York.

Gilliland, S.E., Morelli, L. & Reid, G. (2001) Health and nutritional properties of probiotics in food including powder milk with live lactic acid bacteria. *Evaluation of Health and Nutritional Properties of Probiotics in Food Including Powder Milk with Live Lactic Acid Bacteria* – Report of joint FAO/WHO of expert consultation, pp. 1–34, Food and Agriculture Organization of the United Nations, Rome. http://www.fao.org/es/ESN/probio/report.pdf

Gobbetti, M., Corsetti, A., Smacchi, E., Zocchetti, A. & de Angelis, M. (1998) Production of Crescenza cheese by incorporation of bifidobacteria. *Journal of Dairy Science*, **81**, 37–47.

Godward, G. & Kailasapathy, K. (2000) Viability and survival of free, encapsulated and co-encapsulated probiotic bacteria in ice-cream. *Milchwissenschaft*, **58**, 161–164.

Godward, G. & Kailasapathy, K. (2003a) Viability and survival of free and encapsulated probiotic bacteria in Cheddar cheese. *Milchwissenschaft*, **58**, 624–627.

Godward, G. & Kailasapathy, K. (2003b) Viability and survival of free, encapsulated and co-encapsulated probiotic bacteria in ice-cream. *Milchwissenschaft*, **58**, 161–164.

Godward, G. & Kailasapathy, K. (2003c) Viability and survival of free, encapsulated and co-encapsulated probiotic bacteria in yoghurt. *Milchwissenschaft*, **58**, 396–399.

Godward, G., Sultana, K., Kailasapathy, K., Peiris, P., Arumugaswamy, R. & Reynolds, N. (2000) The importance of strain selection on the viability and survival or probiotic bacteria in dairy foods. *Milchwissenschaft*, **55**, 441–445.

Gomes, A.M.P. & Malcata, F.X. (1998) Development of probiotic cheese manufactured from goat milk: response surface analysis via technological manipulation. *Journal of Dairy Science*, **81**, 1492–1507.

Gomes, A.M.P. & Malcata, F.X. (1999) *Bifidobacterium* spp. and *Lactobacillus acidophilus*: biological, biochemical, technological and therapeutical properties relevant for use as probiotics. *Trends in Food Science & Technology*, **10**, 139–157.

Gomes, A.M.P., Malcata, F.X., Klaver, F.A.M. & Grande, H.J. (1998) Incorporation and survival of *Bifidobacterium* spp. strain Bo and *Lactobacillus acidophilus* strain Ki in a cheese product. *Netherlands Milk and Dairy Journal*, **49**, 71–95.

Gomes, A.M.P., Malcata, F.X. & Klaver, F.A.M. (1995) Growth enhancement of *Bifidobacterium lactis* Bo and *Lactobacillus acidophilus* Ki by milk hydrolysate. *Journal of Dairy Science*, **81**, 2811–2825.

Goni-Cambrodon, I. & Gudiel-Urbano, M. (2001) Oligosaccharide supplementation in infant formula: natural prebiotics. *Alimentaria*, **38**, 77–81.

Gudmundsson, B. (1987) Skyr. *Scandinavian Dairy Industry*, **1**(4), 240–242.

Hagen, M. & Narvhus, J.A. (1999) Production of ice-cream containing probiotic bacteria. *Milchwissenschaft*, **54**, 265–268.

Halpin-Dohnalek, M.I., Hilty, M.D. & Bynum, D.G. (1999) Method and formula for the prevention of diarrhea. *U.S. Patent* No. 5,902,578.

Hamed, A.I., Zedan, M.A., Salem, O.M., Moussa, A.M. & Yousef, E.T.A. (2004) Impact of frozen yoghurt ingredients on its quality and survival of bifidobacteria: III Effect of milk solids not fat sources.. *9th Egyptian Conference for Dairy Science and Technology, Cairo 9–11 October 2004, Research Papers I*, 227–241.

Hansen, S.L. (1997) Protective cultures for cheese. *Meieriposten*, **86**, 261–262.

Hansen, L.T., Allan-Wojtas, P.M., Jin, Y.-L. & Paulson, A.T. (2002) Survival of Ca-alginate microencapsulated *Bifidobacterium* spp. in milk and simulated gastrointestinal conditions. *Journal of Food Microbiology*, **19**, 35–45.

Hawrelak, J. (2002) Probiotics: Are supplements really better than yoghurt? *Journal of the Australian Traditional-Medicine*, **8**, 11–23.

Haynes, I.N. & Playne, M.J. (2002) Survival of probiotic cultures in low-fat ice-cream. *Australian Journal of Dairy Technology*, **57**, 10–14.

Helland, M.H., Wicklund, T. & Narvhus, J.A. (2004) Growth and metabolism of selected strains of probiotic bacteria in milk- and water-based cereal puddings. *International Dairy Journal*, **14**, 957–965.

Heller, K.J. (2001) Probiotic bacteria in fermented foods: product characteristics and starter organisms. *American Journal of Clinical Nutrition*, **73**(S supplement), 374S–379S.

Heller, K.J., Bockelmann, W., Schrezenmeir, J. & de Vrese, M. (2003) Cheese and its Potential as a Probiotic Food. *Handbook of Fermented Functional Foods*, (ed. E.R. Farnworth), pp. 203–225, CRC Press, Boca Raton.

Holm, F. (2003) Gut health and diet: the benefits of probiotic and prebiotics on human health. *The World of Ingredients*, **2**, 52–55.

Hsiao, H.-C., Lian, W.-C. & Chou, C.-C. (2004) Effect of packaging conditions and temperature on viability of microencapsulated bifidobacteria during storage. *Journal of the Science of Food and Agriculture*, **84**, 134–139.
Hughes, D.B. & Hoover, D.G. (1991) Bifidobacteria: Their potential for use in American dairy products. *Food Technology*, **45**(4), 74–75, 78–80, 83.
HungChi, H., WenChian, L. & ChengChun, C. (2004) Effect of packaging conditions and temperature on viability of microencapsulated bifidobacteria during storage. *Journal of the Science of Food and Agriculture*, **84**, 134–139.
IDF (2002) *World Dairy Situation 2002*, Document No. 378, pp. 60, International Dairy Federation, Brussels.
Iwana, H., Masuda, H., Fujisawa, T., Suzuki, H. & Mitsuoka, T. (1993) Isolation and identification of *Bifidobacterium* spp. in commercial yoghurts sold in Europe. *Bifidobacteria Microflora*, **12**, 39–45.
Jan, G., Leverrier, P., Proudy, I. & Roland, N. (2002) Survival and beneficial effects of propionibacteria in the human gut: *in vivo* and *in vitro* investigations. *Lait*, **82**, 131–144.
Jiang, T., Mustapha, A. & Saviano, D.A. (1996) Improvement of lactose digestion in humans by ingestion of unfermented milk containing *Bifidobacterium longum*. *Journal of Dairy Science*, **79**, 750–757.
Kailasapathy, K. (2002) Microencapsulation of probiotic bacteria: technology and potential applications. *Current Issues in Intestinal Microbiology*, **3**(2), 39–48.
Kailasapathy, K. & Rybka, S. (1997) *L. acidophilus* and *Bifidobacterium* spp. – their therapeutic potential and survival in yogurt. *Australian Journal of Dairy Technology*, **52**, 28–35.
Kailasapathy, K. & Sultana, K. (2003) Survival of β-D-galactosidase activity of encapsulated and free *Lactobacillus acidophilus* and *Bifidobacterium lactis* in ice-cream. *Australian Journal of Dairy Technology*, **58**, 223–227.
Kaneko, T. (1999) A novel bifidogenic growth stimulator produced by *Propionibacterium freudenreichii*. *Bioscience and Microflora*, **18**, 73–80.
Kasimoğlu, A., Göncüoğlu, M. & Akgün, S. (2004) Probiotic white cheese with *Lactobacillus acidophilus*. *International Dairy Journal*, **14**, 1067–1073.
Katla, A.K., Kruse, H. Johnsen, G. & Herikstad, H. (2001) Antimicrobial susceptibility of starter culture bacteria used in Norwegian dairy products. *International Journal of Food Microbiology*, **67**, 147–152.
Kaur, I.P., Chopra, K. & Saini, A. (2002) Probiotics: potential pharmaceutical applications, *European Journal of Pharmaceutical Sciences*, **15**(1), 1–9.
Kebary, K.M.K., Hussein, S.A. & Badawi, R.M. (1998) Improving the viability of bifidobacteria and their effect on frozen ice milk. *Egyptian Journal of Dairy Science*, **26**, 319–337.
Kebary, K.M.K, Hamed, A.I., Salem, O.M. & Yousef, E.T.A. (2004) Impact of frozen yoghurt ingredients on its quality and survival of bifidobacteria:- I Effect of sweeteners. *9th Egyptian Conference for Dairy Science and Technology, Cairo 9–11 October 2004, Research Papers I*, 165–163.
Klaver, F.A.M., Kingma, F. & Weerkamp, A.H. (1993). Growth and survival of bifidobacteria in milk. *Netherlands Milk and Dairy Journal*, **47**, 151–164.
Krasaekoopt, W., Bhandari, B. & Deeth, H. (2003) Evaluation of encapsulation techniques of probiotics for yoghurt. *International Dairy Journal*, **13**, 3–13.
Krasaekoopt, W., Bhandari, B. & Deeth, H. (2004) The influence of coating materials on some properties of alginate beads and survival of microencapsulated probiotic bacteria. *International Dairy Journal*, **14**, 737–743.
Krishnakumar, V. & Gordon, I.R. (2001) Probiotics: challenges and opportunities. *Dairy Industries International*, 66(2), 38–40.
Kunz, C. & Rudolff, S. (2002) Health benefits of milk-derived carbohydrates. Fresh Perspectives on Bioactive Dairy Products, Document No. 375, pp. 72–79, International Dairy Federation, Brussels.
Kurmann, J.A., Rašić, J.Lj. & Kroger, M. (1992) *Encyclopedia of Fermented Milk Products*. Van Nostrand Reinhold, New York.
Lankaputhra, W.E.V. & Shah, N.P. (1995) Survival of *Lactobacillus acidophilus* and *Bifidobacterium* spp. in the presence of acid and bile salts. *Cultured Dairy Products Journal*, **30**(3), 2–7.
Lankaputhra, W.E.V. & Shah, N.P. (1996) A simple method for selective enumeration of *Lactobacillus acidophilus* in yogurt supplemented with *L. acidophilus* and *Bifidobacterium* spp. *Milchwissenschaft*, **51**, 446–451.
Lankaputhra, W.E.V. & Shah, N.P. (1997) Improving viability of *L. acidophilus* and bifidobacteria in yoghurt using two step fermentation and neutralized mix. *Food Australia*, **49**, 363–369.
La Torre, L., Tamime, A.Y. & Muir, D.D. (2003) Rheology and sensory profiling of set-type fermented milks made with different commercial probiotic and yoghurt starter cultures. *International Journal of Dairy Technology*, **56**, 163–170.
Leporanta, K. (2001) The world of fermented milks – 1: drinking type products originating from Finland and Russia. *LGG Action*, **2**, 10–12.

Leporanta, K. (2003) The world of fermented milks – 4: viili and långfil – exotic fermented products from Scandinavia. *Valio Foods & Functionals*, **2**, 3–5.

Leroy, F. & de Vuyst, L. (2004) Lactic acid bacteria as functional starter cultures for the food fermentation industry. *Trends in Food Science & Technology*, **15**, 67–78.

Lian, W.-C., Hsiao, H.-C. & Chou, C.-C. (2002) Survival of bifidobacteria after spray-drying. *International Journal of Food Microbiology*, **74**, 79–86.

Lonnerdal, B. (2003) Breast milk and breastfed infants: implications for improving infant formula. *American Journal of Clinical Nutrition*, **77**, 1537S–1543S.

Lourens-Hattingh, A. & Viljoen, B.C. (2001a) Yogurt as probiotic carrier food. *International Dairy Journal*, **11**, 1–17.

Lourens-Hattingh, A. & Viljoen, B.C. (2001b) Growth and survival of a probiotic yeast in dairy products. *Food Research International*, **34**, 791–796.

Lourens-Hattingh, A. & Viljoen, B.C. (2002) Survival of probiotic bacteria in South African commercial bio-yogurt. *South African Journal of Science*, **98**, 298–300.

Lucas, A., Sodini, I., Monnet, C., Jolivet, P. & Corrieu, G. (2004) Probiotic cell counts and acidification in fermented milks supplemented with milk protein hydrolysates. *International Dairy Journal*, **14**, 47–53.

Ma, G. (1995) An ice-cream containing *Lactobacillus bifidus*. *Journal of Food Industry*, **1**, 15–16.

Mahdi, H.A., Tamime, A.Y. & Davies, G. (1990) Some aspects of the production of "Labneh" by ultrafiltration using cow's, sheep's and goat's milk. *Egyptian Journal of Dairy Science*, **18**, 345–367.

Maiocchi, G. (2001) YOMO ABC: Functional food for consumers well-being and satisfaction. *Industria del Latte*, **XXXVII**(1-2), 94–98.

Mantere-Alhonen, S. (1995) Propionibacteria used as probiotics – a review. *Lait*, **75**, 447–452.

Mantere-Alhonen, S. & Forsen, R. (1990) [Microbes in fermented milk products]. [*Microbes in Milk and Milk Products*], (ed. S. Mantere-Alhonen), pp. 129–148, VAPK-Kustannus, Helsinki.

Marshall, V.M. (1987) Fermented milks and their future trends - I. microbiological aspects. *Journal of Dairy Research*, **54**, 559–574.

Marshall, V.M.E. & Tamime, A.Y. (1997) Physiology and biochemistry of fermented milks. *Microbiology and Biochemistry of Cheese and Fermented Milk*, (ed. B.A. Law), 2nd edn., pp. 153–192, Blackie Academic & Professional, London.

Martín-Diana, A.B., Janer, C., Peláez, C. & Requena, T. (2003) Development of a fermented goat's milk containing probiotic bacteria. *International Dairy Journal*, **13**, 827–833.

Masco, L., Ventura, M., Zink, R., Huys, G. & Swings, J. (2004) Polyphasic taxonomic analysis of *Bifidobacterium animalis* and *Bifidobacterium lactis* reveals relatedness at the subspecies level: reclassification *Bifidobacterium animalis* as *Bifidobacterium animalis* subsp. *animalis* subsp. nov. and *Bifidobacterium lactis* as *Bifidobacterium animalis* subsp. *lactis* subsp. nov. *International Journal of Systematic and Evolutionary Microbiology*, **54**, 1137–1143.

Matsuzaki, T. (2003) Health properties of fermented milk with *Lactobacillus casei* strain Shirota (LcS). *Handbook of Fermented Functional Foods*, (ed. E.R. Farnworth), pp. 145–175, CRC Press LLC, Boca Raton.

Mattila-Sandholm, T., Myllärinen, P., Crittenden, R., Mogensen, G., Fondén, R. & Saarela, M. (2002). Technological challenges for future probiotic foods. *International Dairy Journal*, **12**, 173–182.

Mättö, J., Malinen, E., Suihko, M.-L., Alander, M., Palva, A. & Saarela, M. (2004) Genetic heterogeneity and technological properties of intestinal bifidobacteria. *Journal of Applied Microbiology*, **97**, 459–470.

McBrearty, S., Ross, R.P., Fitzgerald, G.F., Collins, J.K., Wallace, J.M. & Stanton, C. (2001) Influence of two commercially available bifidobacteria cultures on Cheddar cheese quality. *International Dairy Journal*, **11**, 599–610.

Medici, M., Vinderola, C.G. & Perdigón, G. (2004) Gut mucosal immunomodulation by probiotic fresh cheese. *International Dairy Journal*, **14**, 611–618.

Mehanna, N.S., Sharaf, O.M., Ibrahim, G.A. & Tawfik, N.F. (2002) Incorporation and viability of some probiotic bacteria in functional dairy food – 1. soft cheese. *Egyptian Journal of Dairy Science*, **30**, 217–229.

Milanovic, S.D., Panic, M.D. & Caric, M.D. (2004) Quality of quarg produced by probiotics application. *Acta Periodica Technologica*, **35**, 37–48.

Minelli, E.B., Benini, A, Marzotto, M., Sbarbati, A., Ruzzenete, O., Ferrario, R., Hendriks, H. & Dellaglio, F. (2004) Assessment of a novel probiotic *Lactobacillus casei* strain for production of functional dairy foods. *International Dairy Journal*, **14**, 723–736.

Mogensen, G. (1980) Production and properties of yoghurt and ymer made from ultrafiltrated milk. *Desalination*, **35**, 213–222.

Muir, D.D., Tamime, A.Y. & Wszolek, M. (1999) Comparison of the sensory profiles of kefir, buttermilk and yogurt. *International Journal of Dairy Technology*, **52**, 129–134.

Murad, H.A., Sadek, Z.I. & Fathy, F.A. (1998) Production of bifidus Kariesh cheese. *Deutsche Lebensmittel-Rhundschau*, **94**, 409–412.

Oberman, H. & Libudzisk, Z. (1998) Fermented milks. *Microbiology of Fermented Products*, (ed. B.J.B. Wood), Vol. 1, 2nd edn., pp. 308–350, Blackie Academic & Professional, London.

Oliviera, M.N. & Damin, M.R. (2003) Efeito do teor de sólidos e da concentrção de sacarose na acidificaçã, fermeza e viabilidade de bactérias do iogurte e probióticas em leite fermentado. *Ciência e Technologia de Alimentos*, **23**(Supplement), 172–176.

Ordonez, G.A., Fung, D.Y.C. & Jeon, I.J. (2000a) Effect of Oxyrase™ on the metabolic process of lactic acid bacteria in frozen yogurt mix. *Journal of Rapid Methods and Automation in Microbiology*, **8**, 71–81.

Ordonez, G.A., Jeon, I.J. & Roberts, H.A. (2000b) Manufacture of frozen yogurt with ultrafiltered milk and probiotic lactic acid bacteria. *Journal of Food Processing and Preservation*, **24**, 163–176.

O'Riordan, K. & Fitzgerald, G.F. (1998) Evaluation of bifidobacteria for the production of antimicrobial compounds and assessment of performance in Cottage cheese at refrigerated temperature. *Journal of Applied Microbiology*, **85**, 103–114.

Osman, M.M. & Abbas, F.M. (2001) Fate of *Lactobacillus acidophilus* La-5 and *Bifidobacterium lactis* Bb-12 in 'probiotic' Ras cheese. *8th Egyptian Conference for Dairy Science and Technology, Cairo 3–5 November 2001, Research Papers II*, 653–664.

Ostlie, H.M., Helland, M.H. & Narvhus, J.A. (2003) Growth and metabolism of selected strains of probiotic bacteria in milk. *International Journal of Food Microbiology*, **87**, 17–27.

Ouwehand, A.C., Isolauri, E., Kirjavainen, P.V., Tölkkö, S. & Salminen, S.J. (2000) The mucus binding of *Bifidobacterium lactis* Bb12 is enhanced in the presence of *Lactobacillus* GG and *Lactobacillus delbrueckii* subsp. *bulgaricus*. *Letters in Applied Microbiology*, **30**, 10–13.

Perko, B., Matijasic, B.B. & Rogelj, I. (2002) Production of probiotic cheese with addition of *Lactobacillus gasseri* LF221 (Rif$^T$) and K7 (Rif$^T$). *Kmetijstvo, Zootehnika*, **80**, 61–70.

Picot, A. & Lacroix, C. (2003) Optimization of dynamic loop mixer operating conditions for production of o/w emulsion for cell microencapsulation. *Lait*, **83**, 237–250.

Picot, A. & Lacroix, C. (2004) Encapsulation of bifidobacteria in whey protein-based microcapsules and survival in simulated gastrointestinal conditions and in yoghurt. *International Dairy Journal*, **14**, 505–515.

Playne, M.J., Bennet, L.E. & Smithers, G.W. (2003) Functional dairy foods and ingredients. *Australian Journal of Dairy Technology*, **58**, 242–264.

Prajapati, J.B., Shah, R.K. & Dave, J.M. (1986) Nutritional and therapeutic benefits of a blended spray dried acidophilus preparation. *Australian Journal of Dairy Technology*, **42**, 17–21.

Psomas, E., Andrighetto, C., Litopoulou-Tzanetaki, E., Lombardi, A. & Tzanetakis, N. (2001) Some probiotic properties of yeast isolates from infant faeces and Feta cheese. *International Journal of Food Microbiology*, **69**, 125–133.

Rao, H.G.R. & Prakash, A.S. (2004) Development of probiotic *kulfi* (Indian ice-cream). *Indian Dairyman*, **56**(6), 57–64.

Rao, A.V., Shivnarain, N. & Maharaj, I. (1989) Survival of microencapsulated *Bifidobacterium pseudolongum* in simulated gastric and intestinal juices. *Journal of Canadian Institute of Food Science and Technology*, **22**, 345–349.

Rašić, J.Li. & Kurmann, J.A. (1978) *Yoghurt – Scientific Grounds, Technology, Manufacture and Preparations*, pp. 117, Technical Dairy Publishing House, Copenhagen.

Rašić, J.Li. & Kurmann, J.A. (1983) *Bifidobacteria and their Role*, Experienta Supplementum Vol. 39, pp. 102–133, Birkhäuser Verlag, Basel.

Ravula, R.R. & Shah, N.P. (1998) Viability of probiotic bacteria in fermented frozen dairy desserts. *Food Australia*, **50**, 136–139.

Rogelj, I., Matijasic, B.B., Majhenic, A.C. & Stojkovic, S. (2002) The survival and persistence of *Lactobacillus acidophilus* LF221 in different ecosystems. *International Journal of Food Microbiology*, **76**, 83–91.

Ross, R.P., Fitzgerald, G., Collins, K., Stanton, C., Deeth, H.C., Smithers, G.W. & Hourigan, J. (2002) Cheese delivering biocultures – probiotic cheese. *Australian Journal of Dairy Technology*, **57**, 71–78.

Rothschild, P. (1995) Internal defenses. *Dairy Industries International*, **60**(2), 24–25

Ryhänen, E.L., Pihlanto-Leppälä, A. & Pahkala, E. (2001) A new type of ripened, low-fat cheese with bioactive properties. *International Dairy Journal*, **11**, 441–447.

Saarela, M., Mogensen, G., Fonden, G., Mättö, J. & Mattila-Sandholm, T. (2000) Probiotic bacteria: safety, functional and technological properties. *Journal of Biotechnology*, **84**, 197–215.

Saavedra, J.M., Abi-Hanna, A., Moore, N. & Yolken, R.H. (2004) Long term consumption of infant formulas containing live probiotic bacteria: tolerance and safety. *American Journal of Clinical Nutrition*, **79**, 261–267.

Sabikhi, L. & Mathur, B.N. (2000) Selection of suitable strain of *Bifidobacterium bifidum* for the manufacture of probiotic Edam cheese. *Indian Journal of Dairy Science*, **53**, 112–122.

Sabikhi, L. & Mathur, B.N. (2002) Technology and quality attributes of probiotic Edam cheese containing *Bifidobacterium bifidum*. *Indian Journal of Dairy Science*, **55**, 331–337.

Saito, T. (2004) Selection of useful probiotic lactic acid bacteria from *Lactobacillus acidophilus* group and their application to functional foods. *Animal Science Journal*, **75**, 1–13.

Salama, F.M.M. (2002) Production of therapeutic and diabetic stirred yoghurt-like fermented milk products. *Egyptian Journal of Dairy Science*, **30**, 177–190.

Salji, J. (1992) Acidophilus milk products: foods with a third dimension. *Food Science and Technology Today*, **6**, 142–147.

Santos, A., San Mauro, M., Sanchez, A., Torres, J.M. & Marquina, D. (2003) The antimicrobial properties of different strains of *Lactobacillus* spp. isolated from kefir. *Systematic and Applied Microbiology*, **26**, 434–437.

Saxelin, M., Grenov, B., Svensson, U., Fonden, R., Reniero, R. & Mattila-Sandholm, T. (1999) The technology of probiotics. *Trends in Food Science and Technology*, **10**, 387–392.

Schillinger, U. (1999) Isolation and identification of lactobacilli from novel-type probiotic and mild yoghurts and their stability during refrigerated storage. *International Journal of Food Microbiology*, **47**, 79–87.

Schoevers, A. & Britz, T.J. (2003) Influence of different culturing conditions on kefir grains increase. *International Journal of Dairy Technology*, **56**, 183–187.

Shah, N.P. (2000) Probiotic bacteria: Selective enumeration and survival in dairy foods. *Journal of Dairy Science*, **83**, 894–907.

Shah, N.P. (2004) Probiotics and prebiotics. *Agro Food Industry Hi-Tech*, **15**(1), 13–16.

Shah, N.P. & Ravula, R. (2000a) Influence of water activity on fermentation, organic acids production and viability of yoghurt and probiotic bacteria. *Australian Journal of Dairy Technology*, **55**, 127–131.

Shah, N.P. & Ravula, R. (2000b) Microencapsulation of probiotic bacteria and their survival in frozen fermented dairy desserts. *Australian Journal of Dairy Technology*, **55**, 139–144.

Shah, N.P. & Ravula, R. (2004) Selling the cells in desserts. *Dairy Industries International*, **69**(1), 31–32.

Shah, N. P., Lankaputhra, W. E. V., Britz, M. L. & Kyle, W. S. A. (1995) Survival of *Lactobacillus acidophilus* and *Bifidobacterium bifidum* in commercial yoghurt during refrigerated storage. *International Dairy Journal*, **5**, 515–521.

Shehata, A.E., El-Nawawy, M.A., El-Kenany, Y.M. & Aumara, I.E.M. (2001) Production of soft cheese with health benefits. *8th Egyptian Conference for Dairy Science and Technology, Cairo 3–5 November 2001, Research Papers II*, 635–651.

Shehata, A.E., El-Nawawy, M.A., El-Kenany, Y.M. & Aumara, I.E.M. (2004a) Use of bifidobacteria in Ras cheese production: I Chemical and organoleptic properties. *9th Egyptian Conference for Dairy Science and Technology, Cairo 9–11 October 2004, Research Papers II*, 533–562.

Shehata, A.E., El-Nawawy, M.A., El-Kenany, Y.M. & Aumara, I.E.M. (2004b) Use of bifidobacteria in Ras cheese production: II Microbiological properties. *9th Egyptian Conference for Dairy Science and Technology, Cairo 9–11 October 2004, Research Papers II*, 563–585.

Shehata, A.E., El-Nawawy, M.A., El-Kenany, Y.M. & Aumara, I.E.M. (2004c) Use of bifidobacteria in Ras cheese production: III Changes in protein and microstructure during ripening. *9th Egyptian Conference for Dairy Science and Technology, Cairo 9–11 October 2004, Research Papers II*, 587–594.

Sheu, T.Y. & Marshall, R.T. (1993) Microentrapment of lactobacilli in calcium alginate gels. *Journal of Food Science*, **54**, 557–561.

Shihata, A. & Shah, N.P. (2000) Proteolytic profiles of yoghurt and probiotic bacteria. *International Dairy Journal*, **10**, 401–408.

Shihata, A. & Shah, N.P. (2002) Influence of addition of proteolytic strains of *Lactobacillus delbrueckii* subsp. *bulgaricus* to commercial ABT starter cultures on texture of yoghurt, exopolysaccharides production and survival of bacteria. *International Dairy Journal*, **12**, 765–772.

Shortt, C. (1999) The probiotic century: Historical and current prospective. *Trends in Food Science & Technology*, **10**, 411–417.

Simova, E., Beshkova, D., Angelov, A., Hristozova, T., Frengova, G. & Spasov, Z. (2002) Lactic acid bacteria and yeasts in kefir grains and kefir made from them. *Journal of Industrial Microbiology and Biotechnology*, **28**, 1–6.

Sridar, R., Nguyen, M. & Kailasapathy, K. (2003) Studies on the effect of encapsulation on the survival of probiotic microorganisms under high acid and bile conditions. *Journal of Food science and Technology (Mysore)*, **40**, 4[illegible] 460.

Stanton, [illegible] ardiner, G., Lynch, P.B., Collins, J.K., Fitzgerald [illegible] Ross, R.P. (1998) Probiotic cheese. *Int[illegible] al Dairy Journal*, **8**, 491–496

Stant[illegible] Coakley [illegible] R.P. (2002) Development of dairy-bas[illegible]

Stroeh[illegible] *atment Options in Gastroenterology*, **7**, 2[illegible]

Strohm[illegible] ical aspects. *Proceedings of Food Ingre[illegible]*

Suarez-[illegible] argoso, L. (2002) Probiotic fresh chees[illegible]

Surawic[illegible] *um difficile* diarrhoea in humans. *Best P[illegible]*

Sultana, [illegible] apathy, K. (2000) Encapsulation of prob[illegible] l gastrointestinal conditions and yoghurt[illegible]

Taha, S.H., [illegible] fat substitution with vegetable oils on t[illegible] 193–203.

Takahashi, [illegible], Kokubo, S. & Hosono, A. (2004) Se[illegible] ble acid tolerance response in *Bifidobac[illegible]*

Talwalkar, A[illegible] oxicity in probiotic bacteria. *Australian[illegible]*

Talwalkar, A. [illegible] robiotic bacteria to oxygen. *Journal of [illegible]*

Talwalkar, A. [illegible] c bacteria with reference to *L. acidophil[illegible]* **5**, 1–8.

Talwalker, A., [illegible] ng materials and dissolved oxygen on th[illegible] *od Science & Technology*, **39**, 605–611.

Tamime, A. Y. (2[illegible] eview. *European Journal of Clinical Nu[illegible]*

Tamime, A.Y. (ed[illegible]

Tamime, A.Y. & [illegible] head Publishing Ltd., Cambridge.

Tamime, A.Y. & [illegible] ology of fermented milks. *Microbiology and Biochemistry of* [illegible] *Milk*, (ed. B.A. Law), 2nd edn., pp. 57–152, Blackie Academic & Professional, London.

Tamime, A.Y., Marshall, V.M.E. & Robinson, R.K. (1995) Microbiological and technological aspects of fermented milks by bifidobacteria. *Journal of Dairy Research*, **62**, 151–187.

Tamime, A.Y., Skriver, A. & Nilsson, L.-E. (2005) Starter cultures. *Fermented Milks*, (ed. A.Y. Tamime), Blackwell Publishing, Oxford (in press).

Tharmaraj, N. & Shah, N.P. (2004) Survival of *Lactobacillus acidophilus*, *Lactobacillus paracasei* subsp. *paracasei*, *Lactobacillus rhamnosus*, *Bifidobacteria animalis* and *Propionibacterium* in cheese-based dips and the suitability of dips as effective carriers of probiotic bacteria. *International Dairy Journal*, **14**, 1055–1066.

Thorball, J., Skolling, O. & Casas, I.A. (2001) Dispensing Tube. *Patent Cooperation Treaty (PCT) / World Intellectual Property Organization*, **WO 01/15985 A1**.

Tratnik, L., Suskovic, J., Bozanic, R. & Kos, B. (2000) Creamed Cottage cheese enriched with *Lactobacillus* GG. *Mljekarstvo*, **50**, 113–123.

Tungjaroenchai, W., White, C.H., Holmes, W.E. & Drake, M.A. (2004) Influence of adjunct cultures on volatile free fatty acids in reduced-fat edam cheeses. *Journal of Dairy Science*, **87**, 3224–3234.

Ulrich, P. (1980) [Manufacture of ymer by means of ultrafiltration (UF)]. *Maelkeritidende*, **93**, 67–73.
Van de Water, J. (2003) Yogurt and immunity: the health benefits of fermented milk products that contain lactic acid bacteria. *Handbook of Fermented Functional Foods*, (ed. E.R. Farnworth), pp. 113–144, CRC Press LLC, Boca Raton.
Varga, L., Szigeti, J. & Csengeri, É. (2003) Effect of oligofructose on the microflora of an ABT-type fermented milk during refrigerated storage. *Milchwissenschaft*, **58**, 55–58.
Vinderola, C.G, Bailo, N. & Reinheimer, J.A. (2000a) Survival of probiotic microflora in Argentinian yoghurts during refrigerated storage. *Food Research International*, **33**, 97–102.
Vinderola, C.G., Prosello, W., Ghiberto, D. & Reinheimer, J.A. (2000b) Viability of probiotic (*Bifidobacterium* spp., *Lactobacillus acidophilus* and *Lactobacillus casei*) and nonprobiotic microflora in Argentinian Fresco cheese. *Journal of Dairy Science*, **83**, 1905–1911.
Vinderola, C.G., Mocchiutti, P. & Reinheimer, J.A. (2002) Interactions among lactic acid starter and probiotic bacteria used for fermented dairy products. *Journal of Dairy Science*, **85**, 721–729.
de Vrese M. (2003) Probiotiische Käse: Kaum Produkte am Markt. *Deutche Milchwirtschaft*, **54**(24), 1066–1070.
Weinrichter, B., Ginzinger, W., Sollberger, H. & Rohm, H. (2004a) Production of Emmental cheese with adjunct starters with varying degree of autolysis. *Milchwissenschaft*, **59**, 515–519.
Weinrichter, B., Sollberger, H., Ginzinger, W., Jaros, D. & Rohm, H. (2004b) Adjunct starter properties affect characteristic features of Swiss-type cheeses. *Nahrung*, **48**, 73–79.
WenChian, L., HungChi, H. & ChengChun, C. (2003) Viability of microencapsulated bifidobacteria in simulated gastric and bile solution. *International Journal of Food Microbiology*, **86**, 293–301.
WenRong, S. & Griffiths, M.W. (2000) Survival of bifidobacteria in yogurt and simulated gastric juice following immobilization in gellan-xanthan beads. *International Journal of Food Microbiology*, **61**, 17–25.
Witthuhn, R.C., Schoeman, T. & Britz, T.J. (2004) Isolation and characterization of the microbial population of different South African kefir grains. *International Journal of Dairy Technology*, **57**, 33–37.
Wolpert, V.M. (1988) Skyr and skyrwhey. *Dairy Industries International*, **90**(3), 29.
Wszolek, M., Tamime, A.Y., Muir, D.D. & Barclay, M.N.I. (2001) Properties of kefir made in Scotland and Poland using bovine, caprine and ovine milk with different starter cultures. *Lebensmittel-Wissenschaft und -Technologie*, **34**, 251–261.
Yilmaztekin, M., Özer, B.H. & Atasoy, F. (2004) Survival of *Lactobacillus acidophilus* LA-5 and *Bifidobacterium bifidum* BB-02 in white-brined cheese. *International Journal of Food Science and Nutrition*, **55**, 53–60.
Young, C.K. & Nelson, F.E. (1978) Survival of *L. acidophilus* in "sweet acidophilus milk" during refrigerated storage. *Journal of Food Protection*, **41**, 248–250.

# 4 Current Legislation of Probiotic Products

M. Hickey

## 4.1 Introduction

For centuries it was believed that the consumption of cultured products was beneficial to human health. However, the first scientific claims for this were made by the Russian-born Nobel Laureate Elie (Ilya) Metchnikoff, who was working at the Pasteur Institute in Paris, and attributed the longevity of Bulgarians to their regular consumption of fermented milk products (Metchnikoff, 1907). In addition, Metchnikoff believed that there was potential to replace harmful bacteria in our bodies with beneficial ones. At about the same time a French physician, Henry Tissier, observed that the stools of young children with diarrhoea were characterised by low numbers of particular y- or bifid-shaped bacteria, while those of healthy children had high numbers of the same type of organisms. Tissier suggested that there was a possibility of administering such bacteria to the ill children, and a Japanese doctor, Minoru Shirota, was the first to follow up on these ideas. During the 1930s, Shirota succeeded in isolating one such bacterium, *Lactobacillus casei* strain Shirota, which he believed could help to maintain the balance of beneficial bacteria in the gut. Over time, Shirota cultivated the beneficial micro-organism to survive digestion, and went on to incorporate it into a fermented milk beverage known as Yakult. Initially, Yakult was sold on the Japanese market from the early 1950s, but today it is sold in over 25 countries worldwide.

Since that time, many researchers have isolated other micro-organisms that have health benefits associated with humans, and these have been commercialised. Recently, a scientific paper listed 17 strains of lactobacilli and five strains of bifidobacteria that are used by internationally known food manufacturers, such as Nestlé, Danone, Arla, Valio, Yakult and Fonterra, and major starter culture suppliers, such as Chr. Hansen and Danisco (Sanders, 2003).

The total safety of any product cannot be guaranteed, but some of the *Lactobacillus* species have been used in the manufacture of fermented milk products and cheese for centuries, and have had a very good safety record during that time. On very rare occasions, there have been reports of infections linked with the consumption of commercial probiotic fermented products by individuals with underlying medical conditions, but the lactobacilli used were not necessarily causal (Mackay *et al.*, 1999; Rautio *et al.*, 1999). A recent review on the safety of certain micro-organisms, such as lactobacilli and bifidobacteria, concluded that there is essentially no risk involved in their oral consumption by healthy individuals (Salminen *et al.*, 1998). Nonetheless, the increasing trend of incorporating into foods specific strains, which have been isolated from humans but have not traditionally been used as starter cultures, is likely to increase the need, requirements and rigour of safety assessment. Also, the increasing numbers of immunocompromised individuals in the population can only contribute to this need. Furthermore, *Enterococcus*, *Bacillus* and *Saccharomyces* species are also used in fermented food and dietary supplements throughout the world. Safety assessments of such strains need to be quite rigorous as some pathogenic strains are also found in these genera.

From a legal perspective, while the concept of using beneficial bacteria originated early in the last century, the term 'probiotic' only appeared in the 1960s. A number of scientific definitions have appeared in the literature since then but, as yet, there is no internationally recognised legal definition. This has led to a situation where there is no clear and legally consistent way in which the term is, or can be, used in product labelling.

In addition, as an increasing number of scientific reports appear in the literature detailing the specific health benefits of probiotic micro-organisms, there is an increasing clamour to allow the use of specific claims in the labelling, marketing and advertising of products containing probiotic microflora. These considerations arise not only with probiotic products, but also encompass the overall area of functional foods, which fall into the grey area between foods and medicines. This has led to some countries developing their own legislative systems to ensure food safety and consumer protection and to address the issues raised by this relatively new category of products.

Inevitably, differences arise between the approaches used in different jurisdictions, and this chapter will cover three particular areas in the world (e.g. the European Union, with some particular reference to the United Kingdom, the USA and Japan). It will then go on to describe the progress that has taken place at an international level and, in particular, the role and status of Codex Alimentarius will be examined. Finally, some possible future developments will also be considered.

## 4.2 The situation in the European Union (EU)

At the outset, it should be stated that, at present, in the EU there is no legal definition of the term probiotic, no specific legislation governing them, or the broader category of functional foods to which they belong. Nonetheless, a number of horizontal legislation considerations need to be taken into account, such as aspects of food safety, food labelling and product claims.

### 4.2.1 *Relevant food safety legislation*

Regulation 178/2002 lays down some general principles and requirements of EU food law (EU, 2002). Article 14.1 of this regulation requires that food put on the market should be safe. The Regulation goes on to specify that safety, in this context, encompasses short-term, intermediate and long-term effects on consumers, and also any possible effects on subsequent generations. Prior to its adoption, food manufacturers were required to ensure that their products met the specific requirements of EU legislation but, unlike the laws of an increasing number of Member States, the law did not include the broad requirement that food put on the market should be safe. While it could be held that the Product Liability Directive 85/374 (EU, 1985, 1999) made it a requirement to have safe products by imposing strict liability on manufacturers whose products caused harm to individuals, it did not allow action by the authorities before the products actually caused a problem. Furthermore, the scope of the Liability Directive excluded primary agricultural products. By way of contrast, the General Food Law Regulation allows pre-emptive action and also removes the exemption for primary produce.

The Novel Foods Regulation 258/97 (EU, 1997a), which came into force on 15 May 1997, requires novel foods and novel food ingredients that have not been previously used to a significant degree within the EU to undergo a safety assessment before being placed on the market. One could have an interesting discussion on the precise legal meaning of the words 'significant degree' in the definition, but doubtless this will emerge over time. The categories of foods listed in the scope of the regulation include food ingredients consisting of micro-organisms and, hence could possibly include new probiotics. It is interesting to note that between May 1997 and May 2004 a total of 53 applications were made under this regulation, but none related to a probiotic dairy or food product. (Applications under Regulation (EC) No. 258/97 of the European Parliament and of the Council, as at June 2004 (www.europa.eu.int/comm/food/food/biotechnology/novelfood/app_list_en.pdf).) The likely reason for this is that many of the probiotic products were on the market prior to the regulation coming into force. Commercial foods in use, in at least one member state before regulation EU (1997a) came into force, are allowed on the EU market under the 'principle of mutual recognition'. One example of such prior approval of a probiotic is *Lactobacillus rhamnosus* GG (sometimes referred to simply as LGG), which was approved by the UK Advisory Committee for Novel Foods and Processes (ACNFP) in 1992 (Anonymous, 1993). It should be noted that its approval was as a novel starter culture and not as a novel probiotic. This was the first micro-organism to undergo a formal novel food evaluation by the ACNFP, and there was discussion in the Committee as to whether the proposed use was actually novel. The conclusion was that, while the organism was already present in the UK diet at low levels, the proposed use could significantly increase the level of consumption and that justified its assessment as a novel food. Furthermore, the conclusion of the assessment was that there were no food safety concerns regarding its consumption in foods. In addition, Salminen (1996) outlines some of the selection criteria and clinical evaluation applicable to *Lactobacillus* GG, and the specific techniques used in the selection process for the organism are described fully in a United States Patent (Gorbach & Goldin, 1989) (see http://patft.uspto.gov/netahtml/srchnum.htm).

### 4.2.2 *The EU novel food application procedure*

When a company is planning to launch a novel food or food ingredient for the first time in the EU, it must submit an application to the competent authority in the relevant Member State where the food is to be first marketed. A copy of this application, including a summary, should also be submitted to the EU Commission at the same time. The European Commission Recommendation 97/618/EC (EU, 1997b) lays down the requirements on how this application should be made. This is a long and detailed document, and is indicative that the requirements for an application dossier are quite detailed and onerous.

An initial assessment of the application is carried out by the competent national authority, with the help of any relevant and appropriate committees of scientific experts, and a report, which may be favourable or unfavourable, is produced within 3 months of receipt of the application. In the UK, the Food Standards Agency is the competent authority for novel foods and works in consultation with its ACNFP.

A copy of the initial assessment is forwarded to the Commission. The summary of the application dossier and the initial assessment are circulated by the Commission to all the Member States for comments within the following 60 days; this summary may also include some reasoned objections. Where the initial assessment is favourable and no reasoned objections are raised in the subsequent procedure, the applicant is notified by the member state that the product can be marketed. Where the initial assessment is not favourable, the applicant company has the opportunity to submit additional information, which the Commission will bring to the attention of the Member States and the European Food Safety Authority (EFSA) for a scientific opinion.

Where an objection is raised by a Member State or further assessment is requested, the application is transferred to the Standing Committee for the Food Chain and Animal Health (SCFCAH) with a scientific opinion from the EFSA if required. If the SCFCAH fails to make a qualified majority decision, the application is referred to the Council of Ministers to decide. If within 3 months the Council fails to act, or fails to reach a qualified majority decision, the Commission may adopt the proposal. An authorisation decision defines its scope and specifies, as appropriate, the conditions of use, the designation of the food or food ingredient, its specification, and the specific labelling requirements. Such authorisation decisions are published in the Official Journal of the European Commission (see www.europa.eu.int/comm/food/food/biotechnology/novelfood/app_list_en.pdf).

### 4.2.3 *Simplified procedure/notification*

Where an applicant considers their food or ingredient to be substantially equivalent to a similar product already on the EU market, they may apply to the Commission directly with supporting scientific evidence or alternatively seek the opinion of a competent authority of a member state. This is termed the simplified procedure or a notification. Up to July 2004, there have been 27 notifications received by the European Commission (European Commission – Notifications pursuant to Article 5 of Regulation (EC) No. 258/97 of the European Parliament and of the Council as at July 2004; this is available at www.europa.eu.int/comm/food/food/biotechnology/novelfood/notif_list_en.pdf). Again, none of the notifications relates to a probiotic.

### 4.2.4 *Genetic modification*

Genetically modified organisms (GMOs) are organisms, such as plants, animals and microorganisms, the genetic characteristics of which have been modified artificially in order to give them new or improved properties. To ensure the safety of GMOs, the EU has established a detailed legal framework covering the area.

Given the controversy in recent years regarding GMOs, it is likely to be some time before genetically modified probiotics will appear on the EU market. Nonetheless, as part of the selection and evaluation of probiotics, improvements of the performance of some of the strains are made. If any of these procedures involve genetic modification, as defined by the regulation, then the EU legislation on GMOs must be respected. This establishes the conditions under which GMOs, or foods containing or derived from GMOs, may be developed, used, or marketed.

The definition of a GMO, given in Article 2 of Directive 2001/18 (EU, 2001a), is an organism in which the genetic material has been altered in a way that does not occur naturally by mating and/or by natural recombination. The particular GMO legislation that applies in any situation depends on the stage of the process involved, but the following is the list of the regulations that apply:

- Directive 90/219 is on the contained use of genetically modified micro-organisms in the case of, for example, the confined environment of laboratory research (EU, 1990, 2001b, 2001c).
- Directive 2001/18 on the deliberate release into the environment of genetically modified organisms, for the placing on the market of GMOs, for cultivation, import or processing into industrial products (EU, 2001a).
- Regulation 1829/2003 is on genetically modified food and feed, for the placing on the market of GMOs intended for food and of food products containing, consisting of, or produced from GMOs (EU, 2003a).

Where a food product contains or consists of GMOs, the applicant has a choice: either the application as a whole is subject solely to Regulation (EC) 1829/2003 or the application, or part of it, is subject to Directive 2001/18/EC and to Regulation (EC) 1829/2003 (EU, 2001a, 2003a). In addition, GMOs and food products derived from GMOs placed on the market must also comply with labelling and traceability requirements. These requirements are found in Regulation 1829/2003 and in Regulation 1830/2003 on the traceability and labelling of GMOs and of food products produced from GMOs (EU, 2003a, 2003b).

### *Approval of GMOs*

As regards the approval of GMOs, foods containing GMOs and foods containing ingredients derived from GMOs are regarded as a particular category of Novel Foods and are treated as outlined above for Novel Foods as a whole.

### *Claims and labelling*

Food labelling in the EU is governed by Directive 2000/13 (EU, 2000, 2001d). The general principle underpinning food labelling is that purchasers must not be misled. Consequently, the information on food labels must be clear and unambiguous, and must not be such as could mislead the consumer to a material degree. The text on food labels regarding the characteristics of the food and, in particular, its nature, identity, properties, composition, quantity, durability, origin, method of manufacture or production should avoid the following claims:

- Attributions to a particular food, effects or properties which it does not possess.
- Suggestions that a particular food possesses special characteristics when, in fact, all similar foodstuffs possess such characteristics.
- Attributing to a foodstuff the property of preventing, treating or curing a human disease or refer to such properties.

These principles and requirements must be respected when companies wish to make claims for their food products. A claim on a food label may be regarded as a declaration regarding the properties of that food above and beyond those that are required by legislation.

The regulation of claims in the EU has proved very difficult. Efforts were made during the late 1980s and early 1990s, at a time when nine of the then 12 member states had either national legislation or guidelines on claims, and this situation had the potential to create barriers to the completion of the open market. The EU Commission produced draft proposals for a regulation, which was worked on for some years, but this effort came to nought, foundering on the contentious issue of health claims. Some countries wanted a total prohibition on health claims, while others were prepared to accept them provided they were subject to strict criteria.

Following further internal discussions and, in light of developments in Codex Alimentarius, in July 2003 the EU Commission published a proposal for a regulation on nutrition and health claims used in the labelling, presentation and advertising of foods (EU, 2003c; see also http://europa.eu.int/prelex/detail_dossier_real.cfm?CL=en&DosId=184390). By June 2004, the outgoing European Parliament had not authorised its first reading opinion on the draft, and the Council has not yet adopted a common position. Nonetheless, it is worth considering its provisions in some detail.

The proposal defines a health claim as '*any claim that states, suggests or implies that a relationship exists between a food category, a food, or one of its constituents and health*'. The proposal envisages that only claims that comply with the provisions of the proposed regulation would be allowed. It is proposed that health claims should only be authorised for use on the Community market after a scientific assessment of the highest possible standard. In order to ensure harmonised scientific assessment of these claims, the EFSA should carry out such assessments.

In addition to claims relating to nutrients, the proposal also covers other substances with nutritional or physiological effects. Probiotic bacteria are specifically mentioned in this context. It recognises that many such claims are already used on products on the market; therefore, failure to include them would result in claims that either would be outside any legislation or would continue to be regulated by differing national provisions, which would have the potential to impede the free movement of goods or fail to ensure an equal and high level of consumer protection throughout the EU. It is also envisaged that nutritional labelling should be mandatory for all foods bearing nutrition and health claims.

A number of specific requirements are identified in the proposal that are particularly relevant to probiotics. In order to ensure that the claims made are truthful, it is proposed as a requirement that the substance, that is the subject of the claim, be present in the final product in quantities that are sufficient to produce the effect claimed. The substance should also be available to be used by the body. In addition, a significant amount of the substance, producing the claimed nutritional or physiological effect, should be provided by a quantity of the food that can reasonably be expected to be consumed.

To ensure that health claims are truthful, clear, reliable and useful to the consumer in choosing a healthy diet, the wording and the presentation of health claims should be taken into account in the opinion of the EFSA, and in the subsequent authorisation procedure. It is envisaged that a register of assessed claims shall be established to avoid multiple applications and, for transparency purposes, the register shall be revised regularly to keep it up-to-date in the light of scientific and technological developments. The register should also include both accepted and rejected health claims.

### 4.2.5 *Proposed health claim application procedure*

The proposal envisages that applications to use health claims should be made to the EFSA, and should be accompanied by the following particulars and documents:

- the name and address of the applicant;

- the food, or the category of food, in respect of which the health claim is to be made and its particular characteristics;
- a copy of the studies which have been carried out with regard to the health claim including, where available, independent, peer-reviewed studies that have been carried out and any other material which is available to demonstrate that it complies with the criteria provided for in the regulation;
- a copy of other scientific studies which are relevant to that health claim;
- a proposal for the wording, in all Community languages, of the health claim for which authorisation is sought including, as the case may be, specific conditions for use;
- a summary of the dossier.

It is stated in the proposal that the EFSA shall publish detailed guidance to assist applicants in the preparation and the presentation of applications prior to the entry into force of the regulation. This has not yet been published. Furthermore, the proposal commits the EFSA to give an opinion within three months from the date of receipt of a valid application. The time limit could be extended where the authority seeks supplementary information from the applicant. The EFSA opinion shall include the following particulars:

- the name and address of the applicant;
- the designation of the food or category of food in respect of which a claim is to be used and its particular characteristics;
- the recommended wording, in all Community languages, of the proposed health claim;
- where necessary, conditions of use of the food and/or an additional statement or warning that should accompany the health claim on the label and advertising.

Within 3 months of receipt of the opinion of the EFSA, the Commission shall submit a draft of the decision to be taken in respect of the application, taking into account the opinion of the Authority, any relevant provisions of Community law, and any other legitimate factors relevant to the matter under consideration. No doubt all interested parties are studying the implications of this. Where the draft decision is not in accordance with the opinion of the Authority, the Commission shall provide an explanation for the differences.

### **4.2.6** *EU Commission proposal for a regulation on yoghurt and yoghurt-like products*

As mentioned elsewhere, at present, there is no legal definition of the term probiotic, and no specific legislation governing their use. However, this position may change if and when a Commission proposal for a regulation on yoghurt and yoghurt-like proposals progresses and is adopted (EU, 2003d). The preamble to the proposal refers to the fact that products containing probiotics are becoming more and more popular in Europe, and there is a risk that consumers may be misled by claims relating to the health effects of such products if they are not supported by scientific evidence of the effects; also if the probiotic bacteria are not present in sufficient quantities. Article 4 of the third draft of the proposed regulation defines probiotic bacteria as '*live food supplements which benefit the health of consumers if consumed regularly in sufficient quantities*'. It further proposes that health claims relating to such probiotic bacteria shall be based on, and supported by, generally accepted scientific data, and would require the following information to be given on the label:

- The quantity of the food and the required consumption pattern to obtain the claimed health effect;
- Where appropriate, a statement addressed to persons who should avoid the product;
- Where appropriate, a statement warning not to exceed quantities of the product that might represent a health risk.

Claims that would make reference to general non-specific benefits, advice of doctors or other health professionals, or which suggest that health could be affected by not consuming the food, would not be allowed.

The proposal also contains a requirement that the level of probiotic bacteria in such products should be a minimum of $10^7$ colony forming units (cfu) $mL^{-1}$ or a sufficient number in cfu $mL^{-1}$ to obtain the declared health effect. This requirement would have to be met from the date of production through to the end of the shelf-life (date of minimum durability) specified on the product, and taking into account the storage temperature specified on the label.

### 4.2.7 *Use of the term 'Bio'*

At present, many probiotic products on the market use designations such as 'Bio-Yoghurts' to distinguish products containing additional cultures other than the normal starter bacteria in yoghurt (*Streptococcus thermophilus* and *Lactobacillus delbrueckii* subsp. *bulgaricus*). This now conflicts with the provisions of the latest regulation (EU, 2004) that limits the use of the terms such as 'bio' and 'eco' to organic products, which meet the requirements of Regulation 2092/91 (EU, 1991). The basis of this restriction is that the word for organic in nine of the 11 official languages begins with these, or very similar, diminutives. The two language exceptions are English and Finnish. The consequence is that, in future, the term Bio-Yoghurt may be applied only to organic yoghurts.

### 4.2.8 *The UK market*

As a member state of the EU, the UK government is expected to implement Community legislation where such exists. However, in areas where there is no specific Community legislation, the UK can, and does, develop its own legislation. In such cases it is obliged to inform the Commission of its plans and proposed texts.

In the past, the UK had a governmental advisory committee (the ACNFP) in place, reporting initially to the Ministry of Agriculture Fisheries and Food (MAFF) and the Department of Health to assess the safety of novel foods prior to the elaboration of the EU regulation (Anonymous, 1993). In addition, another advisory body, the Food Advisory Committee (Anonymous, 1991), issued a report in the early 1990s, which contained a definition of a health claim. It recommended that health claims be allowed only if they can be justified by recommendations made, or supported, by the Chief Medical Officer and laid down some general principles for such claims. However, the government, because of the ongoing work at EU level at that time, did not adopt that report.

Nonetheless, the need to prevent the use of false, exaggerated, misleading and prohibited health claims became a priority for the food industry, law enforcement officers and consumer groups, not only to protect consumers but also to promote fair trade. MAFF initiated a Functional Foods Initiative in 1994 to get an overview of the requirements of the marketplace. From this the Joint Health Claims Initiative (JHCI) was established in June 1997 involving consumer groups, enforcement authorities and industry representatives, with the aim of drawing up a code of practice for the use of health claims on foods. Key motivating factors were an increased appreciation of the role of diet in maintaining good health and the anticipated growth in the market for functional foods. The need to clarify and strengthen the requirements for evidence to substantiate health claims was accepted and, when this scientific basis was established, claims should be able to express clearly and more directly the increasing variety of relationships between foods or food constituents and human health which was being documented by research.

The JHCI (Code of Practice – see www.food.gov.uk/multimedia/pdfs/jhci_healthreport.pdf) defines a health claim as '*a direct, indirect or implied claim in food labelling, advertising and promotion that consumption of a food carries a specific health benefit or avoids a specific health detriment*', and further divides such claims into three main types:

- Generic Health Claims: health claims based on generally accepted scientific knowledge from literature or the recommendations of national or international public bodies; such claims would not require specific substantiation, and the Code Administration Body will draw up a list of generic food claims.
- Innovative Health Claims: health claims that are not generic claims, but are based on scientific evidence applied to existing or new foods; these claims would require specific substantiation.
- Medicinal Claims: health claims which state or imply that a food has the property of treating, preventing, or curing human disease, or makes any reference to such a property; the prohibition on medicinal claims for foods would remain.

It was recognised that a voluntary Code, such as this, cannot lay down a definitive interpretation of the law; this is the role of the courts. However, provided that this Code is followed in its entirety, including seeking pre-market advice at an early stage and taking into account the outcomes of an assessment by the Code Administration Body, disputes should not arise over the legality or scientific justification of innovative health claims. To date, the list of generic health claims has not been published, although a number are being considered, and there have been no applications made for innovative health claims on probiotic products.

It is hoped that the Code will also provide guidance to the Independent Television Commission (ITC), the Radio Authority (RA) and the Committee of Advertising Practice (CAP) (for enforcement by the Advertising Standards Authority) and, thus, the Code would apply to all health claims made in the UK.

Despite all that has been discussed, the success of foods containing probiotic micro-organisms on the UK market indicates that the situation with labelling claims has not proved a major obstacle to marketing. Table 4.1 lists some products that are available on the market at present, together with the probiotic organisms they contain, and claim statements on the label. It would appear that these claims are acceptable and have not caused problems to the relevant enforcement authorities. Of course, some individual dialogue may well have taken place between manufacturers and regulators, and such contacts are welcomed and encouraged.

**Table 4.1** Examples of claims used on labels of some probiotic fermented milk products on the UK and Irish markets.

| Product | Probiotic micro-organism | Claim on label |
|---|---|---|
| Danone Actimel | *Lactobacillus casei* strain imunitass | Helps support your body's natural defences |
| Danone Activia | *Bifidobacterium* spp. strain essensis or digestivum | *Bifidobacterium* spp. strain essensis or digestivum is a natural culture unique to Danone Activia, has been proven to help our digestion work better as it supplements and supports the essential cultures in our intestinal flora. A healthy digestion is essential to a healthy life. |
| Yoplait Every Body Probiotic Yogurt Drink + 15 Vitamins and Minerals | *Lactobacillus rhamnosus* GG (LGG) | + LGG, the most clinically researched probiotic in the world proven to enhance your natural resistance and help you maintain a healthy digestive system.<br>+ Balance from within LGG<br>+ Healthy digestive system LGG |
| Müller Vitality | *Lactobacillus acidophilus* LA-5 and *Bifidobacterium animalis* subsp. *lactis* BB-12 (see Masco *et al*., 2004) | The good bacteria in Müller Vitality can help maintain the balance of "good" and "bad" bacteria in your digestive system. Vitality is packed with millions of "good" probiotic bacteria *Lb. acidophilus* LA-5 and *B. animalis* subsp. *lactis* BB-12. Great at any time of day as often as you like. Contains prebiotic inulin. |
| Ocean Spray Probiotic Yoghurt | Not listed | Contains millions of friendly bacteria that help to maintain the balance of natural flora in your body, which in turn may aid digestion and general well-being. |

## 4.3 The US situation

In the USA, the Food and Drugs Administration (FDA) has defined four categories of foods:

- Conventional Foods.
- Dietary Supplements (i.e. intended to supplement the diet and marketed as a conventional food, they also have to be labelled as dietary supplements).
- Foods for Special Dietary Use (i.e. intended to supply particular dietary needs for physiological conditions, overweight, food allergies, infant formula).
- Medicinal Foods (i.e. intended for dietary management of a specific disease, under the supervision of a doctor or other health professional).

In theory, probiotics could fit into any of the above-mentioned categories but, to date, none would seem to be used in medical foods and there are very few applications for their use in foods for special dietary purposes. A number of conventional foods contain probiotics, and are mainly dairy products, such as yoghurts, cultured milk, milk and cottage cheese (see Chapter 3 for further details). The US standard of identity for yoghurt requires the use of the conventional starter culture organisms (*S. thermophilus* and *Lb. delbrueckii* subsp. *bulgaricus*) to be present, but no levels are proscribed. Certain species of lactobacilli and bifidobacteria may also be added; indeed, it is estimated that approximately 80% of yoghurts sold in the US market contain *Lactobacillus acidophilus* (Sanders, 2003). However, The National Yogurt Association has introduced a voluntary 'Live Active Culture' seal for products that contain live starter cultures, and this requires refrigerated yoghurt to contain $10^8$ cfu $g^{-1}$ and frozen yoghurt $10^7$ cfu $g^{-1}$ at the time of manufacture. These requirements do not apply to the levels of other cultures named on the label (Sanders, 2003).

The main market for probiotics in the US is in dietary supplements, where they are sold in the form of pills, capsules, powders and drinks. The reason that so many probiotics are sold as dietary supplements would appear to be that prior approval of structure/function claims is not required for this category – this aspect will be further discussed in a subsequent section.

As is the case with the EU, the US does not have any specific legislation on probiotics or functional foods. However, The US Institute of Medicine of the Academy of Sciences has defined functional foods as '*any modified food or food ingredient that may provide a health benefit beyond the traditional nutrients it contains*'. Incidentally, this is not a legal definition.

### 4.3.1 *Food safety*

The US food safety requirements depend on the category of foods to which a product belongs, but for foods and food supplements the onus for food safety is primarily on the manufacturers. Dietary supplements that were on the market prior to 1994 are assumed to be safe, and new products may be marketed 75 days after the manufacturer provides notice and substantiation to the FDA that the product can reasonably be expected to be safe. This is much shorter and easier than the prior approval needed for food additives, where Generally Recognised As Safe (GRAS) status cannot be demonstrated by manufacturers; however, such prior approval is expensive and lengthy. Of course, the FDA can act at any time if it has concerns about the safety of any product on the market, and it can also challenge a manufacturer's notice and substantiation.

### 4.3.2 *Claims and labelling*

The Nutrition Labeling and Education Act (NLEA) of 1990 allowed health and disease prevention claims on food labels, such as '*any claim made on a label or in labelling of a food that expressly or by implication characterises the relationship of any substance to a disease or health-related condition*'. Such claims on foods are subject to prior authorisation by the FDA.

The Dietary Supplement Health Education Act (DSHEA), which was enacted in 1994, created another category of statements, generally referred to as 'structure/function' claims that may be made for dietary supplements. These statements may claim a benefit related to a nutrient deficiency disease or describe the role of a nutrient or dietary ingredient intended to affect a structure or function in humans that characterise the means by which a nutrient or dietary ingredient acts to maintain such structure or function or describe general well-being from consumption of a nutrient or dietary ingredient. The manufacturer is responsible for ensuring the accuracy and truthfulness of these claims; the FDA does not approve them. For this reason, the law stipulates that if a dietary supplement label includes such a claim, it must state in a 'disclaimer' that the FDA has not evaluated the claim. The disclaimer must also state that the dietary supplement product is not intended to 'diagnose, treat, cure or prevent any disease' because only a drug can legally make such a claim. Some report that this requirement is not always followed in practice (Berner & O'Donnell, 1998).

In 1997, the Food and Drug Administration Modernization Act (FDAMA) provided a second way for a health claim to be used on foods. This allows certain health claims to be made as a result of a successful submission of a notification based on an 'authoritative statement' from a scientific body of the US Government or the National Academy of Sciences. The government did not include dietary supplements in the provisions for health claims based on authoritative statements. Consequently, this method of managing health claims cannot be used for dietary supplements at this time. Qualified health claims may be made for dietary supplements only and came about as a result of the US Court of Appeals in the case of Pearson *versus* Shalala. This court decision requires the FDA to allow appropriately qualified health claims that would be misleading without such qualification. These qualified claims are based on the weight of the scientific evidence – that is, there is more evidence for than against the relationship, but it falls short of the validity standard required for foods under NLEA.

At present, the FDA has approved 12 health claims for foods (see Table 4.2), and some of these apply to dietary supplements as well as conventional foods. As well as approved claims, the FDA has given the requirements for the food making the claim, the food claim requirements, and model claim statements. Full details of these claims can be found in the Code of Federal Regulations and in Appendix C of the Food Labelling Guide, all of which are accessible on the FDA website (www.cfsan.fda.gov).

As can be seen, none of the FDA-approved claims for food applies to the use of probiotics. Indeed, in US legislations there is no explicit recognition of any benefits of probiotics or foods with added cultures. Berner and O'Donnell (1998) contend that statements that a probiotic helps proper digestive function would be a structure function statement and not necessarily a claim, but if the claim was that it helped to reduce the risk of cancer it would be a health-related claim and, thus, subject to FDA approval. Many probiotics sold as dietary supplements make structural/function claims, such as '*when taken daily, helps fortify your body's natural defenses and helps keep your body at its best*' (Actimel, Danone), and '*Helps create a favourable environment for the growth of beneficial flora, which dramatically influences metabolism and physical well being*' (Acidophilus, Cell Tech) (Sanders, 2003).

**Table 4.2** Health claims approved by the FDA.

| Claims |
|---|
| Calcium and osteoporosis |
| Dietary lipids and cancer |
| Sodium and hypertension |
| Dietary saturated fat, cholesterol and coronary heart disease |
| Fibre-containing grain products, fruit and vegetables and cancer |
| Fruits, vegetables, and grain products that contain fibre, particularly soluble fibre, and risk of coronary heart disease |
| Fruits and vegetables and cancer |
| Folate and neural tube defects |
| Dietary sugar alcohols and dental caries |
| Soluble fibre from certain foods and the risk of coronary heart disease |
| Soy protein and the risk of coronary heart disease |
| Plant sterol/stanol esters and the risk of coronary heart disease |

## 4.4 The Japanese model

Whenever legislation on the subject of functional foods is discussed, reference is made to the Japanese system of Foods for Specified Health Use (FOSHU). How did this come about? The concept of functional foods and the consequential legislation on FOSHU have their origins in two main factors:

- The severe malnutrition in certain sections of the population in the aftermath of World War II, leading to the schools lunch programme and permitting the addition of certain nutrients to certain staple foods, such as bread and rice; and
- Concerns regarding the ageing of the population and the burden that this was likely to place on the state's finances in the future.

In 1984, the Ministry of Education, Science and Culture carried out a study on statistical analysis and an outlook on food nutrition (Ichikawa, 1994) and, in 1986, a similar report of this study identified three main functions of food: first, the food's nutritive value; second, the food's organoleptic appeal; and third, the food's physiological factors, which include the regulation of bodily functions, disease prevention, promotion of recovery and good health.

The concept of functional foods grew out of the third function, and a Functional Foods Forum was established comprising of experts from the food and nutrition departments of Japanese universities. In addition, The Ministry of Health and Welfare was promoting functional foods, and the industry was keen to market such products using health claims. In 1989, the Nutrition Law was enacted, and this included the manufacture of functional foods, which could make health claims. A Functional Foods Liaison Board was established to work with the industry, and various Functional Food Sub-Committees were set up. Since 1991, FOSHU has become the official term for such foods, and they are defined as '*processed foods containing ingredients that aid specific bodily functions in addition to being nutritious*'. The law also aimed at preventing misleading and ill-defined health claims.

Hence, approved FOSHU are specific foods for which the manufacturers or sellers have provided sufficient scientific justification of specific function(s) to the authorities and, based on this justification, can use approved health claims and the FOSHU logo on their labels (see Fig. 4.1). The approval procedure differs from other countries, such as the US, where only generic claims can be made on certain foods.

**Fig. 4.1** The FOSHU logo.

### 4.4.1 *The process for obtaining FOSHU approval*

In order to gain approval, the manufacturer or seller has to provide the following information:

- Name of the applicant company and the name of a representative.
- Name and address of the head office and factory.
- Name of the product.
- Expiration date or validity date.
- Content amount.
- Reason for seeking permission or approval.
- Health claim and nutrition information on the label of the product.
- List of ingredients and composition as percentages.
- Details of manufacturing process.
- Certificate of nutrient and energy analysis.
- Explanation of how the product contributes to the improvement of one's diet and the maintenance/enhancement of health of the entire population.
- Daily intake amount recommended.
- Considerations and precautions regarding intake.
- Instructions for preparation, storage, or intake of the product.
- Other materials as appropriate.

In addition, attachments must include the following information:

- A sample of the product packaging with labels and claims.
- Documentation that demonstrates clinical and nutritional evidence relating to the health claim of the product or its functional components.
- Documentation regarding safety of the product or its functional components.
- Documentation in relation to the physicochemical properties of the product and the analytical methods for the product's functional components.
- Results of the quantitative and qualitative analyses of its functional components as nutrient constituents of the food, and the analytical methods.
- Results of the nutrient analysis and energy calculation.
- A copy of the applicant's organisational statutes or certificate of contribution.
- Description of the production method and factory equipment and details of the quality control system.
- A copy of the consignment contract, if the applicant is not the manufacturer.
- Reasons for not attaching any of the above requirements.

The labelling of FOSHU-approved foods must provide the following information:

- The name of the food.
- The date of manufacture.
- The name and address of the manufacturer and the distributor of the food.
- The net contents in weight or volume.
- The reason for the approval (i.e. the effect on health) and general guidelines on health.
- A table of nutritional and energy values.

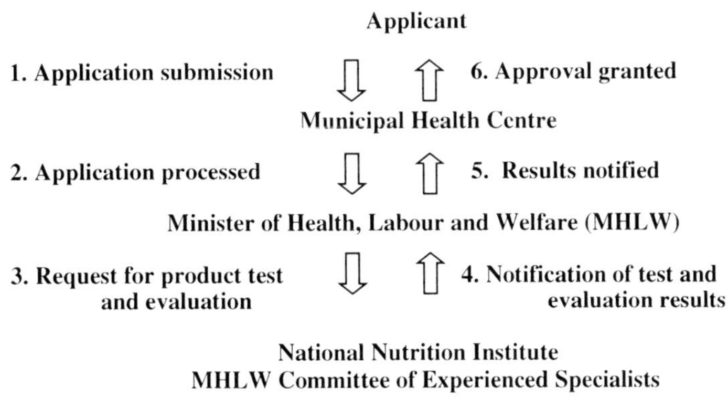


**Fig. 4.2** Flowchart for the process of granting FOSHU approval. Note: see also Shortt (2004) regarding the perspectives on foods for specific health uses (FOSHU) (http://www.foodsciencecentral.com/library.html#ifis/12686).

- A list of ingredients of the food in descending order.
- The date of minimum durability under specified storage conditions.
- An indication that it is a FOSHU (this could be the FOSHU Logo), the recommended intake, and a warning against excessive intake under the heading of 'Cautions'; any special cautions relating to intake, cooking or storage where necessary.
- The name and address of the company (or person) to whom approval was granted if the that company (or person) is not the manufacturer.

This application is then submitted to the local or regional health authority as outlined in Fig. 4.2. It generally takes about one year from submission to obtain approval. Initially, the rate of applications was slow because it was attributed to the length of time taken to obtain approval (in the earlier years it was reputed to take some years), and also to the costs involved. From 1991 to 1998, only 126 products received FOSHU approval; however, amendments to the FOSHU Law in 1998 reduced the requirements of: (a) the amount of scientific documentation that manufacturers must submit; (b) a certificate confirming that all submitted scientific documentation had been reviewed by outside scientific experts was removed and replaced by a requirement that studies had been published in a scientific journal; an industry-sponsored journal is deemed acceptable in this respect; (c) that products be tested by the National Health and Nutrition Laboratory was removed, as the manufacturer's own analytical tests are now acceptable, and (d) the responsibility for reviewing applications has been transferred from the Ministry of Health and Welfare to the local prefectures.

As a result, the rate of approvals has increased (see Fig. 4.3) and, by the end of 2003, a total of 396 products had been granted FOSHU status (see Food Industry Bulletin at www.japanscan.com/). Approved product claims fit into seven main categories relating to:

- Gastrointestinal health
- Cardiovascular health
- Hypertension control
- Blood sugar control
- Mineral absorption

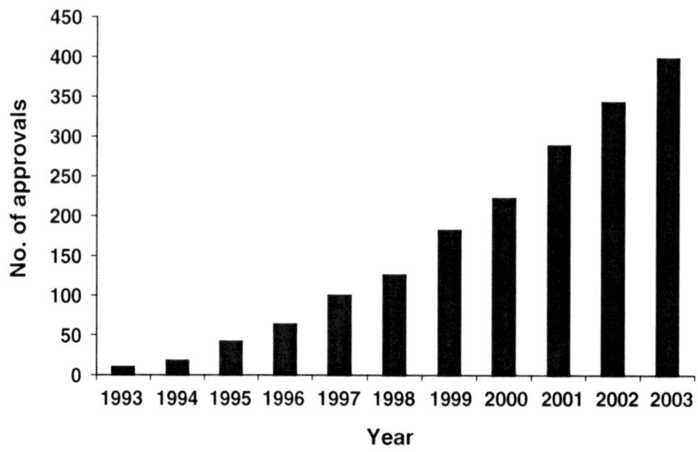


**Fig. 4.3** FOSHU products approved between 1993 and 2003. Note: see Shortt (2004) regarding the perspectives on foods for specific health uses (FOSHU) (http://www.foodsciencecentral.com/library.html#ifis/12686), and Toyoshima (2004) (http://websrv2.tekes.fi/opencms/opencms/OhjelmaPortaali/Kaynnissa/ELITE/fi/Dokumenttiarkisto/Viestinta_ja_aktivointi/Julkaisut/Functional_Food_in_Japan_- Status_and_Trend.doc).

- Dental health
- Triglyceride health (including cholesterol control)

The largest category, in terms of product approvals, is that of gastrointestinal health, which consists of probiotics, prebiotics and dietary fibre. By the end of 2001, such products accounted for about 60% of the 289 approved products (Arai *et al.*, 2002), and probiotics accounted for about one-third of the gastrointestinal health category. Some examples of health claims that have been approved for certain probiotic-type FOSHU products on the Japanese market are listed in Table 4.3.

**Table 4.3** Examples of health claims approved on FOSHU products containing probiotic micro-organisms.

| Commercial product | Health claims |
|---|---|
| Meiji Milk Products (Bulgarian Yoghurt) | Due to the effects of *Lactobacillus* LB 81, this yoghurt regulates the balance of intestinal bacteria that lead to and maintain a good intestinal condition. |
| Yakult | Due to the effects of the Yakult strain (*Lactobacillus casei* strain Shirota), which can reach the intestine alive. Yakult maintains the intestine in good health by increasing beneficial bacteria, decreasing harmful bacteria and by improving the intestinal environment. |
| Morinaga Milk Industry (Bifidus Yoghurt) | This yoghurt contains living bifidobacteria (*Bifidobacterium longum* BB536). It helps to increase intestinal bifidobacteria, improve the intestinal environment and regulate the intestinal condition. |
| Takanashi Milk Products (Onaka-He-GG) | Due to the effects of *Lactobacillus rhamnosus* GG, this organism can reach the intestine alive. This yoghurt increases beneficial bacteria and decreases harmful bacteria. It improves the intestinal environment and regulates the intestinal condition. |

See also Shortt (2004) regarding the perspectives on foods for specific health uses (FOSHU) (http://www.foodsciencecentral.com/library.html#ifis/ 12686).

### 4.4.2 *Costs of approval*

A recent estimate of the typical cost of obtaining approval is about $830 000, but most of these costs are accounted for by the scientific data on safety and efficacy; the direct costs and fees associated with an application are only $1600 (see P. Yamaguchi at www.npicenter.com/newscontrol/).

In summary, the Japanese FOSHU system has a number of unique features. First, it is a voluntary procedure, but it has government approval. Second, specific health claims are approved prior to use. Third, approval is based on documented scientific safety and effective considerations. Fourth, approved products may use the FOSHU logo on the label. However, the fact that the system is voluntary has led to a situation that many non-FOSHU functional foods are also on the market, considerably outnumbering FOSHU products. These products are marketed as health foods and are often sold in health stores, but cannot legally make specific health claims.

## 4.5 Codex Alimentarius

### 4.5.1 *Background*

At the 11th Session of the Conference of the Food and Agriculture Organization (FAO) of the United Nations in November 1961, a resolution was passed to establish a Codex Alimentarius Commission (CAC), and a request was made for the World Health Organisation (WHO) to support the establishment of a joint FAO/WHO food standards programme. In 1962, the Joint FAO/WHO Food Standards Conference asked the CAC to implement the joint FAO/WHO foods standards programme, and to create the Codex Alimentarius. In May 1963, the 16th World Health Assembly approved this, and adopted the statutes of the CAC.

The CAC is an international intergovernmental body. Its membership is open to member nations and associate members of the FAO and/or the WHO. As of July 2003, it had 169 member countries from all continents and regions. In general, the CAC meets every two years, and the venue alternates between the FAO headquarters in Rome and WHO headquarters in Geneva. Nominated senior officials represent member governments at Codex meetings. National delegations may also include representatives of the industry, consumers and academia. A significant number of other international governmental organisations, for example, the International Office of Epizootics (OIE), the World Trade Organisation (WTO), and international Non-governmental Organisations (NGOs), such as the International Dairy Federation (IDF) and the Confederation of the Food and Drink Industries of the EU (i.e. Confédération des Industries Agro-Alimentaires – CIAA), also attend in an observer capacity. Such observers are allowed to contribute to meetings at all stages except in final decisions. This is the exclusive prerogative of member governments.

The CAC has established two types of subsidiary committees: (a) Codex Committees; and (b) Co-ordinating Committees. The former type committee is subdivided into General Subject Committees (currently nine in number) and it is so-called because of the horizontal nature of their work, and Commodity Committees (currently 16 in number), which develop the standards for specific foods or classes of foods. There are five Regional Co-ordinating Committees whose role is to ensure that the CAC is responsive to regional interests and the needs of developing countries. The CAC also establishes *ad hoc* Intergovernmental Task Forces given stated tasks on specific topics. Currently, there are three such Task Forces on foods derived from biotechnology, animal feeding, and fruit and vegetables. The Codex Alimentarius structure is shown in Fig. 4.4.

Codex Alimentarius can be translated, from the Latin, as the Food Code, and its three main aims are:

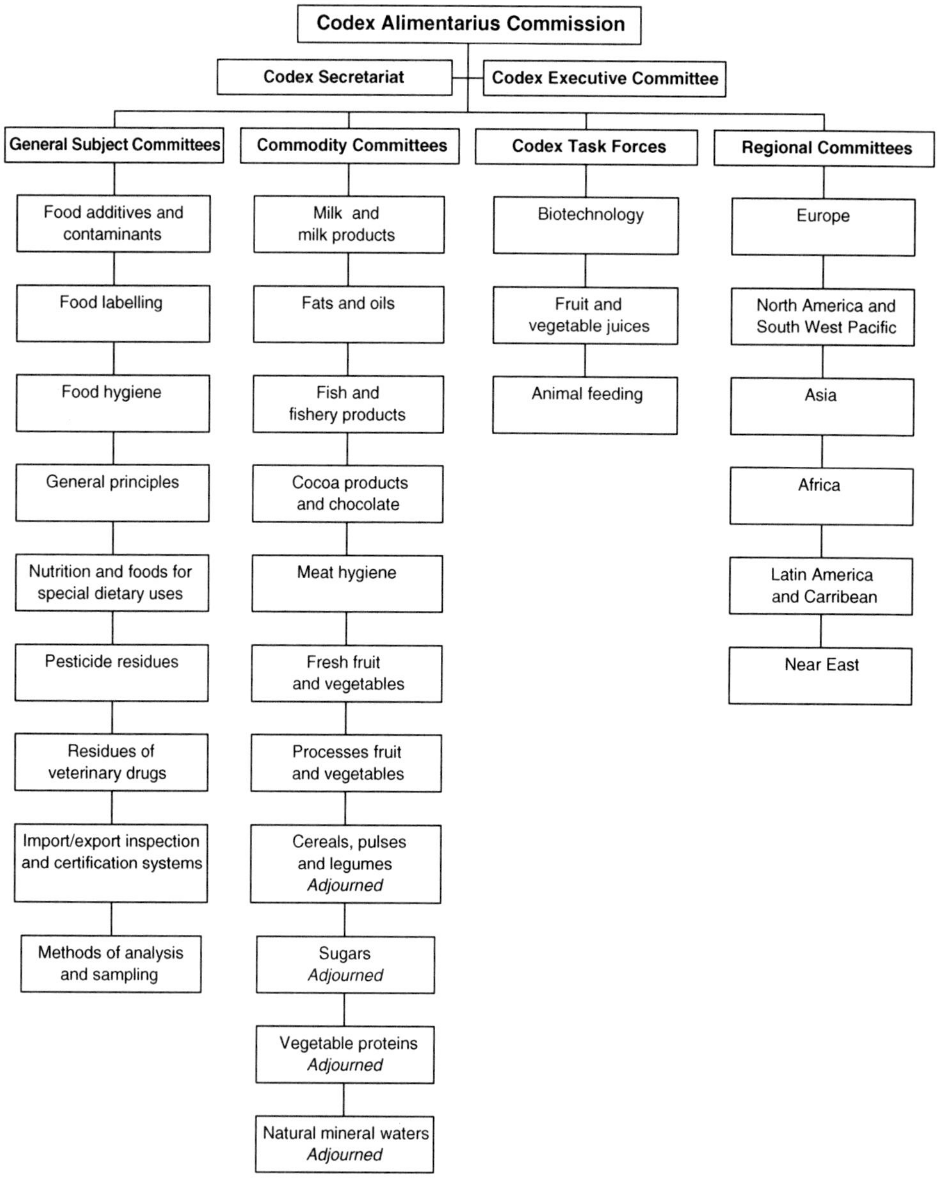


**Fig. 4.4** Structure of the Codex Alimentarius Commission. Note: not all the commodity committees are included.

- The protection of consumer health.
- Ensuring fair trading practices.
- Facilitating international trade.

Furthermore, the Codex Alimentarius currently comprises:

- 237 food standards;
- 41 Codes of hygienic or technological practice;
- 185 pesticide evaluations;
- 3 274 limits for pesticide residues;
- 25 guidelines for contaminants;
- 1 005 food additive evaluations;
- 54 veterinary drug evaluations.

Food standards cover specific commodities and also general issues, which have horizontal application. For example, they encompass food labelling, food additives, food hygiene, contaminants, nutrition and foods for special dietary uses, and methods of analysis and sampling.

The CAC has established a number of principles on the scientific basis for its decision-making (Randel & Race, 1996). These principles ensure that the quality and food safety provisions shall be based on sound science and, that in establishing food standards, other legitimate factors relevant to consumers' health may be considered and in the promotion of fair trade. Nevertheless, the Codex Alimentarius consists of 13 volumes, which contain general principles, general standards, commodity standards, definitions, codes, methods and recommendations (see www.codexalimentarius.net/). The standards and related texts are subject to revision, as and when deemed necessary by the CAC and its subsidiary bodies, to ensure that they are consistent with, and reflect current scientific knowledge. Any member of Codex may identify and present new scientific or other information to the relevant body that may warrant a revision

## **4.5.2** *Acceptance of Codex Standards and their role in the WTO*

Once adopted, the Codex Standards are proposed to member countries for acceptance, which may fall into one of these categories: (a) full acceptance; (b) acceptance with specified deviations; and (c) free distribution. These different forms of acceptance are outlined in the General Principles of Codex.

The Uruguay round of multilateral trade negotiations held under the General Agreement on Tariffs and Trade (GATT), which took place between 1986 and 1994, led to the formation of the WTO on the 1st of January 1995. For the first time, GATT agreements included agriculture and food in its scope; however, the Marrakesh agreement of 1994 also included the agreements on sanitary and phytosanitary (SPS) measures and on technical barriers to trade (TBT). These agreements acknowledge the need for the harmonisation of international standards to minimise the risk of sanitary, phytosanitary and other technical standards becoming barriers to international trade. Thus, the SPS and TBT agreements gave formal recognition to international standards, guidelines and recommendations of international organisations, including the CAC, as reference points for facilitating international trade and resolving disputes. Hence, the role of Codex Alimentarius in this regard is now well recognised.

It should be noted, however, that consumer groups have expressed some criticism of Codex standards on the basis of the time taken to elaborate standards, and sound science; the latter basis may not necessarily take into account other considerations, such as consumer concerns (O'Rourke, 1999).

### 4.5.3 *Codex and the issue of claims*

The Codex General Standard for the Labelling of Prepackaged Foods (GSLPF) was adopted by the CAC in 1981; it was subsequently revised in 1985 and 1991, and amended in 1999 and 2001 (FAO/WHO, 1985a, 1990, 1991a, 2000). In addition, The Codex General Standard for the Labeling of and Claims for Prepackaged Foods for Special Dietary Uses was adopted by the Codex Alimentarius Commission in 1985 (FAO/WHO, 1985b), and it defines a claim as '*any representation which states, suggests or implies that a food has particular qualities relating to its origin, nutritional properties, nature, processing, composition or any other quality*'.

The Codex General Guidelines on Claims was originally adopted in 1979, with a revised version adopted in 1991 (FAO/WHO, 1979, 1991b). The text contains the definition given above under the GSLPF, and also gives a list of claims which should be prohibited:

- Claims which state that a given food will provide an adequate source of all nutrients, except in certain well-defined products where a Codex standard regulates this claim as admissible or appropriate authorities have accepted that the specific food does so.
- Claims that imply that a balanced diet or ordinary foods cannot supply adequate amounts of all ingredients.
- Claims that cannot be substantiated.
- Claims that a food is suitable for use in the prevention, alleviation, treatment or cure of a disease, disorder or particular condition unless those are in accordance with and follow the principles of Codex standards or guidelines on foods for special dietary uses or, in the absence of Codex standards and guidelines, are permitted by the laws of the country of sale (these are usually called medicinal claims).

The task of developing guidelines on the use of nutrition and health claims comes under the remit of the Codex Committee on Food Labelling (CCFL). This has proved a difficult task; for example, in written comments submitted to the CCFL meeting in 1994, some countries (Denmark and Finland) opposed both nutrient function and health claims, whilst others (New Zealand and Switzerland) opposed health claims, but could accept nutrient function claims. However, Australia, Sweden and USA were prepared to accept both claims provided that they were subject to strict criteria (Pascal, 1996). At its meeting in 1996, the CCFL agreed to delete all references to health claims in the guidelines and sent the Guidelines for Use of Nutrition Claims to the CAC, and these were adopted in 1997. Work continued addressing the more contentious issue of health claims and, finally at its 23rd Session in 2004, the CCFL agreed Draft Guidelines on the use of Nutrition and Health Claims and forwarded them to the CAC for adoption at its 27th session (FAO/WHO, 2004). These guidelines are long and detailed and, at the outset, they lay down a number of principles concerning health claims, such as:

- They should be consistent with national health policy, nutrition policy and support same as applicable.
- They should be supported by a sound and sufficient body of scientific evidence to substantiate the claim.
- They should provide truthful and non-misleading information to aid consumers in choosing healthy diets.
- They should be supported by specific consumer information.
- Their impact on consumers' eating habits and buying habits should be monitored.
- The prohibition of claims as stated in Section 3.4 of the Codex General Guidelines for the use of claims should remain.

The Codex definition of a health claim is given as '*any representation that states, suggests, or implies that a relationship exists between a food or a constituent of that food and health*', and three types of claims are also defined: (a) Nutrient Function Claims; (b) Other Function Claims; and (c) Reduction of Disease Risk Claims.

There is a requirement that any health claim must be accepted by, or acceptable to, the competent authorities of the country where the product is sold. As regards substantiation of claims, reference is made to a parallel text being developed by the Codex Committee on Nutrition and Foods for Special Dietary Uses (CCNFSDU). At its 25th Session in November 2003, the CCNFSDU considered a text with the working title of a Proposed Draft Recommendations on the Scientific Basis for Health Claims (FAO/WHO, 2003a). While this text is still at a relatively early stage (Step 3 of the Codex procedure), it is worth considering, at least, the principles therein. They are intended for governments to facilitate their own evaluation of health claims made by industry, and as a reference for industry in preparing dossiers aimed at providing substantiation of such claims. They cover the following aspects:

- The nature of the evidence provided on the characteristics of the product on which the claim is based [e.g. its origin, nature, food safety, specifications, bioavailability, shelf-life stability, validated analytical methods, interactions with other foods, and Acceptable Daily Intake (ADI)].
- Scientific requirements for the claimed effect should be provided, and a high level of scientific justification is obligatory. The evidence should be supported by data from experimental trials, such as *in vivo* or *in vitro*, epidemiological or clinical human studies including specific subpopulation data if relevant to the claim. The trials should be large enough, continued for long periods, and in an amount necessary to demonstrate the effect.
- The claims submitted should be evaluated scientifically by a group of qualified experts, consistent with the principles of risk analysis, taking all available data into account, and using up-to-date scientific methods.
- There should be periodic review of health claims to study new knowledge of the benefits, to look at consumption patterns of products with health claims, to ensure they are in line with the claims, and to look at positive effects and possible negative effects after long-term consumption.

When combined together, the above guidelines and recommendations should provide a firmer legal basis for health claims at an international level, but leave the actual approval or acceptance of such claims to individual governments to ensure that they are in line with national dietary policies and guidelines.

### 4.5.4 *Codex Standard for fermented milks*

At its 26th Session in July 2003, the CAC adopted a new Codex Standard for Fermented Milks (FAO/WHO, 2003b), which replaced two former standards: (a) for Yoghurt and Sweetened Yoghurt; and (b) for Flavoured Yoghurt and Products Heat-Treated after Fermentation. However, the new standard is also expanded to encompass a broader range of fermented milks, such as kefir, acidophilus milk and koumiss. While not specifically aimed at probiotics, this new standard includes compositional requirements for the minimum level of starter culture organisms ($1 \times 10^7$ cfu $g^{-1}$) and, where a content labelling claim is made for specific micro-organisms other than the normal starter cultures, a minimum of $1 \times 10^6$ cfu $g^{-1}$ is required. The earlier standards required only that micro-organisms should be viable and abundant without setting specific minima.

## 4.6 Conclusions and pointers to the future

It is evident from the current review that many aspects pertaining to the regulation of the category of so-called functional foods, and specific subcategories, such as probiotics, still exist in a 'grey' area. As international trade in these products continues to grow, many regulators express a need to develop an appropriate regulatory framework for the control of such products to ensure the on-going need to protect human health, to prevent consumer deception and to facilitate fair trade. This is reflected in developments at Codex level. One such initiative was the joint FAO/WHO expert consultation on health and nutritional properties of probiotics in food, and was held in October 2001 at Cordoba in Argentina (www.fao.org/es/ESN/food/foodandfoo_probio_en.stm). Eleven experts from ten countries attended, and they evaluated the latest information (i.e. scientific evidence available on functional foods, safety aspects of probiotics, and the methodologies to assess such aspects). The report of this consultation lists a number of conclusions including that the regulatory status of probiotics is not well established on an international basis, and that only in a few countries are regulatory procedures in place or sufficiently developed to permit probiotic products to describe specific health benefits. In addition to the scientific recommendations, the international experts also made a number of recommendations pertaining to regulatory matters:

- In order to be termed a probiotic, the micro-organism must be able to confer defined health benefits on the host, in the actual product that will be made available to humans.
- Good manufacturing practices must be applied in terms of quality control, shelf-life conditions established, and the labelling is made clear to include minimum dosage and verifiable health claims.
- The regulatory status of probiotics as a component of foods should be established on an international level.
- The regulatory framework to be established should include issues related to probiotics, such as efficacy, safety, labelling, fraud and claims.
- Probiotic products shown to confer defined health benefits should be permitted to describe these health benefits.

- Surveillance systems should be put in place to record and analyse any adverse events associated with probiotics in food; these systems should include trace-back and post-marketing surveillance, and they should also monitor the long-term health benefits of probiotic strains.

To follow on from the consultation, a Working Group was established to draft guidelines for the evaluation of probiotics in food. This group met in London, Ontario during 2002 and, in the report of their discussions, they concluded that, from the current state of evidence, probiotic effects are strain-specific (ftp://ftp.fao.org/es/esn/food/wgreport2.pdf). The same Working Group proposed a definition for probiotics as follows: '*Live micro-organisms which when administered in adequate amounts confer a health benefit on the host*'. Furthermore, they also proposed that adoption of the guidelines of their report should be a prerequisite for calling a strain a probiotic.

It must be stressed that the conclusions and recommendations of these groups are those of the participants, and do not imply any opinion on the part of the organisers. In addition, the participants were all scientists and not legislators. The needs, expectations and priorities of the two groups do not necessarily concur; however, they do identify gaps and shortcomings which, if addressed, could help to clarify and remove at least some of the 'greyness' and uncertainty from the current regulatory status of probiotics.

## References


Anonymous (1991) *Food Advisory Committee – Report of its Review of Food Labelling and Advertising 1990*, Report Fd AC/REP/10, Her Majesty's Stationery Office.

Anonymous (1993) *Annual Report 1992*, Advisory Committee for Novel Foods and Processes, Ministry of Agriculture Fisheries and Foods Publications.

Arai, S., Moringa, Y., Yoshikawa, T., Ichiishi, E., Kiso, Y., Yamazaki, M., Morotimo, M., Shimizu, M., Kuwata, T. & Kaminogawa, S. (2002) Recent trends in functional food science and industry in Japan. *Bioscience Biotechnology & Biochemistry*, **66**, 2017–2029.

Berner, L.A. & O'Donnell, J.A. (1998) Functional foods and health claims legislation: applications to dairy foods. *International Dairy Journal*, **8** 355–362.

EU (1985) Council Directive 85/374/EEC of 25 July 1985 on the approximation of the laws, regulations and administrative provisions of the member states concerning liability for defective products. *Official Journal of European Commission*, **L210**: 29–33.

EU (1990) Council Directive 90/219/EEC of 23 April 1990 on the contained use of genetically modified micro-organisms. *Official Journal of the European Commission*, **L117**, 1–14.

EU (1991) Council Regulation (EEC) No. 2092/91 of 24 June 1991 on organic production of agricultural products and indications referring thereto in agricultural products and foodstuffs. *Official Journal of the European Commission*, **L198**, 1–15 as amended.

EU (1997a) Regulation (EC) No. 258/97/EC of 27 January 1997 concerning novel foods and novel food ingredients. *Official Journal of the European Commission*, **L43**, 1–6.

EU (1997b) Commission Recommendation 97/618/EC of 29 July 1997 concerning the scientific aspects and the presentation of information necessary to support applications for the placing on the market of novel foods and novel food ingredients and the preparation of initial assessment reports under Regulation (EC) No 258/97 of the European Parliament and of the Council. *Official Journal of the European Commission*, **L253**, 1–36.

EU (1999) Council Directive 1999/34/EC of 10 May 1999 on the approximation of the laws, regulations and administration provisions of the Member States concerning liability for defective products. *Official Journal of the European Commission*, **L141**, 20–21 (this Directive is an amendment of EU, 1985).

EU (2000) Council Directive 2000/13/EC of 20 March on the approximation of laws of the Member States relating to the labelling, presentation and advertising of foodstuffs. *Official Journal of the European Commission*, **L109**, 29–42.

EU (2001a) Council Directive 2001/18/EC on the deliberate release into the environment of genetically modified organisms and repealing Council Directive 90/220/EC. *Official Journal of the European Commission*, **L106**, 1–39.

EU (2001b) Council Directive 98/81/EC of 26 October 1998 on the contained use of genetically modified micro-organisms. *Official Journal of the European Commission*, **L330**, 13–31 (this Directive is an amendment of EU, 1990).

EU (2001c) Council Decision 2001/204/EC of 8 March 2001 supplementing Directive 90/219/EEC as regards the criteria for establishing the safety, for human health and the environment, of types of genetically modified micro-organisms. *Official Journal of the European Commission*, **L73**, 32–34 (this Decision is a supplement for EU, 2001b).

EU (2001d) Commission Directive 2001/101/EC of 26 November 2001. *Official Journal of the European Commission*, **L310**, 19–21 (this Directive is an amendment of EU, 2000).

EU (2002) Regulation (EC) No. 178/2002 of the European Parliament and of the Council of 28 January 2002 laying down the general principles and requirements of food law, establishing the European Food Safety Authority and laying down procedures in matters of food safety. *Official Journal of the European Commission*, **L31**, 1–24.

EU (2003a) Regulation (EC) No. 1829/2003 of the European Parliament and of the Council of 22 September 2003 on genetically modified food and feed. *Official Journal of the European Commission*, **L268**, 1–23.

EU (2003b) Regulation (EC) No. 1830/2003 concerning the traceability and labelling of genetically modified organisms and the traceability of food and feed products produced from genetically modified organisms and amending Directive 2001/18/EC. *Official Journal of the European Commission*, **L268**, 24–28.

EU (2003c) Proposal for a Regulation of the European parliament and of the Council on nutrition and health claims made on foods, COM (2003) 424 Final of 16.7.2003.

EU (2003d) Proposal for a Council Regulation laying down quality characteristics and requirements for the composition, labelling and control of yoghurt and yoghurt-like products, AGRI/38743/2003, Rev. 3 of 19 August 2003.

EU (2004) Council Regulation (EC) No. 392/2004 of 24 February 2004 amending Regulation (EC) No. 2092/91 on organic production of agricultural products and indications referring thereto in agricultural products and foodstuffs. *Official Journal of the European Commission*, **L65**, 1–3.

FAO/WHO (1979) *Codex General Guidelines on Claims*. CAC/GL 1

FAO/WHO (1985a) *Codex General Standard for the Labelling of Prepackaged Foods*. CODEX STAN 1.

FAO/WHO (1985b) *Codex General Standard for the Labelling of and Claims for Prepackaged Foods for Special Dietary Uses*. CODEX STAN 146.

FAO/WHO (1990) *Codex Alimentarius-Abridged Version*, Joint FAO/WHO Food Standards Programme-Codex Alimentarius Commission, Ed. by Smith, B.L., Food and Agricultural Organization of the United Nations, Rome.

FAO/WHO (1991a) *Codex General Standard for the Labelling of Prepackaged Foods*, (Revised of FAO/WHO, 1985a).

FAO/WHO (1991b) *Codex General Guidelines on Claims*. CAC/GL 1, (Revised of FAO/WHO, 1979).

FAO/WHO (2000) *Codex Alimentarius-Abridged Version*, Joint FAO/WHO Food Standards Programme-Codex Alimentarius Commission, (2nd edn. of FAO/WHO, 1990), Food and Agricultural Organization of the United Nations, Rome.

FAO/WHO (2003a) Codex Committee on Nutrition and Foods for Special Dietary Uses. *Proposed Draft Recommendations on the Scientific Basis for Health Claims at Step 3 of the Procedure*. Agenda Item for the 25th Session of the Codex Committee Nutrition and Foods for Special Dietary Uses August 2003 CX/CCNFSDU 03/9.

FAO/WHO (2003b) *Standard for Fermented Milks*. CODEX STAN 243.

FAO/WHO (2004) *Draft Guidelines for use of Nutrition and Health Claims (At Step 8 of the Procedure)*. Report of the Thirty-second Session of the Codex Committee on Food Labelling July 2004 Alinorm 04/27/22 Appendix III.

Gorbach, S.L. & Goldin, B.R. (1989) *Lactobacillus* strains and methods of selection, US Patent No. 4 839 281, (http://patft.uspto.gov/netahtml/srchnum.htm).

Ichikawa, T. (1994) Functional Foods in Japan. *Functional Foods Designer Foods, Pharmafoods, Nutraceuticals*, (ed. Goldberg, I.), pp. 453–467, Chapman & Hall, London.

Mackay, A.D., Taylor, M.B., Kibbler, C.C. & Hamilton-Miller, J.M.T. (1999) *Lactobacillus* endocarditis caused by a probiotic organism. *Clinical Microbiology Infectious Diseases*, **5**, 290–292.

Masco, L., Ventura, M., Zink, R., Huys, G. & Swings, J. (2004) Polyphasic taxonomic analysis of *Bifidobacterium animalis* and *Bifidobacterium lactis* reveals relatedness at the subspecies level: reclassification *Bifidobacterium animalis* as *Bifidobacterium animalis* subsp. *animalis* subsp. nov. and *Bifidobacterium lactis* as *Bifidobacterium animalis* subsp. *lactis* subsp. nov. *International Journal of Systematic and Evolutionary Microbiology*, **54**, 1137–1143.

Metchnikoff, E. (1907) Lactic acid as inhibiting intestinal putrefaction. *The Prolongation of Life: Optimistic Studies*, pp. 163–183, W. Heinemann, London.

O'Rourke, R. (1999) *European Food Law with 1999 Update*, pp. 148–160, Palladian Law Publishing Ltd., London.

Pascal, G. (1996) Functional foods – the future: how to regulate these foods. *Nutrition Reviews*, **54**, 199–201.

Randell, A. & Race, J. (1996) Regulatory and legal aspects of functional foods: an international perspective. *Nutrition Reviews*, **54**(11), 152–155.

Rautio, M., Jousimes-Somer, H., Kauma, H., Pirtarinen, I., Saxelin, M., Tynkkynen, S. & Koskela, M. (1999) Liver abscess due to a *Lactobacillus rhamnosus* strain indistinguishable from *Lactobacillus rhamnosus* strain GG. *Clinical Infectious Diseases*, **29**, 1159–1160.

Salminen, S. (1996) Functional dairy foods with *Lactobacillus* strain GG. *Nutrition Reviews*, **54**(11), 99–101.

Salminen, S., von Wright, A., Morelli, L., Marteau, P., Brassart, D., de Vos, W.M., Fonden, R., Saxelin, M., Collins, K., Mogensen, G., Birkeland, S.E. & Mattila-Sondholm, T. (1998) Demonstration of safety of probiotics – a review. *International Journal of Food Microbiology*, **44**, 93–106.

Sanders, M.E. (2003) Probiotics: considerations for human health. *Nutrition Reviews*, **61**(3), 91–99.

Shortt, C. (2004) Perspectives on foods for specific health uses (FOSHU), *Food Science and Technology Bulletin*, 13th July, 2004, (http://www.foodsciencecentral.com/library.html#ifis/ 12686).

Toyoshima, T. (2004) Functional Food in Japan – Status and Trends (http://websrv2.tekes.fi/opencms/opencms/OhjelmaPortaali/Kaynnissa/ELITE/fi/Dokumenttiarkisto/Viestinta_ja_aktivointi/Julkaisut/Functional_Food_in_Japan_-Status_and_Trend.doc for 2002).

# 5 Enumeration and Identification of Mixed Probiotic and Lactic Acid Bacteria Starter Cultures

A.L. McCartney

## 5.1 Introduction

Lactic acid bacteria (LAB) encompass a range of taxonomically distinct genera that are extremely important to the food industry, and traditionally used in a wide range of fermented foods (including cheese, fermented milk, meat and plant products) (Klein *et al.*, 1998; Giraffa & Neviani, 2000; Holzapfel *et al.*, 2001). Historically, LAB were incorporated into foods to reduce spoilage and to prolong the edible viability of food products, thus affording storage for leaner seasons. More recently, health-promoting properties attributed to certain LAB (particularly *Lactobacillus* and *Bifidobacterium* species) have expanded their application into health foods, namely probiotics (Salminen & von Wright, 1993; Heller, 2001). As such, certain LAB have significant commercial interest, either as starter cultures for industrial fermentations (e.g. for production of yoghurt, kefir, cheese, fermented-milk drinks, fermented sausages, sauerkraut, sour dough, wine or beer) or as probiotic cultures (in either foods or supplements). Different LAB species and/or subspecies can confer distinctive characteristics upon the food product (Heller, 2001). Indeed, the metabolic end-products of certain strains, or combinations of strains, can render the product organoleptically undesirable. Similarly, it does not follow that all bacterial strains or species within a genetically related group are probiotics just because health-promoting activity has been identified for one strain or species. Taken together with the legal ramifications of patents, and the modern-day consumers' desire for transparency in the food industry with accurate labelling of food stuffs, it is increasingly important to distinguish LAB used in the food industry and health care market. Furthermore, along with the advent of improved methods for monitoring and characterising mixed LAB cultures (probiotics, starters or otherwise) has come the ability to examine the diversity and dynamics of LAB throughout the fermentation process, ripening and even storage of products (Pintado *et al.*, 2003; Ercolini, 2004). This affords the opportunity to better understand and technologically exploit the interplay of LAB, and to improve control during production, thus ensuring consistency in the final product delivered to the consumer.

This chapter will discuss recent developments in microbiology that relate to the selective detection, enumeration and differentiation of LAB. Greatest attention will be given to modern molecular methods, for which the advantages, disadvantages and biasing of the different techniques will be identified. Particular interest is afforded to probiotic LAB, namely lactobacilli and bifidobacteria.

## 5.2 Classic approaches to LAB enumeration and differentiation

Prior to the advent of molecular-based methodologies to monitor bacterial numerics and composition in samples, cultivation methods were the mainstay for identifying microbial content (Charteris *et al.*, 1997). These relied on physiological and/or phenotypic characteristics of the organisms. For example, selective media and/or incubation conditions were employed in order to electively stimulate isolation of the bacterial group of interest. Such approaches exploited the growth characteristics of said organisms, including their nutritional requirements, preferred incubation conditions (temperature, pH, redox potential) and susceptibility to, or prerequisite for, certain compounds (such as antibiotics, vitamins and blood). Multiple agar types and liquid media were needed to examine the full diversity of micro-organisms, usually employing a range of incubation strategies.

Traditional identification of the isolated bacteria similarly relied heavily on their phenotypic characteristics. Indeed, any feature that afforded a descriptive end-point was useful in expanding the catalogue of information on any given isolate. Classically, growth characteristics (e.g. anaerobic, facultatively anaerobic, microaerophilic or aerobic), growth requirements, morphology (both cellular and colonial), Gram's reaction, biochemical-fermentation profiles, enzymatic profiles, antibiotic susceptibility, serology and cell-wall-protein profiles were the means for differentiation and identification of microbes (Charteris *et al.*, 1997; Klein *et al.*, 1998; McCartney, 2002).

The major drawback of phenotypic characterisation is the plasticity of bacterial growth and the fact that phenotype does not necessarily provide accurate information of the genotypic relationship of organisms (Giraffa & Neviani, 2000; Dalezios & Siebert, 2001). For example, the taxonomically diverse components of the LAB group share the common phenotypic trait of lactic acid production. Different genera within the LAB group can also be subsequently subdivided (to varying levels) according to other phenotypic and metabolic traits. However, the ambiguous nature of phenotypically derived classification and reproducibility concerns limit the power of such strategies for the accurate identification of LAB. Furthermore, such techniques provide no insight into those bacterial components that are unculturable in the laboratory, either due to the fastidious nature of the organisms or the limitations of available culture techniques.

Modern identification methods rely less on carbohydrate-fermentation patterns and more on genotypic analysis and/or phenotypic characteristics based on the molecular rather than the physiological level (Charteris *et al.*, 1997; Lick, 2003). Indeed, certain phenotypic assays (revealed at the molecular level) have been shown to be useful for some members of the LAB. For example, *Bifidobacterium longum*, *Bifidobacterium infantis* and *Bifidobacterium suis* can be identified on the basis of their cell-wall-peptidoglycan composition (Holzapfel *et al.*, 2001); cell-wall-protein fingerprinting affords good differentiation of *Lactobacillus* species and/or subspecies (including reliable discrimination of the *Lactobacillus acidophilus* group, and rapid identification of thermophilic lactobacilli) (Gatti *et al.*, 1997; Pérez *et al.*, 2000); and lectin typing can differentially group lactobacilli isolates (Annuk *et al.*, 2001).

The advance of molecular technology, which affords rapid and reliable characterisation and/or differentiation of isolates (including closely related species and/or strains), has resulted in an abundance of genotyping assays for a broad range of functions (Olive & Bean 1999; McCartney, 2002). Formerly, such methodologies were developed and employed to discriminate bacterial isolates. Accordingly, a number of genetic fingerprinting assays became available that provided more sensitive and accurate differentiation of cultured strains than traditional phenotypic approaches (Prasad *et al.*, 1998; Gancheva *et al.*, 1999). Probing and sequencing strategies were soon developed in an effort to provide definitive identification of each biotype. With increased understanding of the molecular biology of LAB came the emergence of species-specific amplification techniques (via the polymerase chain reaction;

PCR) and expansion of probing approaches. As such, a number of direct genetic profiling techniques are now available. Although more robust, rapid and reliable than classic cultivation and phenotyping assays, genotyping and genetic profiling methodologies are not without their limitations. Thus, a polyphasic approach is wise in most cases.

## 5.3 Current approaches to LAB enumeration

### 5.3.1 *Differential plating methods*

By definition, probiotic LAB and starter cultures are viable, and thus classic plating strategies are applicable. A number of differential plating methods are available for monitoring such strains. The desired deliverable normally dictates which assay is most suitable (Charteris *et al.*, 1997). For example, different media and incubation conditions (temperature, time, atmosphere) will afford different coverage of the microbial content of samples. In general, 'selective' or differential media are used when monitoring specific probiotic or starter cultures. As mentioned previously, so-called selective media promote enhancement of the bacterial group of interest (e.g. Beerens' agar is often used to isolate bifidobacteria from mixed populations) (Beerens, 1990). However, there are very few truly selective media, and confirmation of the identity of isolates is required. Differential media enable discrimination of different bacterial groups based on their growth characteristics. In some cases, differential media may even allow the tentative classification of organisms. A large number of agar types have been proposed for the specific enumeration of LAB probiotic and starter cultures (Charteris *et al.*, 1997). Briefly, the International Dairy Federation (IDF) suggests the standard protocol of MRS (de Man–Rogosa–Sharpe) and M17 media usage for enumeration of the starter cultures *Lactobacillus delbrueckii* subsp. *bulgaricus* and *Streptococcus thermophilus*, respectively. The list of media useful for enumeration of probiotic LAB (*Bifidobacterium* and *Lactobacillus* species) is extensive, and selection often reflects the preference of the user (be it industry, academia or a government standards agency). Modified media have also been formulated to afford good recovery of bifidobacteria from fermented dairy products. Charteris *et al.* (1997) provide a fairly extensive review of differential plating techniques for the isolation of potentially probiotic LAB species. However, it is important to remember that growth on 'selective' media does not even provide presumptive identification of isolates, and further characterisation is essential. Therefore, in order to obtain accurate viable counts of particular LAB cultures, colony hybridisation should be performed and probing strategies used.

In general practice, to enumerate the bacterial contents of samples using plating strategies, a dilution series is first prepared and aliquots of each dilution plated and incubated accordingly. Anaerobic conditions are essential to recover bifidobacterial cultures. As such, enumeration of lactobacilli from mixed LAB cultures is often performed aerobically, to minimise bifidobacterial growth. Incubation times are also important, as some LAB grow more rapidly than others. It is often sensible to check the growth on plates on a number of consecutive days, to ensure small colonies are not missed. The dilution plate containing between 25 and 250 discrete colonies is (generally) used for plate counts. In practice, multiple plating of each dilution in the series is performed, to enable a more accurate calculation of the cultiviable component of the original sample.

## 5.3.2 *Probing strategies*

A number of probing strategies are applicable for monitoring the bacterial composition of mixed LAB populations. Three assays are commonly employed: colony hybridisations, dot-blot hybridisations and in-situ hybridisations (Charteris *et al.*, 1997; O'Sullivan, 1999; Giraffa & Neviani, 2000; Lick *et al.*, 2000). Colonies grown on agar plates can be transferred to nylon membranes, the cells lysed, and the released DNA fixed to the membrane for hybridisation with selected probes. Provided that appropriate conditions are used to afford efficient cell lysis of all colonies, this assay provides accurate enumeration of the viable levels of specific organisms. Initially, colony hybridisations were performed to obtain accurate identification of individual bacterial colonies. With the ever-increasing library of probes (especially for commercially and clinically important species), it is feasible to select appropriate probes to enable differential counts of LAB within mixed probiotic cultures. Furthermore, the development of gene sequencing and the ever-expanding ribosomal database have led to means of identifying isolates that are faster and more effective than colony hybridisation.

Dot-blot assays can be prepared for DNA extracts of isolates or samples directly (i.e. either culture-dependent or culture-independent). Total nucleic acids are prepared from pure cultures, mixed cultures or samples themselves. Purity and yields are usually checked against standard DNA ladders prior to setting up the dot-blot membrane. Depending on the application, either (a) a standard amount of DNA is applied (e.g. 1 μg) or (b) a dilution series is prepared from the crude DNA extract and a standard aliquot of each used. Denatured DNA is applied to the membrane using a dot-blot apparatus. Multiple samples can be assayed at the same time, and probe grids (or checkerboard hybridisations) also allow examination of multiple entities concurrently (using a different probe for each) – provided that similar hybridisation conditions apply (O'Sullivan, 1999). During checkerboard hybridisations, DNA extracts from different samples are usually applied along one axis and probes along the other. In general, quantification of target rRNA is measured as a percentage of total rRNA. Such quantification strategies overcome the problems relating to the ribosomal copy numbers of different bacterial cultures.

An adaptation of the dot-blot procedure – a reverse dot-blot technique – was shown to be useful for identifying LAB in fermented foods (Ehrmann *et al.*, 1994). *Lactobacillus* species-specific probes were bound to filter membranes. Non-specific PCR was used to amplify the rRNA genes from samples and the labelled amplification products were added to hybridisation buffer and applied to filter membranes. More recent developments in probing strategies, such as well assays and microchip assays, similarly utilise a grid system. In these cases, labelled oligonucleotide probes are bound to the wells or microchip and denatured DNA samples applied (Satokari *et al.*, 2003). *Bifidobacterium* species-specific probes have been used in such a manner (microtitre plates) to monitor the faecal bifidobacterial populations during a synbiotic feeding trial in humans (Malinen *et al.*, 2002). In this case, genus-specific PCR primers were employed for rRNA gene amplification. Microchip arrays (also called DNA chips or DNA microarrays) afford the ability to include many hundreds and/or thousands of probes on a single glass slide (Wilson *et al.*, 2002). Such technological advances are most exciting for the community analysis of complex microbial ecosystems, such as the microflora of the gastrointestinal (GI) tract. However, the current cost of such technologies is rather prohibitive.

Direct probing strategies, such as fluorescence in-situ hybridisation (FISH), are also culture-independent. In general, bacterial cells are fixed using paraformaldehyde prior to hybridisation with fluorescently labelled probes. In some cases, additional permeabilisation steps (such as enzymatic treatment with lysozyme and lipases) are necessary during the fixation procedure. Enzymatic permeabilisation is employed when preparing faecal samples for FISH analysis of the *Lactobacillus/Enterococcus* group using the probe LAB158 (Franks *et al.*, 1998). Furthermore, some hybridisations require more stringent conditions than others and formamide is added to the hybridisation buffer (Harmsen *et al.*, 2002). Hybridised cells are then enumerated by epifluorescence microscopy or flow cytometry (Langendijk *et al.*, 1995; Harmsen *et al.*, 2002).

Modern technology allows the manufacture of synthetic oligonucleotides. The target sequence of the oligonucleotide probe and the stringency of hybridisation conditions define the probe's specificity, be it universal, group-specific, genus-specific, species-specific or even strain-specific (O'Sullivan, 1999; McCartney, 2002). Due to the extensive nature of the ribosomal database, the majority of probing strategies have concentrated on regions within the bacterial ribosome, with highly variable regions being most applicable for species-specific and strain-specific probes. Of course, each probe must be validated and the hybridisation conditions optimised for specificity and sensitivity. Furthermore, it is important to optimise sample preparation (e.g. homogenisation, dilution factors, washing steps) according to the application. Certain matrices present more challenges for direct analysis than others. Essential to studies investigating the dynamics of microbial composition is the thorough homogenisation of samples and removal of any inhibitors. In many cases, bacterial concentrations also need to be considered, and may warrant dilution, concentration or even amplification to enable adequate results. One further consideration is the detection threshold of probing strategies. For example, FISH is generally considered to be poor for determining the levels of subdominant members of complex populations (McCartney, 2002).

### **5.3.3** *Quantitative PCR*

The advantages of PCR, once optimised, are its simplicity and speed, added to the capacity to investigate different phylogenetic levels (such as bacterial groups, species or strains). Universal primers rely on conserved regions of house-keeping genes common to all bacteria, whereas specific primers (at whatever level) utilise regions conserved only within the group of interest. Conventional PCR does not enable quantitative analysis of the target sequence in the template DNA (or, therefore, quantitative detection of bacterial groups within the initial sample). Real-time PCR incorporates chemistries such as fluorescently labelled probes or non-specific DNA-binding fluoropores (such as SYBR green) to measure the PCR product yield after each cycle (Mackay, 2004; McKillip & Drake, 2004). Using a calibration curve for amplification yields from controls of known concentration, specific quantitative detection of target bacteria in the initial sample can be determined.

To date, the application of real-time PCR has largely involved pathogenic organisms and diagnostic tools. However, the potential of the technique is rapidly becoming apparent and the next few years will undoubtedly see a major increase in real-time PCR and its application across microbiology. The abundance of species-specific PCR assays developed for rapid identification of probiotic and LAB starter cultures (for a discussion, see Section 5.4.6) suggests it is only a matter of time before optimised real-time PCR is available for numerous LAB species and/or strains of importance to the food industry. Indeed, published reports have already begun to filter out in this regard. Vitali *et al.* (2003) demonstrated the ability of real-time PCR to quantify three probiotic bifidobacteria strains (*B. infantis* Y1, *Bifidobacterium breve* Y8 and *B. longum* Y10) in the mixed probiotic product VSL-3. The development of real-time PCR assays for quantitative analysis of the LAB content of fermented milk products has also been published recently (Furet *et al.*, 2004). This study demonstrated the ability of real-time PCR methodologies to couple species-specific identification and accurate quantification of LAB strains in fermented milk products. Furthermore, the detection threshold of the assay was shown to be $10^3$ cells per mL product (Furet *et al.*, 2004).

Overall, PCR is a powerful tool in molecular studies. However, like all assays, it is not without its limitations. The main concern with PCR involves inherent biasing, from sample collection and processing to PCR amplification (Wintzingerode *et al.*, 1997). Such issues must be considered, and are often overcome by appropriate optimisation of the protocol. An additional limitation to PCR-based assays is the detection limit. The microbial complexity and nature of the sample of interest (be it food, environmental or clinical) have a profound impact on efficiency of both DNA extraction and PCR amplification, and thus the detection limit. Differences in the *rrn* operon copy numbers of different organisms further complicate detection limits and accurate quantification using PCR (Wintzingerode *et al.*, 1997). In general, incorporating multiple DNA extraction protocols and PCR amplification strategies, in conjunction with replication, are considered the most effective manner to obtain accurate coverage of the predominant microbial community (McCartney, 2002). An additional drawback of PCR-based assays is that they can detect DNA from dead cells as well as living cells, and thus do not provide accurate information on the viability of the microbial constituents they detect (Satokari *et al.*, 2003; Temmerman *et al.*, 2003).

## 5.4 Modern genetic approaches to LAB differentiation

### 5.4.1 *Background*

Genotypic fingerprinting and profiling systems fall into two categories: culture-dependent and culture-independent. Initial genotyping concentrated on developing assays to differentiate bacterial strains. This was a natural progression from phenotypic characterisation of isolates and afforded more robust methods for monitoring bacterial strains of interest (for example, probiotics) or investigating the diversity and dynamics of the culturable component beyond species level (McCartney, 2002). In general, such genetic fingerprinting provides subspecies discrimination, although certain assays also enable species identification.

Analyses of restriction fragment length polymorphisms (RFLPs) provide the platform for most forms of genetic fingerprinting (Holzapfel *et al.*, 2001). A number of these profiling systems have been investigated in recent years, largely involving restriction endonuclease analysis (REA) and/or modifications thereof. The selection and number of restriction enzymes is critical, as are the interpretation tools used. Nowadays, numerous software packages are available to enable comparison of complex banding profiles and provide similarity indices. The inclusion of standard ladders in each gel allows normalisation across different gels (provided that the electrophoresis conditions are the same). Formerly, such analysis was dependent on visual differentiation by the user. Due to the complexity of standard REA, comprising more than 100 DNA fragments of varying size, discrimination by the naked eye was limited. Methods to simplify the banding profile were thus examined, including: (a) the use of rare-cutting enzymes (pulsed-field gel electrophoresis; PFGE) (McCartney *et al.*, 1996); (b) highlighting specific fragments from the complex fingerprint through probing strategies (ribotyping) (Rodtong & Tannock, 1993; McCartney & Tannock, 1995) or selective amplification (amplified fragment length polymorphism; AFLP) (Vos *et al.*, 1995); or (c) preparing REA from discrete regions of the genome (amplified ribosomal DNA restriction analysis; ARDRA) (Roy *et al.*, 2001).

## 5.4.2 *Pulsed-field gel electrophoresis (PFGE)*

The selection of restriction endonucleases for PFGE is made with respect to the G+C content of the DNA of the bacteria of interest. For example, an enzyme targeting a sequence rich in A and T nucleotides would be employed for bacteria with a high G+C content (Finney, 1995). Conversely, enzymes containing G+C-rich recognition sites are more effective for low-G+C bacterial groups. In addition, enzymes containing the tetranucleotide CTAG (such as *Xba*I, *Spe*I and *Nhe*I) are useful for certain genera since this sequence is extremely rare in bacterial genomes with a G+C content of greater than 45 mol% (Bourget *et al.*, 1993). The resulting large DNA fragments cannot be separated by conventional gel electrophoresis, and an alternating field is thus employed with increasing pulse times across the electrophoresis run (hence the name pulsed-field gel electrophoresis). The larger DNA macromolecules spend more of the run adjusting their conformation to the alternating fields, whilst smaller macromolecules migrate in a zigzag fashion down the gel.

Numerous studies have demonstrated the discriminatory power of PFGE for LAB (Fig. 5.1), including studies monitoring the LAB flora of humans (McCartney *et al.*, 1996; Kimura *et al.*, 1997) or tracking probiotic strains during human feeding trials (O'Sullivan & Kullen, 1998). PFGE has been successfully applied for subspecies differentiation of LAB, grouping bacterial strains within a species in some instances, and can also be used to distinguish specific strains of interest. Multiple studies have successfully used PFGE to differentiate probiotic strains, including within the *Lb. acidophilus* complex and three of the four *Lactobacillus delbrueckii* subspecies (*delbrueckii*, *bulgaricus* and *lactis – indicus* subsp. nov.), amongst others (Klein *et al.*, 1998; Giraffa & Neviani, 2000; Lick, 2003; Dellaglio *et al.*, 2005).

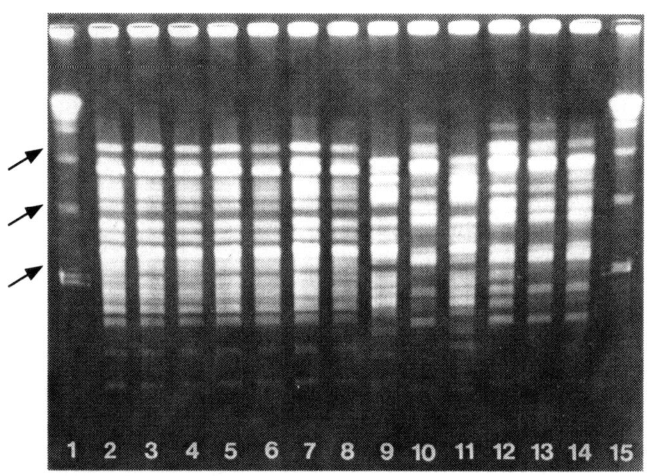


**Fig. 5.1** PFGE profiles of *Apa*I-digested DNA from 13 *Lactobacillus* isolates which were indistinguishable by ribotyping using *Eco*RI and *Hin*dIII. Note: Lanes 1 and 15: λ ladder PFGE marker #340 (arrows: 145.5, 97.0 and 48.5 kb, respectively; New England BioLabs); lanes 2–8: isolates from subject A; lanes 9–14, isolates from subject B.

Essential to PFGE is intact chromosomal DNA, which necessitates embedding bacterial cells in low-melting-point agarose prior to DNA extraction, digestion and electrophoresis. Taken together with the need to extend incubation times, this ensures that PFGE is a relatively time-consuming method. However, the main advantage of PFGE is its discriminatory power, which most likely reflects the representation of the whole genome in the profile, rather than selected regions (as is the case with the majority of other RFLP fingerprinting assays). The superior differentiation power of PFGE for LAB has been shown in a number of studies comparing different genetic fingerprinting techniques (McCartney *et al.*, 1996; O'Sullivan, 1999; Tynkkynen *et al.*, 1999). Indeed, PFGE is generally accepted to be the 'gold standard' genotyping technique. Due to its extensive discriminatory power for strain differentiation, PFGE genotyping is particularly useful for quality assurance purposes.

### **5.4.3** *Ribotyping and amplified rDNA restriction analysis (ARDRA)*

Ribotyping and ARDRA are modified RFLP techniques that enable simplification of the resulting banding profile. In ribotyping, the DNA fragments from conventional REA are transferred to a nylon membrane (by Southern blotting) and hybridised with an oligonucleotide probe of ribosomal origin. As such, the DNA fragments which contain the rRNA operon (*rrn*) are highlighted from the RFLP profile (Fig. 5.2) (Grimont & Grimont, 1986; Popovic *et al.*, 1993; McCartney, 1996). Indeed, due to the sensitivity of radioactively labelled probes, some of the bands obtained in the ribotype may not have been clear in the original REA profile. As such, visualisation strategies and their sensitivity can greatly impact the resultant

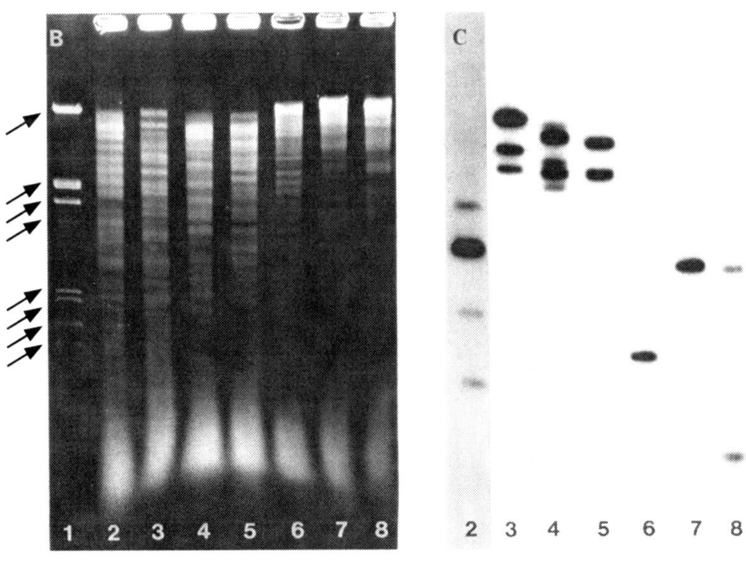


**Fig. 5.2** Ribotyping profiles of *Bifidobacterium infantis* ATCC 15697 generated using seven different restriction endonucleases. Note: Agarose gel electrophoresis (panel B) of DNA digests, and autoradiographs (panel C) produced by hybridisation of the subsequent membrane (Southern blotting) with $^{32}$P-labelled *B. infantis* ATCC 15697 16S rDNA. Lanes: 1, molecular weight marker III (arrows: 21.226, 5.178 and 4.973, 4.277, 3.530, 2.027, 1.907, 1.584, 1.330, 0.983 and 0.831 kb, respectively; Boehringer-Mannheim, Germany); 2, *Kspl*; 3, *PvuII*; 4, *NarI*; 5, *SacI*; 6, *SmaI*; 7, *HindIII*; and 8, *BamHI*. A longer exposure time was required for autoradiography of the *KspI* digest, in order to develop fainter bands.

biotypes. This is a particularly important aspect to consider for inter-laboratory comparison and/or reproducibility. Additional considerations for ribotyping include the choice of restriction endonucleases, which are usually selected to afford ribotypes comprising fewer than 20 bands whilst enabling sufficient discrimination for the application. As well as being dependent on the restriction enzyme used, the number of bands in a ribotype is dependent on the *rrn* copy number of the bacterial chromosome. Comparison of the ribotypes generated by several enzymes is advised, as minor variations in the bacterial DNA sequence may result in a change in the ribotype that is detected by a single enzyme (Rodtong & Tannock, 1993). The inclusion of a second ribotype, generated using a second restriction endonuclease with a distinctly different recognition site, increases the accuracy of the technique. Furthermore, the use of multiple enzymes to generate an overall ribotyping profile can often afford greater discriminatory power (McCartney & Tannock, 1995). Good subspecies differentiation has been shown with ribotyping for both *Lactobacillus* and *Bifidobacterium* species, and for examining the dynamics and diversity of LAB strains. However, PFGE has been shown to have greater discriminatory power than ribotyping (Fig. 5.3), with *Lactobacillus* isolates of the same ribotype displaying distinctive PFGE banding profiles.

ARDRA essentially reverses the mechanics of ribotyping, by first amplifying the rRNA gene (usually the 16S rRNA gene) using PCR and then performing REA on the resultant amplicons. ARDRA is a rapid and reliable fingerprinting technique that has been successfully applied to distinguish LAB, most notably *Lb. acidophilus*, *Lactobacillus helveticus* and *Lb. delbrueckii* subspecies (*delbrueckii*, *lactis* and *bulgaricus*), at the species level (Giraffa *et al.*, 1998; Roy *et al.*, 2001; Delley & Germond, 2002). Similar to ribotyping, the discriminatory power of ARDRA relies heavily on the selection of the restriction endonuclease. Furthermore, the size of the amplification product (i.e. partial or complete 16S rRNA gene) has an impact on the differentiation power (Satokari *et al.*, 2003). ARDRA profiles are, however, highly reproducible and afford easy inter-laboratory comparisons. To date, most

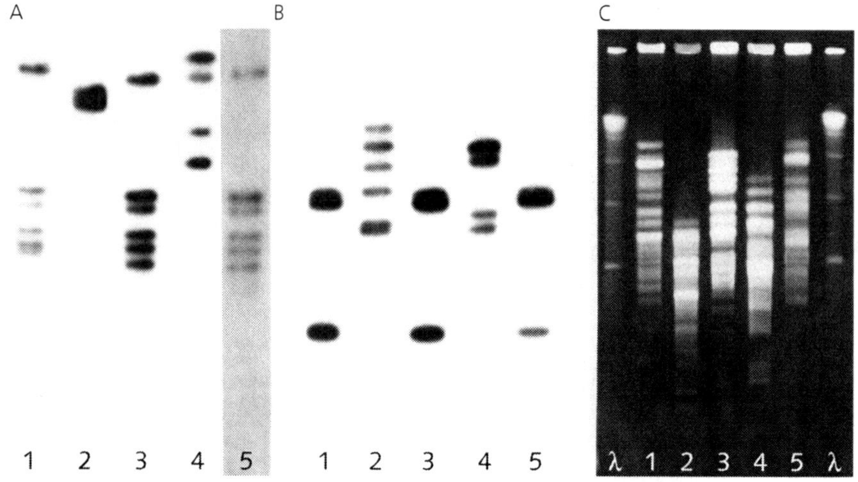


**Fig. 5.3** Differentiation of representative *Lactobacillus* isolates by ribotyping and PFGE. Note: Autoradiographs displaying ribotypes of *Eco*RI digests (panel A) and *Hin*dIII digests (panel B), and PFGE of *Apa*I-digested DNA (panel C). Lanes 1–5 represent lactobacilli isolates numbers 1 to 5, respectively; the lambda ladder (λ) was run as a marker in the PFGE.

applications of ARDRA have employed single-enzyme assays. Theory would suggest that inclusion of additional restriction endonucleases, with dissimilar target sequences, should increase the discriminatory power of ARDRA (as seen with ribotyping). However, this is dependent on the selection of enzymes the recognition sites of which are contained within the amplified region. Since ARDRA profiles do not include the flanking regions of the *rrn* operon, this technique displays less discriminatory power than ribotyping. Both techniques are inferior to methods such as PFGE and randomly amplified polymorphic DNA (RAPD) analysis, which encompass the total bacterial chromosome.

ARDRA patterns generated by *Cfo*I, *Hin*fI, *Tru*9I and *Scr*FI were shown to provide an effective identification tool for LAB species within the *Lb. acidophilus* and *Lactobacillus casei* complex (Roy *et al.*, 2001). Furthermore, *Cfo*I ARDRA patterns afforded reliable differentiation of *Lb. delbrueckii* subsp. *lactis* and subsp. *bulgaricus*. Overall, the work of Roy and colleagues demonstrated that ARDRA was a useful identification method for homofermentative lactobacilli commonly used as starter or probiotic cultures, the only exception being *Lactobacillus amylovorus*.

### **5.4.4** *Amplified fragment length polymorphism (AFLP) analysis*

As for PFGE, the selection of the restriction endonucleases for AFLP analysis is made with respect to the G+C content of the DNA of the bacteria of interest. The AFLP fingerprinting technique was originally designed to allow plant researchers to identify cultivars, to localise desirable genes, and to develop genetic maps. In recent years, the technique has been increasingly used in taxonomic and epidemiological studies for bacteria, because of the reproducibility, the requirement for small amounts of DNA, and the resolution of multiple polymorphic bands per reaction (Huys *et al.*, 1996; Janssen *et al.*, 1996; Dijkshoorn *et al.*, 1996; Gibson *et al.*, 1998; Koeleman *et al.*, 1998). Genomic DNA is cut with two restriction enzymes, and double-stranded adaptors are ligated to the 'sticky ends' of the resulting DNA fragments to generate template DNA for amplification (Fig. 5.4). The sequences of the adaptors and the adjacent restriction sites serve as primer-binding sites for a subsequent preselective amplification of the restriction fragments. A selective amplification is then carried out using PCR primers containing sequences homologous to the adaptors with additional selective bases at the 3' end. Thus, a subset of restriction fragments is selected – that is, only those fragments containing the complementary flanking sequences to the restriction sites of the selective primers (Vos *et al.*, 1995; Olive & Bean, 1999). The complexity and, therefore, discriminatory power of the AFLP banding profile is dependent on the number of selective nucleotides and the complexity of the genomic DNA. Polyacrylamide gel electrophoresis (PAGE) is used to separate the selectively amplified fragments, generating species-/subspecies-/strain-specific banding patterns (depending on the primers used and the level of discrimination required). As mentioned above, AFLP analysis has largely been applied to epidemiological analyses, and has demonstrated good reproducibility and discriminatory power greater than ribotyping and ARDRA (Olive & Bean, 1999). The few studies published to date using AFLP for LAB have similarly shown good reproducibility, providing inter- and intra-specific differentiation. Torriani and colleagues presented a poster demonstrating species identification by AFLP typing for *Lactobacillus plantarum*, *Lactobacillus pseudoplantarum* and *Lactobacillus pentosus* at the 6th Symposium on LAB (Giraffa & Neviani, 2000). In addition, intra-specific discrimination was displayed for cheese and silage isolates. Gancheva *et al.* (1999) similarly showed species identification by AFLP using 98 homofermenta-

1) Digest chromosomal DNA with restriction enzymes (*Eco* RI and *Mse* I)

5'-----GAATTC-----TTAA-----3'
3'-----CTTAAG-----AATT-----5'

2) Ligate *Eco* RI and *Mse* I adaptor sequences

*Eco* RI adaptor *Mse* I adaptor
5' **CTCGTAGACTGCGTACC**AATTC-----T**TACTCAGGACTCAT** 3'
**CTGACGCATGGTTAA**G-----AAT**GAGTCCTGAGTAGCAG**

3) Pre-selective amplification of the *Eco* RI–*Mse* I fragments

*Eco* RI core sequence
5' **GACTGCGTACCAATTC**-----AATGAGTCCTGAGTA 3'
GAGCATCTGACGCATGGTTAAG

5' CTCGTAGACTGCGTACCAATTC-----TTACGAGTCCTGAGTA 3'
**AATGCTCAGGACTCAT**
*Mse* I core sequence

4) Selective amplification of the *Eco* RI–*Mse* I fragments

*Eco* RI primer
5' **GACTGCGTACCAATTCN**-----AATGAGTCCTGAGTA 3'
GAGCATCTGACGCATGGTTAAGN

5' CTCGTAGACTGCGTACCAATTC-----NTTACGAGTCCTGAGTA 3'
**NAATGCTCAGGACTCAT**
*Mse* I primer

5) Separate fragments by polyacrylamide gel electrophoresis

**Fig. 5.4** Outline of the AFLP fingerprinting technique, using the restriction enzymes *Eco*RI and *Mse*I as examples. Note: Arrows indicate the sites of action for restriction enzymes. Sequences for adaptors, core sequences and primers are shown in bold type. *Eco*RI and *Mse*I are suitable enzymes for use with low G+C organisms, and have been used successfully with LAB.

tive LAB species (namely, *Lb. acidophilus*, *Lactobacillus amylolyticus*, *Lb. amylovorus*, *Lactobacillus crispatus*, *Lactobacillus gallinarum*, *Lactobacillus gasseri*, *Lb. helveticus*, *Lactobacillus iners* and *Lactobacillus johnsonii*). Furthermore, subspecies discrimination was also obtained. Both AFLP and RAPD afforded greater intra-specific differentiation of strains from the *L. acidophilus* group than cell-wall-protein profiling (Gancheva *et al.*, 1999). Although it is clear that AFLP allows greater discrimination of bacteria at the subspecies level, the technique is time-consuming and its costs may be prohibitive if a large number of strains is to be examined.

## 5.4.5 *Randomly amplified polymorphic DNA (RAPD)*

RAPD, also termed arbitrarily primed PCR (AP-PCR) or DNA amplification fingerprint (DAF), uses short random sequences (arbitrary primers) coupled with low-stringency PCR conditions (Charteris *et al.*, 1997; Lick, 2003). In general, oligonucleotide primers of 10 bases are employed, and the low-stringency conditions allow annealing to partial or complete complementary sequences on the bacterial genome. The number and position of sites for amplification by these arbitrary primers varies between bacterial strains. The randomly amplified DNA fragments are then separated by conventional gel electrophoresis to provide a genetic fingerprint (Tilsala-Timisjarvi & Alatossava, 1998; Satokari *et al.*, 2003). Single and multiplex RAPD (combining multiple arbitrary primers in a single PCR) assays have been reported. The method is simple and fast, but suffers from poor reproducibility; this is due to the low stringency of conditions and the use of non-specific primers. As such, careful control of conditions is necessary. The purity of the target DNA also impacts the discriminatory power of RAPD (Johansson *et al.*, 1995). One reported adaptation to the method, which improves reproducibility, is tri-RAPD or triplet arbitrary primed PCR (TAP-PCR) (O'Sullivan & Kullen, 1998; O'Sullivan, 1999). Multiple reactions are set up and run at different annealing temperatures, and a consensus RAPD profile is configured comprising the shared bands from each of the reactions.

A large body of work demonstrates the efficacy of RAPD genotyping to differentiate LAB species. These data demonstrate the usefulness of this technique for inter- and intra-species discrimination of lactobacilli, especially those important to the food industry (Cocconcelli *et al.*, 1995; Du Plessis & Dicks, 1995; Johansson *et al.*, 1995; Drake *et al.*, 1996; Van Reenen & Dicks, 1996; Vogel *et al.*, 1996; Andrighetto *et al.*, 1998; Klein *et al.*, 1998; Zapparoli *et al.*, 1998; Gancheva *et al.*, 1999; Tynkkynen *et al.*, 1999; Kwon, 2000; Klein, 2003; Lick, 2003). Furthermore, comparison of RAPD-typing and SDS-PAGE-typing showed better identification of *Lactobacillus* species using the cumulative RAPD profile from patterns obtained from three single-primer reactions (Gancheva *et al.*, 1999). Similar clustering was obtained to that of DNA-DNA hybridisation and rDNA sequencing methods. Furthermore, subspecies discrimination has been demonstrated for LAB using this approach.

Corsetti and colleagues (2003) utilised both RAPD and SDS-PAGE to characterise type strains of eight LAB species and 150 LAB isolates from wheat flours and sourdoughs. RAPD typing successfully differentiated the type strains, but did not provide species-specific profiles. However, intra-species differentiation was possible. Interestingly, LAB isolates from sourdoughs made from similar raw materials or of similar geographical origin displayed similar RAPD profiles. Nine clusters were observed using the combined RAPD fingerprints (combined profiles of primers P1, P4 and P7), and subclusters (59% similarity) comprised isolates of similar origin (either region or manufacturer). *Lb. plantarum* showed the greatest intra-species diversity, with seven of the 17 strains producing unique fingerprints (Corsetti *et al.*, 2003). RAPD was more discriminatory than cell-wall protein profiling. However, species-specific identification was plausible by SDS-PAGE, due to common bands. Overall, analysis combining both RAPD and cell-wall protein profiles provided an effective tool to characterise LAB isolates from sourdough, enabling both species identification and intra-species differentiation which afforded discrimination of strains according to origin (geographical or raw material). As such, the combined assay may be applicable to studies tracing foods based on their microbial content.

Andrighetto *et al.* (2001) had previously shown similar results for enterococcal isolates from traditional Italian cheeses, using RAPD and SDS-PAGE assays. Numerical analysis of the RAPD profiles obtained using four primers and five sets of amplification conditions demonstrated both inter-specific and intra-specific differentiation. Species characterisation was in good agreement with that of SDS-PAGE analysis for 93.5% of the isolates. Most significantly, these studies demonstrated the ability of RAPD profiling to highlight the diversity of enterococcal isolates among Italian cheeses.

### 5.4.6 *PCR-typing*

Along with RAPD-typing, a number of PCR-based fingerprinting methodologies have been developed in recent years. Unlike RAPD, most of these employ specific (rather than arbitrary) primers. Single and multiplex PCR assays are available, depending on the desired output. One such PCR-based fingerprinting technique is repetitive element sequence-based PCR (Rep-PCR) (Versalovic *et al.*, 1991). Three such repetitive elements have been targeted for PCR-typing, namely repetitive extragenic palindromic (REP) elements, enterobacterial repetitive intergenic consensus (ERIC) sequences and the so-called BOX elements (Olive & Bean, 1999). Rep-PCR is a fast and simple method, capable of subspecies level differentiation. Indeed, Rep-PCR is rapidly becoming the assay of choice for DNA-typing in clinical scenarios. A combination of multiple Rep-PCR profiles, for example REP and ERIC patterns, increases differentiation beyond that of either alone.

Rep-PCR, using both REP and ERIC primer sets, was used in a study monitoring the LAB dynamics of cheddar cheeses during ripening (Dasen *et al.*, 2003). A total of 380 *Lactobacillus* isolates were obtained from four cheeses; a control cheese prepared with base milk, a raw milk cheese (cheese prepared with base milk plus 10% retentate) and two adjunct-containing cheeses (two different *Lactobacillus paracasei* subsp. *paracasei* strains). Rep-PCR generated 16 distinctive profiles from the 380 isolates. Representatives of each Rep-PCR profile were identified using species-specific PCR, with eight being shown to be *Lb. paracasei*, five *Lactobacillus curvatus* and three *Lb. plantarum*. The only Rep-PCR profiles seen in the two adjunct-containing cheeses during ripening were those of the respective adjunct strains. However, isolates exhibiting adjunct profiles were also cultured from the retentate cheese, and one of the profiles was also seen in the control cheese. A second Rep-PCR profile (displayed by 5% of isolates) was seen for one of the adjunct cheeses, but only after 12 months ripening. Overall, Rep-PCR typing demonstrated that the *Lactobacillus* species compositions of control cheese and raw milk cheese were similar, but distinctive differences were clearly evident in both subspecies composition and dynamics of these cheeses during ripening (Dasen *et al.*, 2003). Inclusion of *Lb. paracasei* subsp. *paracasei* adjunct-strain in Cheddar cheeses resulted in a predominance of the said strain throughout ripening.

Rep-PCR-typing has been shown to afford greater discriminatory power than other profiling techniques (including genetic fingerprinting) in a number of comparison studies, for different organisms of interest (Olive & Bean, 1999). Few studies have been conducted to date examining the usefulness of Rep-PCR in distinguishing probiotics, food and environmental LAB isolates. Successful Rep-PCR typing has been described for *Lb. curvatus*, *Lb. johnsonii* and *Lactobacillus sakei* (Lick, 2003). In addition, de Urraza *et al.* (2000) differentiated 37 thermophilic LAB at species, subspecies and intra-subspecies level, using BOX Rep-PCR. Andrighetto and colleagues (1998) showed that RAPD analysis could both differentiate and identify the same LAB at species level. However, multiple RAPD banding profiles were required, from more than one primer and/or restriction analysis of the resulting amplification products. Interestingly, wild-type strains of *Lb. delbrueckii* subsp. *lactis* isolated from raw milk provided more complex BOX Rep-PCR fingerprints, with more bands and greater variation, than isolates from commercial products (de Urraza *et al.*, 2000).

These data clearly demonstrate the validity of using PCR-typing protocols to identify and distinguish the microbiota of starter cultures.

Overall, the major application of genetic fingerprinting methods has been differentiation studies of isolated bacterial strains (culture-dependent). Such molecular methods have greatly improved the ability to discriminate LAB, especially closely related species (many of which are indistinguishable phenotypically). Of immense interest is the opportunity to expand the use of molecular techniques into culture-independent applications. The biggest hindrance to this is the food matrix itself, though research and optimisation of sample preparations and DNA extraction protocols is unlocking the door to such investigations. This will afford direct detection of the community profile (complex genetic fingerprint reflecting the entire mixed LAB culture). Such advances are already underway, particularly using denaturing gradient gel electrophoresis (DGGE; discussed below).

A variety of PCR techniques have also been developed for identifying and/or differentiating LAB. Many of these rely on either amplification alone or specified size of amplification product(s), for example, using species-specific PCR primers. A number of target regions have been incorporated for such assays, including 16S rRNA genes, 16S–23S intergenic sequences (ITS), heat-shock protein (*hsp*) genes, *ldh* and *rec*A genes (Lebond-Bourget *et al.*, 1996; Kullen *et al.*, 1997; Tilsala-Timisjarvi & Alatossava, 1997; Tannock *et al.*, 1999; Torriani *et al.*, 1999; Roy & Sirois, 2000; Jian *et al.*, 2001). Furthermore, species- and/or strain-specific oligonucleotide sequences (for use as either primers or probes) have been identified from RAPD-profiling (Lucchinin *et al.*, 1998; Tilsala-Timisjarvi & Alatossava, 1998). The list of LAB species-specific and strain-specific primers is ever-increasing. PCR-based strategies have, indeed, overtaken probing applications as the mode of choice for identifying and differentiating LAB cultures and populations. Furthermore, the tide of real-time PCR applications for LAB is rapidly approaching, and it is only a matter of time before such procedures are applied more extensively.

Numerous species-specific primer sets have been published for *Lactobacillus* and *Bifidobacterium* species to date. In particular, species-specific primers are available for the probiotic and starter LAB cultures: *S. thermophilus*, *Lb. acidophilus*, *Lb. casei*, *Lb. crispatus*, *Lb. curvatus*, *Lb. delbrueckii* subsp. *bulgaricus*, *Lb. delbrueckii* subsp. *lactis*, *Lactobacillus fermentum*, *Lb. gasseri*, *Lb. helveticus*, *Lb. johnsonii*, *Lb. paracasei* subsp. *paracasei*, *Lb. pentosus*, *Lb. plantarum*, *Lactobacillus reuteri*, *Lactobacillus rhamnosus*, *Lactobacillus sanfranciscensis*, *Bifidobacterium adolescentis*, *Bifidobacterium bifidum*, *B. breve*, *B. infantis* and *B. longum* (McCartney, 2002; Satokari *et al.*, 2003). Strain-specific primers are also available. These are of particular interest for quality assurance measures and for monitoring the function and dynamics of starter and probiotic cultures during production, storage and consumption of products.

Multiplex PCR utilises multiple primer sets in a single reaction to simultaneously amplify multiple DNA targets, permitting greater accuracy and reliability. Lucchinin *et al.* (1998) employed multiplex PCR to specifically detect the probiotic *Lb. gasseri* strain 4B2 from faecal isolates grown on selective agar. Optimisation of reaction conditions is the greatest hurdle to multiplex PCR and, similarly to probing strategies combining multiple oligonucleotide probes, is applicable only with primer sets which have similar reaction specifications (especially annealing temperatures).

## 5.4.7 *Gene sequencing*

Perhaps the most accurate tool for identification of bacterial isolates is that of gene sequencing. The extensive ribosomal database and web-based alignment tools allow rapid comparisons of sequences culminating in similarity indices. 16S rRNA gene sequencing is perhaps the best tool available for identifying unknown isolates and/or confirming the identity of cultures (such as probiotic or starter culture strains) (Wintzingerode *et al.*, 1997; Giraffa & Neviani, 2000). Bacterial systematics and phylogenetic classification defines species delineation by DNA-DNA hybridisation of greater than 70%. DNA-DNA homology is not a feasible option for routine laboratory purposes, and the advent of effective nucleic acid sequencing techniques provided an alterative.

Numerous studies have demonstrated the efficacy of 16S rRNA gene sequencing for identification and taxonomy, with the general acceptance of the 97% rule (i.e. a similarity ≥97% indicates that the sequences are from bacteria of the same species) (Wayne *et al.*, 1987). Ribosomal genes have been studied most extensively because they are essential house-keeping genes contained in all bacterial genomes, and they comprise a number of variable and hyper-variable regions which afford differentiation (Woese, 1987; Ludwig & Schleifer, 1994). The information contained in such variable regions provides phylogenetic discrimination at various levels (bacterial group, genus, species or subspecies) and identification of sequences unique to such taxonomic units not only provides the resource for classification, but can also identify targets for specific primers and probes. Other sequences which have proven useful for differentiating LAB include the 16S-23S ITS (Lebond-Bourget *et al.*, 1996; Tannock *et al.*, 1999).

As with all amplification-based strategies, gene sequencing is prone to inherent biasing. Cost can also be prohibitive, with the need for specialised equipment. However, provided that the template is of sufficient quality, gene sequencing is a highly reliable and reproducible method. Furthermore, engineering advances (e.g. capillary sequencers) allow rapid throughput of a large number of samples, overcoming the logistical problems previously encountered. Most importantly, gene sequencing is applicable to both culture-dependent and culture-independent strategies using either bacterial isolates and/or clones as template.

## 5.4.8 *Denaturing gradient gel electrophoresis-PCR (DGGE-PCR)*

DGGE is a powerful profiling method which provides a snapshot (or fingerprint) of the diversity of the microbial composition of samples without requiring cultivation (Muyzer *et al.*, 1993; Ercolini, 2004). Target sequences (usually ribosomal genes) are amplified from DNA extracts by PCR and then separated in polyacrylamide gels containing a linear denaturing gradient (using urea and formamide). Separation of the similarly sized sequences is facilitated by their chemical stability and melting temperatures. The mobility of the DNA fragment in the DGGE gel is inhibited as the double-stranded DNA (dsDNA) denatures or partially melts. As this is a sequence-dependent phenomenon, different amplicons will migrate to distinct regions within the denaturing gradient and thus be separated by DGGE. In general, a GC-clamp is employed (attached to one of the PCR primers) to ensure that the dsDNA amplicons are not completely denatured (Ercolini, 2004).

DGGE has largely been used to study the microbial complexity and dynamics of different samples. However, a number of investigations have shown a further application for DGGE with respect to LAB, that of differentiating LAB species (Walter *et al.*, 2000; Ercolini *et al.*,

2001; Satokari *et al.*, 2001; Heilig *et al.*, 2002; Temmerman *et al.*, 2003). A recent review by Ercolini (2004) discusses PCR-DGGE fingerprinting of LAB in detail and provides a third use for this technique; namely, quality control and microbial assessment of commercial foods, particularly those comprising probiotics and LAB starter cultures.

There are a number of limitations to DGGE, on top of the previously mentioned disadvantages of PCR-based techniques (such as amplification efficiency and inherent biasing). One such limitation is that heterologous sequences do not necessarily migrate distinctively, and thus a single band in a PCR-DGGE profile may contain multiple sequences. Altering the gradient may alleviate this problem, as Temmerman *et al.* (2003) demonstrated by using a narrower denaturing gradient to differentiate *Lb. delbrueckii* subsp. *bulgaricus* and *Lb. acidophilus*.

Other problems which may affect DGGE profiling are the formation of heteroduplexes or chimeric molecules (Muyzer *et al.*, 1993). These may lead to an overestimation of bacterial diversity. Recent investigations have highlighted the impact of template concentration, cycle number and species diversity on heteroduplex frequency (Qiu *et al.*, 2001). The use of PAGE, or T7 endonuclease I to cleave the bubble in heteroduplexes, can potentially eliminate these difficulties. Another problem, which may lead to an overestimation of bacterial diversity, is heterologous rRNA operons within the same organism. As such, the DGGE profile of an organism containing non-homologous copies of the 16S rRNA gene will have multiple bands. Recent studies have highlighted such a problem in *B. adolescentis* (Satokari *et al.*, 2001) and some LAB species of food origin (including *Lactobacillus* and *Streptococcus* species) (Cocolin *et al.*, 2001). The selection of alternate target regions on the 16S rRNA gene may overcome the problem of heterologous copies.

Electrophoresis time has also been shown to impact the PCR-DGGE fingerprint (Sigler *et al.*, 2004). In this study, Sigler and colleagues investigated the banding profiles obtained by running PCR-DGGE under differing applied voltages and running times whilst maintaining the same electrophoresis volt-hours (V·h), namely 1000 V·h. Their results clearly demonstrated that altering the voltage and running time ratio (whilst maintaining the same V·h) significantly affected the banding profile obtained. Shorter run times at a higher voltage afforded better PCR-DGGE fingerprinting, this being attributed to the stability of the denaturing gradient under such conditions (Sigler *et al.*, 2004). Standardised DGGE conditions must therefore be employed to enable comparison of fingerprints from different gels.

The PCR-DGGE of the V3 region successfully differentiated eight out of 21 reference strains of common food LAB (Ercolini *et al.*, 2001), namely *Enterococcus casseliflavus*, *Enterococcus faecium*, *Lb. casei*, *Lb. helveticus*, *Lb. plantarum*, *Lactococcus garvieae*, *Leuconostoc mesenteroides* subsp. *mesenteroides* and *S. thermophilus*. Three *Lactococcus lactis* subspecies (*cremoris*, *hordniae* and *lactis*) generated identical DGGE migratory patterns using the V3 region, although these were distinct from all other reference strains (see Schleifer *et al.*, 1985). Four other migration patterns were identified between the remaining 10 LAB species. Most interestingly, the *Lb. delbrueckii* subspecies (*bulgaricus* and *lactis*) were distinguishable from each other using this PCR-DGGE protocol, although *Lb. delbrueckii* subsp. *bulgaricus* showed an identical profile to *Lb. sakei* and *Lb. delbrueckii* subsp. *lactis* was indistinguishable from *Lb. curvatus* (Ercolini *et al.*, 2001).

Temmerman *et al.* (2003) employed PCR-DGGE of the V3 region to investigate the microbial content of 10 probiotic products. A higher detection sensitivity was provided by DGGE profiling than culture-dependent analysis for four of the products (Aciforce, Actimel, Bacilac and Bififlor). Both methods identified bacterial strains in some products which were not listed on the label. Similarly, for five products (Aciforce, Bacilac, Bififlor, Proflora and Vitamel) some organisms stated on the product labels were not determined by either assay (Temmerman *et al.*, 2003).

An extensive review of PCR-DGGE application in food microbiology has recently been published (Ercoloni, 2004). The greater proportion of studies have utilised the V3 region or sequences including the V3 region, although Ercolini *et al.* (2003) also utilised a sequence across the V4–V5 regions to follow the LAB components in Stilton cheeses. Furthermore, the V1 region has also been used in some studies. Overall, DGGE protocols that provide species-specific differentiation have been identified for certain LAB of interest in the food industry. In addition, assays which afford greater discrimination are also reported, and may be of particular use in traceability studies. To date, PCR-DGGE of LAB cultures has been applied to a range of fermentation products, including cheeses, sausages, malt whisky, sourdoughs and dairy products.

The major application of PCR-DGGE has represented qualitative assessment and comparison analysis of different samples. A recent study, however, investigated the use of multiple competitive PCR-DGGE (cPCR-DGGE) to quantify and profile strain succession during the fermentation process (Pintado *et al.*, 2003). Using this technique, the authors demonstrated the potential to calculate biomass concentration from the DNA extract, although the nature of the biomass (active, inactive or dead) could not be determined. Essential to such procedures is the identification of suitable competitor DNA, which must produce amplicons of the same size with a distinguishable profile (i.e. distinct DGGE mobility). Such techniques would enable bacterial succession during fermentation, ripening and storage of products (Pintado *et al.*, 2003). A further prospective assay that may similarly provide qualitative as well as quantitative data on the microbiota would be to couple real-time PCR with DGGE analysis. To date, however, this has only been alluded to and no published data are available on the potential of this approach.

### 5.4.9 *Probing strategies*

As well as being applicable to the quantification of bacterial groups of interest, probing strategies have been used for qualitative purposes – in particular, the confirmation or identification of bacterial cultures at different phylogenetic levels (genus, species or subspecies). In practice, such assays are useful when a limited number of bacterial strains or species are of interest. One study of 80 LAB strains, including 23 different species, applied *Lb. delbrueckii* species-specific and subspecies-specific oligonucleotide probes and showed good agreement with PCR-typing (one species-specific and two subspecies-specific primer sets) (Lick *et al.*, 2000). Two of the 80 strains generated ambiguous hybridisation results at subspecies level (although the *Lb. delbrueckii* subsp. *bulgaricus* probe reactions were only weakly positive). The PCR reactions for these two strains were negative for *Lb. delbrueckii* subsp. *bulgaricus*. As previously stated, multiple oligonucleotide probes and primers have been developed for LAB and afford identification at different phylogenetic levels, with both dot-blot and in-situ hybridisation methods being effective. However, gene sequencing is preferable to probes for the classification of unknown bacterial cultures, as the use of probes for such would require either a barrage of probes or a 'Russian roulette' approach.

## 5.5 Discussion

In summary, the advent of molecular typing and profiling assays has provided a number of genetic assays for the identification, differentiation or community profiling of lactic acid bacterial cultures. The accuracy, reliability and reproducibility of such techniques are well documented, and the choice of procedure often reflects the desired output/deliverable, availability of necessary equipment (often specialised), cost and throughput (or number of samples to be analysed). Among the available genetic fingerprinting assays, PFGE provides the greatest discriminatory power for LAB cultures. However, RAPD is often the method of choice, due to its superior ease, speed, throughput and the requirement for specialised electrophoresis apparatus for PFGE. PCR-DDGE is a very powerful profiling method which is becoming more commonly used, especially in studies to monitor the bacterial diversity of samples and dynamics over time or across different phases of production. Similarly to PFGE, specialised electrophoresis apparatus is necessary for PCR-DGGE, but this initial cost is rewarded with a high throughput and an effective system to monitor bacterial composition across various phylogenetic levels, without the need for cultivation. Real-time PCR is a method which shows great promise in combining quantitative and qualitative analysis of samples, without the need for cultivation, although as with PFGE and PCR-DGGE, expensive instrumentation is required for this technique. The coupling of real-time PCR and DGGE may provide an excellent strategy to afford reliable qualitative and quantitative assessment of mixed bacterial systems.

The high throughput and accuracy of molecular-based typing and profiling systems is a major advantage, and affords a more extensive examination of the microbial impact of probiotic and LAB starter cultures in food production, storage and human microbiology. However, one of the fundamental aspects of such studies is the viability of said LAB cultures. Function, as well as microbial composition, is also an important aspect to consider; hence polyphasic investigations are necessary. The continued advancement of microbiology (and particularly molecular) methodologies will, no doubt, lead to improved systems which afford both highly discriminatory and functional analyses – potentially affording microbial diversity, dynamics, viability and function to be examined without the need for cultivation. It must not be forgotten, however, that the phenotypic characteristics of probiotic and LAB starter cultures are essential for product development and characteristics. Thus, cultivation studies remain an important aspect in food microbiology.

## References

Andrighetto, C., De Dea, P., Lambardi, A., Neviani, E., Rossetti, L. & Giraffa G. (1998) Molecular identification and cluster analysis of homofermentative thermophilic lactobacilli isolated from dairy products. *Research in Microbiology*, **149**, 631–643.

Andrighetto, C., Knijff, E., Lombardi, A., Torriani, S., Vancanneyt, M., Kersters, K., Swings, J. & Dellaglio, F. (2001) Phenotypic and genetic diversity of enterococci isolated from Italian cheeses. *Journal of Dairy Research*, **68**, 303–316.

Annuk, H., Hynes, S.O., Hirmo, S., Mikelsaar, M. & Wadström, T. (2001) Characterisation and differentiation of lactobacilli by lectin typing. *Journal of Medical Microbiology*, **50**, 1069–1074.

Beerens, H. (1990) An elective and selective isolation medium for *Bifidobacterium* spp. *Letters in Applied Microbiology*, **11**, 155–157.

Bourget, N., Simonet, J-M. & Decaris, B. (1993) Analysis of the genome of five *Bifidobacterium breve* strains: plasmid content, pulsed-field gel electrophoresis genome size estimation and *rrn* loci number. *FEMS Microbiology Letters*, **110**, 11–20.

Charteris, W.P., Kelly, P.M., Morelli, L. & Collins, J.K. (1997) Review article: selective detection, enumeration and identification of potential probiotic *Lactobacillus* and *Bifidobacterium* species in mixed bacterial populations. *International Journal of Food Microbiology*, **35**, 1–27.

Cocconcelli, P.S., Porro, D., Galandini, S. & Senini, L. (1995) Development of RAPD protocol for typing strains of lactic acid bacteria and enterococci. *Letters in Applied Microbiology*, **21**, 376–379.

Cocolin, L., Manzano, M,. Cantoni, C. & Comi, G. (2001) Denaturing gradient gel electrophoresis analysis of the 16S rRNA gene V1 region to monitor dynamic changes in the bacterial population during fermentation of Italian sausages. *Applied and Environmental Microbiology*, **67**, 5113–5121.

Corsetti, A., de Angelis, M., Dellaglio, F., Paparella, A., Fox, P.F., Settanni, L. & Gobbetti, M. (2003) Characterization of sourdough lactic acid bacteria based on genotypic and cell-wall protein analyses. *Journal of Applied Microbiology*, **94**, 641–654.

Dalezios, I. & Siebert, K.J. (2001) Comparison of pattern recognition techniques for the identification of lactic acid bacteria. *Journal of Applied Microbiology*, **91**, 225–236.

Dasen, A., Berthier, F., Grappin, R., Williams, A.G. & Banks, J. (2003) Genotypic and phenotypic characterization of the dynamics of the lactic acid bacterial population of adjunct-containing Cheddar cheese manufactured from raw and microfiltered pasteurised milk. *Journal of Applied Microbiology*, **94**, 595–607.

Dellaglio, F., Felis, G.E., Castioni, A., Torriani, S. & Germond, J.-E. (2005) *Lactobacillus delbrueckii* subsp. *indicus* subsp. nov., isolated from Indian dairy products. *International Journal of Systematic and Evolutionary Microbiology*, **55**, 401–404.

Delley, M. & Germond, J.-E. (2002) Differentiation of *Lactobacillus helveticus*, *Lactobacillus delbrueckii* subsp. *bulgaricus*, subsp. *lactis* and subsp. *delbrueckii* using physiological and genetic tools and reclassification of some strains from the ATCC collection. *Systematic and Applied Microbiology*, **25**, 228–231.

Dijkshoorn, L., Aucken, H., Gerner-Smidt, P., Janssen, P., Kaufmann, M. E., Garaizar, J., Ursing, J. & Pitt, J. L. (1996) Comparison of outbreak and nonoutbreak *Acinetobacter baumannii* strains by genotypic and phenotypic methods. *Journal of Clinical Microbiology*, **34**, 1519–1525.

Drake, M., Small, C.L., Spence, K.D. & Swanson, B.G. (1996) Differentiation of *Lactobacillus helveticus* strains using molecular typing methods. *Food Research International*, **29**, 451–455.

Du Plessis, E. & Dicks, L.M.T. (1995) Evaluation of random amplified polymorphic DNA (RAPD)-PCR as a method to differentiate *Lactobacillus acidophilus*, *Lactobacillus crispatus*, *Lactobacillus amylovorus*, *Lactobacillus gallinarum*, *Lactobacillus gasseri*, and *Lactobacillus johnsonii*. *Current Microbiology*, **31**, 114–118.

Ehrmann, M., Ludwig, W. & Schleifer, K.H. (1994) Reverse dot blot hybridization: a useful method for the direct identification of lactic acid bacteria in fermented food. *FEMS Microbiology Letters*, **117**, 143–150.

Ercolini, D. (2004) PCR-DGGE fingerprinting: novel strategies for detection of microbes in food. *Journal of Microbiological Methods*, **56**, 297–314.

Ercolini, D., Moschetti, G., Blaiotta, G. & Coppola, S. (2001) Behavior of variable V3 region from 16S rDNA of lactic acid bacteria in denaturing gradient gel electrophoresis. *Current Microbiology*, **42**, 199–202.

Ercolini, D., Hill, P.J. & Dodd, C.E.R. (2003) Bacterial community structure and location in Stilton cheese. *Applied and Environmental Microbiology*, **69**, 3540–3548.

Finney, M. (1995) Pulsed-field gel electrophoresis. *Current Protocols in Molecular Biology*, (eds. F.M. Ausubel, R. Brent, R.E. Kingston, D.D. Moore, J.G. Seidman, J.A. Smith and K. Struhl), Vol. 1, pp. 2.5.9–2.5.17 John Wiley and Sons, New York.

Franks, A.H., Harmsen, H.J.M., Raanges, G.C., Jansen, G.J., Schut, F. & Welling, G.W. (1998) Variations of bacterial populations in human feces measured by fluorescent in situ hybridization with group-specific 16S rRNA-targeted oligonucleotide probes. *Applied and Environmental Microbiology*, **64**, 3336–3345.

Furret, J.P., Quence, P. & Tailliez, P. (2004) Molecular quantification of lactic acid bacteria in fermented milk products using real-time quantitative PCR. *International Journal of Food Microbiology*, **15**, 197–207.

Gancheva, A., Pot, B., Vanhonacker, K., Hoste, B. & Kersters, K. (1999) A polyphasic approach towards the identification of strains belonging to *Lactobacillus acidophilus* and related species. *Systematic and Applied Microbiology*, **22**, 573–585.

Gatti, M., Fornassari, E. & Neviani, E. (1997) Cell-wall protein profiles of dairy thermophilic lactobacilli. *Letters in Applied Microbiology*, **25**, 345–348.

Gibson, J.R., Slater, E., Xerry, J., Tompkins, D.S. & Owen, R.J. (1998) Use of an amplified-fragment length polymorphism technique to differentiate isolates of *Helicobacter pylori*. *Journal of Clinical Microbiology*, **36**, 2580–2585.

Giraffa, G. & Neviani, E. (2000) Molecular identification and characterization of food-associated lactobacilli. *Italian Journal of Food Science*, **4**, 403–423.

Giraffa, G., de Vecchi, R. & Rossetti, L. (1998) Identification of *Lactobacillus delbrueckii* subspecies *bulgaricus* and subspecies *lactis* dairy isolates by amplified rDNA restriction analysis. *Journal of Applied Microbiology*, **85**, 918–924.

Grimont, F. & Grimont, P.A.D. (1986) Ribosomal ribonucleic acid gene restriction patterns as potential taxonomic tools. *Annales de l'Institut Pasteur Microbiology*, **137**, 165–175.

Harmsen H.J.M., Raangs, G.C., He, T., Degener, J.E. & Welling, G.W. (2002) Extensive set of 16S rRNA-based probes for detection of bacteria in human feces. *Applied and Environmental Microbiology*, **68**, 2982–2990.

Heilig, G.H.J., Zoetendal, E.G., Vaughan, E.E., Marteau, P., Akkermans, A.D.L. & de Vos, W.M. (2002) Molecular diversity of *Lactobacillus* spp., and other lactic acid bacteria in the human intestine as determined by specific amplification of 16S ribosomal DNA. *Applied and Environmental Microbiology*, **68**, 114–123.

Heller, K.J. (2001) Probiotic bacteria in fermented foods: product characteristics and starter organisms. *American Journal of Clinical Nutrition*, **73**(Suppl.), 374S-379S.

Holzapfel, W.H., Haberer, P., Geisen, R., Bjorkroth, J. & Schillinger, U. (2001) Taxonomy and important features of probiotic microorganisms in food and nutrition. *American Journal of Clinical Nutrition*, **73**(Suppl.), 365S–373S.

Huys, G., Coopman, R., Janssen, P. & Kersters, K. (1996) High resolution genotypic analysis of the genus *Aeromonas* by AFLP fingerprinting. *International Journal of Systematic Bacteriology*, **46**, 572–580.

Janssen, P., Coopman, R., Huys, G., Swings, J., Bleeker, M., Vos, P., Zabeau, M. & Kersters, K. (1996) Evaluation of the DNA fingerprinting method AFLP as a new tool in bacterial taxonomy. *Microbiology*, **142**, 1881–1893.

Jian, W., Zhu, L. & Dong, X. (2001) New approach to phylogenetic analysis of the genus *Bifidobacterium* based on partial HSP60 gene sequences. *International Journal of Systematic and Evolutionary Microbiology*, **51**, 1622–1638.

Johansson, M.-L., Quednau, M., Molin, G. & Ahrené, S. (1995) Randomly amplified polymorphic DNA (RAPD) for rapid typing of *Lactobacillus plantarum* strains. *Letters in Applied Microbiology*, **21**, 155–159.

Kimura, K., McCartney, A.L., McConnell, M.A. & Tannock, G.W. (1997) Analysis of fecal populations of bifidobacteria and lactobacilli and investigation of the immunological responses of their human hosts to the predominant strains. *Applied and Environmental Microbiology*, **63**, 3394–3398.

Klein, G. (2003) Anwendung molekularbiologischer Methoden in der Lebensmittelmikobiologie am Beispiel probiotisch gentzter Laktobazillen. *Berliner and Münchener Tierärztliche Wochenschrift*, **116**, 510–516.

Klein, G., Pack, A., Bonaparte, C. & Reuter, G. (1998) Taxonomy and physiology of probiotic lactic acid bacteria. *International Journal of Food Microbiology*, **41**, 103–125.

Koeleman, J. G. M., Stoof, J., Biesmans, D. J., Savelkoul, P. H. M. & Vandenbroucke-Grauls, C. M. J. E. (1998) Comparison of amplified ribosomal DNA restriction analysis, random amplified polymorphic DNA analysis, and amplified fragment length polymorphism fingerprinting for identification of *Acinetobacter* genomic species and typing of *Acinetobacter baumannii*. *Journal of Clinical Microbiology*, **36**, 2522–2529.

Kullen, M.J., Brady, L.J. & O'Sullivan, D.J. (1997) Evaluation of using a short region of the *rec*A gene for rapid and sensitive speciation of dominant bifidobacteria in the human large intestine. *FEMS Microbiology Letters*, **154**, 377–383.

Kwon, O.-S. (2000) Characterization of isolated *Lactobacillus* spp. and classification by RAPD-PCR analysis. *Journal of Microbiology*, **38**, 137–144.

Langendijk, P.S., Schut, F., Jansen, G.J., Raanges, G.C., Kamphuis, G.R., Wilkinson, M.H.F. & Welling, G.W. (1995) Quantitative fluorescence *in situ* hybridisation of *Bifidobacterium* spp. with genus-specific 16S rRNA-targeted probes and its application in fecal samples. *Applied and Environmental Microbiology*, **61**, 3069–3075.

Lebond-Bourget, N., Philippe, H., Mangin, I. & Decaris, B. (1996) 16S rRNA and 16S to 23S internal transcribed spacer sequence analysis reveal inter- and intraspecific *Bifidobacterium* phylogeny. *International Journal of Systematic Bacteriology*, **46**, 102–111.

Lick, S. (2003) Review: typing systems for lactobacilli. *Milchwissenschaft*, **58**, 256–260.

Lick, S., Brockmann, E. & Heller, K.J. (2000) Identification of *Lactobacillus delbrueckii* and subspecies by hybridization probes and PCR. *Systematic and Applied Microbiology*, **23**, 251–259.

Lucchinin, F., Kmet, V., Cesena, C., Coppi, L., Bottazzi, V. & Morelli, L. (1998) Specific detection of a probiotic *Lactobacillus* strain in faecal samples by using multiplex PCR. *FEMS Microbiology Letters*, **158**, 273–278.

Ludwig, W. & Schleifer, K.H. (1994) Bacterial phylogeny based on 16S and 23S rRNA sequence analysis. *FEMS Microbiology Reviews*, **15**, 155–173.

McCartney, A.L. (1996) *An Investigation of the Composition of the Human Intestinal Microflora, Particularly the* Bifidobacterium *and* Lactobacillus *Populations, Using Ribotyping*. PhD Thesis, University of Otago, Dunedin, New Zealand.

McCartney, A.L. (2002) Application of molecular biological methods for studying probiotics and the gut flora. *British Journal of Nutrition*, **88**(Suppl. 1), S29-S37.

McCartney, A.L. & Tannock, G.W. (1995) Ribotyping of *Bifidobacterium* strains using cells embedded in agarose plugs and a 16S rDNA probe. *Microbial Ecology in Health and Disease*, **8**, 79–84.

McCartney, A.L., Wang, W. & Tannock, G.W. (1996) Molecular analysis of the composition of the bifidobacterial and lactobacillus microflora of humans. *Applied and Environmental Microbiology*, **62**, 4608–4613.

McKillip, J.L. & Drake, M. (2004) Real-time nucleic acid-based detection methods for pathogenic bacteria in food. *Journal of Food Protection*, **67**, 823–832.

Mackay, I.M. (2004) Real-time PCR in the microbiology laboratory. *Clinical Microbiology and Infection*, **10**, 190–212.

Malinen, E., Mättö, J., Salmitie, M., Alander, M., Saarela, M. & Palva, A. (2002) PCR-ELISA II: analysis of *Bifidobacterium* populations in human faecal samples from a consumption trial with *Bifidobacterium lactis* Bb-12 and galactooligosaccharide preparation. *Systematic and Applied Microbiology*, **25**, 249–258.

Muyzer, G., de Waal, E.C. & Uitterlinden, A.G. (1993) Profiling of complex microbial populations by DGGE analysis of PCR-amplified genes coding for 16S rRNA. *Applied and Environmental Microbiology*, **59**, 695–700.

Olive, D.M. & Bean, P. (1999) Minireview: principles and applications of methods for DNA-based typing of microbial organisms. *Journal of Clinical Microbiology*, **37**, 1661–1669.

O'Sullivan, D.J. (1999) Methods for analysis of the intestinal microflora. *Probiotics: A Critical Review*, (ed. G.W. Tannock), pp. 23–44, Horizon Scientific Press, Wymondham.

O'Sullivan, D.J. & Kullen, M.J. (1998) Tracking of probiotic bifidobacteria in the intestine. *International Dairy Journal*, **8**, 513–525.

Perez, G., Cardell, E. & Zarate, V. (2000) Protein fingerprinting as a complementary analysis to classical phenotyping for the identification of lactic acid bacteria from Tenerife cheese. *Lait*, **80**, 589–600.

Pintado, J., Guyot, J.P. & Ampe, F. (2003) Multiple competitive PCR-DGGE as a tool for quantifying and profiling defined mixed cultures of lactic acid bacteria during production of probiotics from complex polysaccharides. *Journal of Applied Microbiology*, **95**, 921–933.

Popovic, D.G., Saunders, N.A. & Olsen, R.J. (1993) Ribotyping in molecular epidemiology. *Diagnostic Molecular Microbiology – Principles and Applications*, (eds. D.H. Persing, T.F. Smith, F.C. Tenover, & T.J. White), pp. 573–583, American Society for Microbiology.

Prasad, J., Gill, H., Smart, J. & Gopal, P.K. (1998) Selection and characterisation of *Lactobacillus* and *Bifidobacterium* strains for use as probiotics. *International Dairy Journal*, **8**, 993–1002.

Qiu, X., Wu, L., Huang, H., McDonel, P.E., Palumbo, A.V., Tiedje, J.M. & Zhou, J. (2001) Evaluation of PCR-generated chimeras, mutations, and heteroduplexes with 16S rRNA gene-based cloning. *Applied and Environmental Microbiology*, **67**, 880–887.

Rodtong, S. & Tannock, G.W. (1993) Differentiation of *Lactobacillus* strains by ribotyping. *Applied and Environmental Microbiology*, **59**, 3480–3484.

Roy, D. & Sirois, S. (2000) Molecular differentiation of *Bifidobacterium* species with amplified ribosomal DNA restriction analysis and alignment of short regions of the *ldh* gene. *FEMS Microbiology Letters*, **191**, 17–24.

Roy, D., Sirois, S. & Vincent, D. (2001) Molecular discrimination of lactobacilli used as starter and probiotic cultures by amplified ribosomal DNA restriction analysis. *Current Microbiology*, **42**, 282–289.

Salminen, S. & von Wright, A. (1993) *Lactic Acid Bacteria*, Marcel Dekker Inc., New York.

Satokari, R.M., Vaughan, E.E., Akkermans, A.D.L., Saarela M. & de Vos, W.M. (2001) Bifidobacterial diversity in human feces detected by genus-specific PCR and denaturing gradient gel electrophoresis. *Applied and Environmental Microbiology*, **67**, 504–513.

Satokari, R.M., Vaughan, E.E., Smidt, H., Saarela, M., Mättö, J. & de Vos, W.M. (2003) Molecular approaches for detection and identification of bifidobacteria and lactobacilli in the human gastrointestinal tract. *Systematic and Applied Microbiology*, **26**, 572–584.

Schleifer, K.H., Kraus, J., Dvorak, C., Kilpper-Bälz, R., Collins, M.D. & Fischer, W. (1985) Transfer of *Streptococcus lactis* and related streptococci to the genus *Lactococcus* gen. nov. *Systematic and Applied Microbiology*, **6**, 183–195.

Sigler, W.V., Miniaci, C. & Zeyer, J. (2004) Electrophoresis time impacts the denaturing gradient gel electrophoresis-based assessment of bacterial community structure. *Journal of Microbiological Methods*, **57**, 17–22.

Tannock, G.W., Tilsala-Timisjarvi, A., Rodtong, S., Ng, J., Munro, K. & Alatossava, T. (1999) Identification of *Lactobacillus* isolates from the gastrointestinal tract, silage, and yoghurt by 16S-23S rRNA gene intergenic spacer region sequence comparisons. *Applied and Environmental Microbiology*, **65**, 4264–4267.

Temmerman, R., Scheirlinck, I., Huys, G. & Swings, J. (2003) Culture-independent analysis of probiotic products by denaturing gradient gel electrophoresis. *Applied and Environmental Microbiology*, **69**, 220–226.

Tilsala-Timisjarvi, A. & Alatossava, T. (1997) Development of oligonucleotide primers from the 16S-23S rRNA intergenic sequences for identifying different dairy and probiotic lactic acid bacteria by PCR. *International Journal of Food Microbiology*, **35**, 49–56.

Tilsala-Timisjarvi, A. & Alatossava, T. (1998) Strain-specific identification of probiotic *Lactobacillus rhamnosus* with randomly amplified polymorphic DNA-derived PCR primers. *Applied and Environmental Microbiology*, **64**, 4816–4819.

Torriani, S., Zapparoli, G. & Dellaglio, F. (1999) Use of PCR-based methods for rapid differentiation of *L. delbrueckii* subsp. *bulgaricus* and *L. delbrueckii* subsp. *lactis*. *Applied and Environmental Microbiology*, **65**, 4351–4356.

Tynkkynen, S., Satokari, R., Saarela, M., Mattila-Sandholm, M. & Saxelin, M. (1999) Comparison of ribotyping, randomly amplified polymorphic DNA analysis, and pulsed-field gel electrophoresis in typing of *Lactobacillus rhamnosus* and *Lactobacillus casei* strains. *Applied and Environmental Microbiology*, **65**, 3908–3914.

de Urraza, P.J., Gómez-Zavaglia, A., Lozano, M.E., Romaowski, V. & de Antoni, G.L. (2000) DNA fingerprinting of thermophilic lactic acid bacteria using repetitive sequence-based polymerase chain reaction. *Journal of Dairy Research*, **67**, 381–392.

Van Reenen, C.A. & Dicks, L.M.T. (1996) Evaluation of numerical analysis of random amplified polymorphic DNA (RAPD)-PCR as a method to differentiate *Lactobacillus plantarum* and *Lactobacillus pentosus*. *Current Microbiology*, **32**, 183–187.

Versalovic, J., Koeuth, T. & Lupski, J.R. (1991) Distribution of repetitive DNA sequences in eubacteria and application to fingerprinting of bacterial genomes. *Nucleic Acids Research*, **19**, 6823–6831.

Vitali, B., Candela, M., Matteuzzi, D. & Brigidi, P. (2003) Quantitative detection of probiotic *Bifidobacterium* strains in bacterial mixtures by using real-time PCR. *Systematic and Applied Microbiology*, **26**, 269–276.

Vogel, R.F., Müller, M., Stolz, P. & Ehrmann, M. (1996) Ecology in sourdoughs produced by traditional and modern technologies. *Advances in Food Sciences*, **18**, 152–159.

Vos, P., Hogers, R., Bleeker, M., Reijans, M., van de Lee, T., Hornes, M., Frijters, A., Pot, J., Peleman, J., Kuiper, M. & Zabeau, M. (1995) AFLP: a new technique for DNA fingerprinting. *Nucleic Acids Research*, **23**, 4407–4414.

Walter, J., Tannock, G.W., Tilsala-Timisjarvi, A., Rodtong, S., Loach, D.M., Munro, K. & Alatossava, T. (2000) Detection and identification of gastrointestinal *Lactobacillus* species by DGGE and species-specific PCR primers. *Applied and Environmental Microbiology*, **66**, 297–303.

Wayne, L.G., Brenner, D.J., Colwell, R.R., Grimont, P.A.D., Kandler, O., Krichevshy, M.I., Moore, L.H., Moore, W.E.C., Murray, R.G.E., Stackerbrandt, E., Starr, M.P. & Truper, H.G. (1987) Report of the ad hoc committee on reconciliation of approaches to bacterial systematics. *International Journal of Systematic Bacteriology*, **37**, 463–464.

Wilson, K.H., Wilson, W.J., Radosevich, J.L., DeSantis, T.Z., Viswanathan, V.S., Kuczmarski, T.A. & Andersen, G.L. (2002) High-density microarray of small-subunit ribosomal DNA probes. *Applied and Environmental Microbiology*, **68**, 2535–2541.

Wintzingerode, F.V., Gobel, U.B. & Stackebrandt, E. (1997) Determination of microbial diversity in environmental samples: pitfalls of PCR-based rRNA analysis. *FEMS Microbiology Reviews*, **21**, 213–229.

Woesce, C.R. (1987) Bacterial evolution. *Microbiology Reviews*, **51**, 221–271.

Zapparoli, G., Torriani, S. & Dellaglio, F. (1998) Differentiation of *Lactobacillus sanfranciscensis* strains by randomly amplified polymorphic DNA and pulsed-field gel electrophoresis. *FEMS Microbiology Letters*, **166**, 325–332.

# 6 Prebiotic Ingredients with Emphasis on Galacto-oligosaccharides and Fructo-oligosaccharides

F. Angus, S. Smart and C. Shortt

## 6.1 Introduction

The gastrointestinal tract is home to around $10^{14}$ colony forming units (cfu) $g^{-1}$ of bacteria, consisting of over 400 different cultivatable species belonging to 190 genera, and known as the gastrointestinal flora. Bacterial numbers and composition vary considerably along the gastrointestinal tract. The density of micro-organisms in the gastrointestinal flora increases dramatically from less than $10^3$ cfu $g^{-1}$ in the stomach due to the harsh acidic conditions to greater than $10^{10}$ cfu $g^{-1}$ in the colon (Table 6.1). The anaerobes, *Bacteroides* spp., *Bifidobacterium* spp. and *Eubacterium* spp. represent greater than 99% of those species present in the colon. The growth and metabolic activity of the flora have a tremendous influence on our physiological and nutritional well-being. A balanced gastrointestinal flora is one in which the health-promoting or beneficial bacteria predominate over the potentially harmful bacteria (Cummings *et al.*, 2004). Many factors such as stress, poor diet, antibiotic therapy, infections, food poisoning and natural ageing process may upset this balance (Holzapfel *et al.*, 1998). The gastrointestinal flora plays an important role in digestion and in maintaining gastrointestinal health by stimulating the immune system (our natural defence system), preventing harmful bacteria from establishing a home on the gastrointestinal wall and promoting acidification – all of which are important in the maintenance of gastrointestinal

**Table 6.1** Type and concentration of bacteria in the large intestine.

| Bacteria | Mean concentration ($\log_{10}$ $g^{-1}$ dry wt of faeces) |
|---|---|
| Bacteroides | 11 |
| Eubacteria | 11 |
| Bifidobacteria | 10 |
| Clostridia | 10 |
| Lactobacilli | 10 |
| Ruminococci | 10 |
| Peptostreptococci | 10 |
| Peptococci | 10 |
| Methanobrevibacteria | 9 |
| Desulfovibrio | 8 |
| Propionibacteria | 9 |
| Actinomyces | 9 |
| Streptococci | 9 |
| Fusobacteria | 8 |
| Escherichia | 9 |

**Table 6.2** Terminology.

| Term | Definition |
|---|---|
| Probiotic | A live microbial food ingredient that is beneficial to health. |
| Prebiotic | A non-digestible food ingredient that beneficially affects the host by selectively stimulating the growth and/or activity of one or a limited number of bacteria in the colon, that have the potential to improve host health. |
| Synbiotic | A mixture of probiotics and prebiotics. |

health. Foods or ingredients that encourage the growth and activity of beneficial bacteria in the gastrointestinal tract help maintain a balanced gastrointestinal flora.

Probiotic micro-organisms and prebiotics (Gibson & Roberfroid, 1995) offer exciting possibilities for the development of innovative foods that can modulate the gastrointestinal microflora (Table 6.2). In addition to the main bifidogenic effect, many prebiotics are reported to be non-cariogenic, provide a good source of dietary fibre, be suitable for inclusion in diabetic and 'light' formulations and may even be used to replace fats.

## 6.2 Classification of prebiotics

Non-digestible oligosaccharides (NDO) contain mixtures of oligomers of different chain lengths, and are characterised by their degree of polymerisation (DP). These carbohydrates are generally composed of 2–20 monosaccharide units that resist hydrolysis by digestive enzymes, but are preferentially utilised in the colon by bifidobacteria and/or lactobacilli. This fermentation produces short-chain fatty acids (SCFA) and gases as well as an increase in metabolic energy, growth and proliferation of these bacteria. NDO can be made commercially from substrates, such as lactose (cows' milk), soluble starch, soybean whey, xylan, chitin and sucrose (beet). About ten NDO are commercially available, with inulin/fructo-oligosaccharides and galacto-oligosaccharides (GOS) having the most scientific support to date (Table 6.3).

The term fructo-oligosaccharide (FOS) is used as a generic name for all malabsorbed oligosaccharides composed mainly of fructose. Inulin and oligofructose are common forms of FOS, which are widely found in nature. Inulin is manufactured by companies such as Orafti, Sensus and Cosucra, and oligofructose is produced on a commercial scale by Orfati, Meiji Seika and Beghin-Meiji Industries.

**Table 6.3** Some examples of prebiotic ingredients.

| |
|---|
| Inulin-type fructans or fructo-oligosaccharides |
| Soybean oligosaccharides |
| Raffinose and stachyose |
| Galacto-oligosaccharides (GOS) |
| Trans-galacto-oligosaccharides |
| Galactosylsucrose |
| Isomalto-oligosaccharides and malto-oligosaccharides |
| Palatinose condensates |
| Xylo-oligosaccharides |
| Chitinoligosaccharides |

Lactulose, GOS and lactosucrose are all prebiotic NDO derived from lactose. Many companies, including Yakult Honsha, Nissin Sugar Manufacturing Company and Snow brand Milk Products in Japan and Borculo Domo Ingredients in Europe, produce GOS. This chapter will focus predominantly on lactose-derived GOS and FOS.

## 6.3 Prominence of prebiotics in FOSHU

Japan is unique in its food for specified health use (FOSHU) system for approving ingredients/foods with a functional claim (Shortt, 2004). The most prominent FOSHU foods are aimed at digestion or gastrointestinal health benefits, and prebiotic-containing foods are most common in this category (Table 6.4). To date, over 300 products have been approved, and gastrointestinal health-orientated products accounted for 60% of FOSHU approvals by the end of 2001 (Arai *et al.*, 2002; Japanscan, 2002; www.japanscan.co.uk). In 2001, the FOSHU-approved oligosaccharide market value was estimated at 6 billion yen. Examples of specific products containing prebiotics that have recently been approved are outlined in Table 6.5 (www.japanscan.co.uk; see also Chapter 4).

## 6.4 Galacto-oligosaccharides as prebiotics

### 6.4.1 *Technical aspects of GOS*

Galacto-oligosaccharides (Fig. 6.1) are NDO derived from lactose. They occur naturally in breast milk at a level of 1 g $L^{-1}$, and are present in trace amounts in cows' milk (Yamashira & Kobata, 1974; Reddy & Bush 1991; Thurl *et al.*, 1991; Miller *et al.*, 1994; Stahl *et al.*, 1994; Kunz *et al.*, 1996; Chaturvedi *et al.*, 1997). In general, GOS have a DP of 2 to 10, are not hydrolysed by digestive enzymes, and beneficially affect the host by selectively

**Table 6.4** The number of FOSHU functional components as of September 1999.

| Health claims | Functional component | Number of products on the market |
|---|---|---|
| Improve gastrointestinal conditions | Oligosaccharides | 47 |
| | Lactic acid bacteria | 35 |
| | Dietary fibre | 31 |
| Helpful for people concerned about high cholesterol and/or neutral fat | Peptide | 8 |
| | Dietary fibre | 5 |
| | Others | 8 |
| Suitable for people with mild hypertension | Glycoside, peptide | 7 |
| Improve mineral absorption | Calcium citrate malate (CCM), casein phosphopeptide (CPP) and/or haem iron | 8 |
| Suitable for people concerned about blood glucose level | Dietary fibre | 2 |
| Non-cariogenic | Sugar alcohol and/or green tea polyphenols | 5 |
| **Total** | | **154** |

**Table 6.5** Recent FOSHU introductions containing prebiotics.

| Claim area | Name | Product detail |
|---|---|---|
| Intestinal health | Como | Croissants with lactosucrose |
| | UCC Ueshima Coffee | Powdered soft drink with lactosucrose |
| | NR[1] | Seasoning vinegar with galacto-oligosaccharides |
| | Nihon Seibutsu Kakagu | Table top galacto-oligosaccharides |
| | Nissin Sugar Manufacturing | Table top galacto-oligosaccharides |
| | Yakult Honsha | Soft drink with galacto-oligosaccharide and polydextrose |
| | | Yoghurt with prebiotics[2] |
| | Snow Brand Milk | |
| Bone health | Nestlé – Milo | Powdered drink with fructo-oligosaccharides |
| | Nestlé – Milo and Milo C | Powdered drink with fructo-oligosaccharides |

[1] NR = not reported.
[2] Contain *Lactobacillus gasseri* and *Bifidobacterium bifidum*.

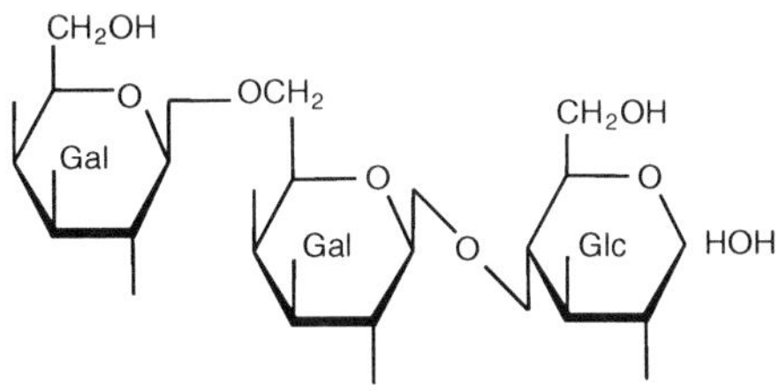


**Fig. 6.1** The chemical structure of galacto-oligosaccharide (GOS), which is composed of galactose, β-1–6 galactose and β-1–4 glucose (6'-galactosyllactose).

stimulating the growth and/or activity of one or a limited number of bacteria in the colon. Human milk GOS are reported to have a protective action against infections of the gastrointestinal tract during the first year of life, and it is thought that they may bear structural homology to cell surface glycoconjugates used as receptors by pathogens (Kunz & Rudloff 1996; Ebrahim 1997; Hamosh 1997; Newberg 1997). In addition to nutritional and physiological characteristics, GOS have interesting physicochemical properties which make them versatile ingredients.

## 6.4.2 *Production of GOS*

A highly concentrated solution of whey-derived lactose is generally used as the primary raw material in the production of GOS (Kwak & Jeon, 1986; Matsumoto *et al.*, 1992; Yanahira *et al.*, 1995). The lactose solution is converted to GOS by the action of β-galactosidases, which have transgalactosylation activity (Fig. 6.2). Generally, the main products are trisaccharides, namely 4'- or 6'-galactosyllactose, and longer oligosaccharides consisting of four or more monosaccharide units; substantial amounts of transgalactosylated disaccharides, which are considered as NDO, are also produced (Matsumoto *et al.*, 1990; Ishikawa *et al.*, 1995). Commercially available GOS are generally mixtures of lactose, glucose, galactose and usually more than 55% of oligosaccharides. The linkage between the galactose units, the efficiency of transgalactosylation, and the components in the final products depend on the enzymes and the specific conditions used in the reaction.

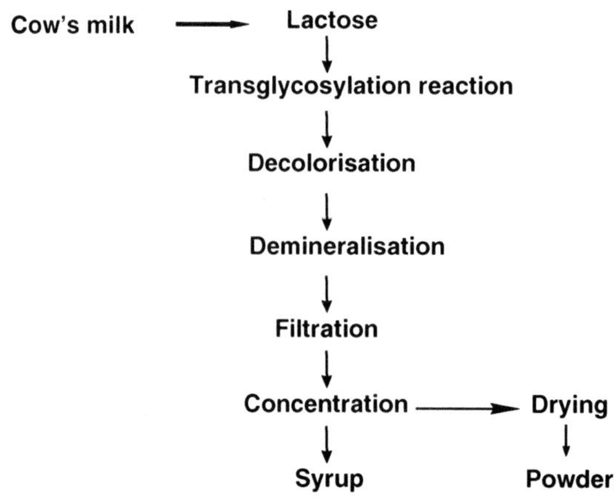


**Fig. 6.2** Production of galacto-oligosaccharide from cows' milk.

### 6.4.3 *Characteristics of GOS*

Galacto-oligosaccharides have a high solubility and a relative sweetness about 35% that of sucrose. They are more viscous than high-fructose corn syrups, decrease the water activity and freezing point, and show good moisture retention capacities. They also have remarkable stability at high temperatures and variable pH levels. In particular, the stability of GOS in acidic and high-temperature conditions enables them to be applied without decomposition in a wider variety of foods. GOS, such as Oligomate 55 (Anonymous, 1999), remain unchanged after treatment at 160°C for 10 min at neutral pH, after treatment at 120°C for 10 min at pH 3 or 100°C for 10 min at pH 2, where about half or more of the sucrose is degraded. Even in acidic conditions at room temperature, GOS tend to be stable during long-term storage.

As GOS can be considered to be dietary fibre, an AOAC method (AOAC, 2003) has been outlined to determine GOS in foods for analytical and labelling purposes. Quemener *et al.* (1994) developed a method to measure trans-GOS in foods and feedstuffs based on high-performance anion-exchange chromatography with pulsed amperometric detection. Subsequently, a collaborative AOAC study of this method was completed and this resulted in the method being approved as an AOAC method – 2001.02 (de Slegte, 2002). In addition, the American Association of Cereal Chemists published the same method for the detection of GOS.

In foods, GOS act as bulk ingredients that contribute to both texture and mouth-feel. They are suitable ingredients for a variety of food formulations and formats. GOS are now used as a source of fibre, as sweeteners by themselves, and for their bifidogenic effects in fermented milk products, confectionery and acidic-beverages. More recently, high value-added products, such as specialised baby foods, foods for geriatrics and for those who are critically ill have been formulated utilising GOS. The calorific value of GOS is calculated to be 1.73 kcal $g^{-1}$ according to the standardised Japanese method (Watanuki *et al.*, 1996; Oku, 1996).

### 6.4.4 *Health effects of GOS*

While many benefits of GOS have been reported (Table 6.6), most research has focused on the demonstration of the bifidogenic effect of GOS consumption (Tanaka *et al.*, 1983; Ito *et al.*, 1990; Bouhnik *et al.*, 1997). GOS have also been shown to alleviate constipation, and to reduce the activity of enzymes that convert pro-carcinogens to carcinogens, albeit mainly in animal studies. Recent studies have also suggested that GOS may have a positive effect on bone mineralisation (Chonan & Watanuki, 1995; Scholz-Arhens *et al.*, 2001).

Since Tissier's discovery during the late 1890s that the administration of bifidobacteria to infants with diarrhoea displaced the pathogens and speeded recovery, the search for 'bifidus factor' candidates derived from human milk has continued (Yazawa & Tamura, 1982). GOS are neither hydrolysed nor absorbed in the small intestinal tract, but are rapidly fermented in the proximal colon, particularly by bifidobacteria (Minami *et al.*, 1985). While GOS are preferentially utilised by many *Bifidobacterium* strains, some strains of *Lactobacillus*, *Bacteroides* and *Clostridium* ferment GOS, with the species that ferment 4'-GOS and 6'-GOS tending to overlap (Table 6.7).

Data from animal and *in vitro* studies have demonstrated decreases in pH, and increases in SCFA and other organic acids in response to GOS (Durand *et al.*, 1992; Kikuchi *et al.*, 1996, 1997). Caecal enlargement and increases in caecal contents are common in animals after consumption of GOS. Bouhnik *et al.* (1997) compared the *in vitro* activity of a batch human faecal culture in relation to the production of adenosine-5-triphosphate (ATP), acid and gas using faeces collected on day 1 (control), 7 days and 14 days after the administration of GOS, and found that in response to added GOS, ATP and acid production were stimulated. In addition, the rate of increase of acetic acid in the batch culture in the presence of GOS was higher than that of the control group. The same authors suggested that this was due to the change in composition of the faecal flora to a bifidobacteria-predominant one. While several authors have detected changes in the faecal microflora of healthy volunteers after ingestion of GOS, changes in the production of SCFA have not generally been detected.

Recently, Boehm *et al.* (2002) showed that GOS-supplemented bovine formula milk stimulated bifidobacteria growth in premature infants. Kanamori *et al.* (2003) also showed that oral administration of a synbiotic containing GOS (3 g $d^{-1}$) in combination with vancomycin helped to eradicate methicillin-resistant *Staphylococcus aureus* (MRSA) and re-established an anaerobic-dominant flora in a 3-month-old infant suffering from MRSA entercolitis. Similarly, earlier studies by Tanaka *et al.* (1983) demonstrated the effects of administration of GOS in healthy adults. After 1 week of intake of 6'-GOS at a dose of 3 or 10 g $d^{-1}$, the faecal count of bifidobacteria increased in dose-dependent manner. It has been suggested that an intake of at least 10 g $d^{-1}$ of 6'-GOS is required to detect a bifidogenic effect. However, a daily intake of 2.5 g 6'-GOS appears to be sufficient to increase the faecal bifidobacteria

**Table 6.6** Potential nutritional attributes of GOS.

| |
|---|
| Stimulate carbohydrate metabolism in colonic bacteria, increase bacterial mass, SCFA and gas production |
| Source of low-calorific sweetener |
| Bifidogenic effect |
| Stool bulking/alleviation of constipation |
| Improved calcium bioavailability |

SCFA: Short-chain fatty acids.

**Table 6.7** Utilisation[1] of 4'-galactosyllactose, lactose and lactosucrose by various intestinal bacteria.

| Micro-organisms | Number of strains | 4'-Galactosyllactose | Lactose | Lactosucrose |
|---|---|---|---|---|
| *Bifidobacterium adolescentis* | 7 | ++ | ++ | ++ |
| *Bifidobacterium bifidum* | 6 | ++ | ++ | ++ |
| *Bifidobacterium breve* | 3 | ++ | ++ | ++ |
| *Bifidobacterium infantis* | 2 | ++ | ++ | ++ |
| *Bifidobacterium longum* | 8 | ++ | ++ | ++ |
| *Lactobacillus acidophilus* | 2 | + | ++ | ++ |
| *Lactobacillus casei* subsp. *casei* | 2 | – | + | ++ |
| *Lactobacillus gasseri* | 1 | – | ++ | ++ |
| *Lactobacillus salivarius* | 2 | – | ++ | ++ |
| *Bacteroides distasonis* | 5 | ++ | ++ | ++ |
| *Bacteroides fragilis* | 2 | ++ | ++ | ++ |
| *Bacteroides ovatus* | 1 | – | ++ | ++ |
| *Bacteroides vulgatus* | 8 | ++ | ++ | ++ |
| *Mitsuokella multiacidus* | 3 | – | ++ | ++ |
| *Rikenella microfusus* | 1 | – | + | – |
| *Megamonas hypermegas* | 1 | ++ | ++ | ++ |
| *Clostridium butyricum* | 1 | – | + | + |
| *Clostridium difficile* | 1 | – | – | – |
| *Clostridium innocuum* | 1 | – | – | – |
| *Clostridium perfringens* | 3 | – | ++ | ++ |
| *Clostridium ramosum* | 2 | – | ++ | ++ |
| *Eubacterium aerofaciens* | 6 | – | ++ | ++ |
| *Eubacterium limosum* | 2 | – | +/– | +/– |
| *Peptostreptococcus anaerobius* | 1 | – | ++ | ++ |
| *Peptostreptococcus prevotii* | 1 | – | – | – |
| *Peptostreptococcus productus* | 1 | – | ++ | ++ |
| *Propionibacterium acnes* | 1 | – | + | + |
| *Fusobacterium varium* | 1 | – | – | +/– |
| *Veillonella alcarescens* subsp. *dispar* | 1 | – | – | – |
| *Megaphaera elsdenii* | 1 | – | – | – |
| *Enterococcus faecalis* | 3 | – | – | + |
| *Enterococcus faecium* | 2 | – | – | + |
| *Escherichia coli* | 6 | – | + | +/– |

[1] Bacterial growth: ++ = same as that on glucose; + = less than that on glucose; +/– = little growth; – = no growth. Adapted from Ohtsuka *et al.* (1989).

count when the initial baseline level is low, which is often the case in elderly people (Ito *et al.*, 1993). Similarly, Ishikawa *et al.* (1995) reported similar increases in bifidobacteria and exchanges in the predominant bacteria after administration of 4'-GOS at doses of 2.5 or 10 g $d^{-1}$. Therefore, in Japanese subjects, daily intakes of 2.5 g $d^{-1}$ of both 6'-GOS and 4'-GOS appear to be sufficient to create a bifidogenic effect in healthy adults.

Teuri *et al.* (1998) showed that daily ingestion of yoghurt containing 15 g GOS increased defecation frequency, but also increased flatulence in healthy subjects. In contrast, while the apparent fermentability of trans-GOS was 100%, no effects were observed in relation to bowel habits (stool composition, concentration of SCFA or bile acids in faecal water, concentration of ammonia, indole, skatoles, pH) or composition of the intestinal flora of healthy Dutch individuals consuming 7.5 and 15 g $d^{-1}$ for 3 weeks (Alles *et al.*, 1999). Several Japanese studies have reported benefits when GOS were administered to constipated individuals, however. Deguchi *et al.* (1997) studied bowel habit, and specifically the frequency of stool formation, in 75 young women after ingestion of GOS for 1 week in a double-blind crossover trial. The results showed that daily ingestion of 5 g GOS improved bowel habit. Likewise, in another study with 50 subjects and conducted by the same group, stool form and abdominal symptoms were shown to have been improved by the daily ingestion of 5 and 10 g GOS, respectively. In a study of diabetic subjects with constipation, a correlation was found between the improvement in constipation and the decrease in faecal Bacteroidaceae after ingestion of GOS (Narimiya *et al.*, 1996).

In a human study with healthy volunteers, the ingestion of 4'-GOS at a dose of 3 g $d^{-1}$ reduced faecal ammonia concentrations significantly (Tamai *et al.*, 1992). Faecal levels of phenol, *p*-cresol and indole were also decreased, indicating that there was a dramatic change in metabolism upon ingestion of 4'-GOS. In patients with hyper-ammonaemia associated with liver cirrhosis, administration of 6'-GOS (30 g $d^{-1}$) together with *Bifidobacterium breve* ($6 \times 10^9$ cfu $d^{-1}$) established an intestinal flora in which *Bifidobacterium* predominated, and the blood ammonia level was gradually decreased from 157 μg $mL^{-1}$ to 88 μg $mL^{-1}$ within several weeks (Matsumoto *et al.*, 1990). The mechanisms underlying these effects are probably due to the increase of bifidobacteria and the decrease of certain indigenous bacteria (e.g. *Bacteroidaceae* strains), which have the potential to produce ammonia (Deguchi *et al.*, 1993).

It has been suggested that several bacterial enzymes, such as β-glucuronidase, β-glucosidase and nitroreductase, may play a role in colon carcinogenesis by converting pre-carcinogens to proximal carcinogens (Gill & Rowland, 2002). Rowland & Tanaka (1993) have shown in human flora-associated rats that GOS (5 g 100 $g^{-1}$) or GOS plus *B. breve* administered for 4 weeks, increased caecal concentrations of total anaerobic bacteria, lactobacilli and bifidobacteria, but decreased the number of Enterobacteriaceae. Caecal pH was significantly reduced through the feeding of GOS, as were the activities of β-glucuronidase and nitrate reductase. However, β-glucosidase activity was increased in GOS-fed rats. GOS was also associated with a decreased conversion of the dietary carcinogen 2-amino-3-methyl-3H-imidazo[4,5-f] quinoline to its genotoxic 7-hydroxy derivative. Similarly, Kikuchi *et al.* (1996) showed that GOS administration (10 g 100 $g^{-1}$) was associated with decreases in β-glucuronidase activity and secondary bile acid concentrations in the faeces of human flora-associated rats. In human studies, Ishikawa *et al.* (1995) showed decreases in secondary bile acids in faeces when subjects consumed 2.5 g and 10 g GOS on a daily basis, while van Dokkum *et al.* (1999) showed that a daily intake of 15 g GOS by young healthy volunteers resulted in a significant increase in faecal acetate and a significant decrease in faecal β-glucuronidase activity.

The effect of GOS on mineral absorption has also received attention. Chonan and Watanuki (1995, 1996) showed that the apparent absorption of calcium in rats was stimulated by feeding GOS. This effect was accompanied by a reduction in caecal pH, and increases in caecal and caecal digest weights. In the presence of NDO in the caecum, caecal bacteria produce a large amount of SCFA and other organic acids, such as lactic acid and succinic acid, and it is likely that this improves the solubility of calcium. This effect of GOS on calcium absorption is detected when calcium is added at a normal dietary concentration, but not at a low concentration (Chonan & Watanuki, 1995). Therefore, the stimulation of calcium absorption seems to be caused by the increased solubility of calcium in the SCFA-rich acidic environment. Chonan *et al.* (1995) analysed the effect of GOS on calcium absorption in ovariectomised (OVX) rats as a model of postmenopausal conditions. In OVX rats, the apparent calcium absorption was stimulated by GOS-feeding until day 20, and the loss of bone mass and calcium content were significantly suppressed by GOS. Rats fed a diet containing GOS (5 g 100 $g^{-1}$) absorbed calcium more efficiently than those on the control diet. The bone (femur and tibia) ash weight and tibia calcium content of OVX rats fed GOS were significantly higher than those of control rats. Although the rat model demonstrates some of the effects of NDO on mineral absorption, there are limited human studies in this area (Chonan *et al.*, 1996). van den Heuvel *et al.* (1998) reported that the consumption of 15 g $d^{-1}$ of GOS, inulin or FOS did not lead to any improvement in calcium or iron absorption in young male subjects. However more recently, van den Heuvel *et al.* (2000) indicated a positive effect of trans-GOS on calcium absorption in postmenopausal women who consumed yoghurt containing GOS (equivalent to 20 g $d^{-1}$). The urinary calcium levels were unchanged, but calcium absorption was increased by 16%. Scholz-Ahrens *et al.* (2001) suggested that the positive effect of GOS on calcium absorption may depend on conditions of increased calcium demand in situations such as the postmenopausal period.

## 6.5 Focus on FOS

### 6.5.1 *Background*

After starch, fructans are the most abundant non-structural natural polysaccharides. In nature, the various fructans are distinguished on the basis of the glycosidic linkages present between the fructose residues, which can be broadly classified into three groups: (a) inulins; (b) levans (or phleins); and (c) the graminan type. FOS is used as a generic term for all malabsorbed oligosaccharides composed mainly of fructose.

FOS are a mixture of short- and medium-length chains of β-D-fructans with fructosyl units bound by a β-2-1 glycosidic linkage. They possess a number of highly desirable attributes, which include a strong bifidus-stimulation; they also provide a selective source of dietary fibre, and enhance the functionality of food products by adding texture and taste (Tungland, 2003). Inulin and oligofructose are common forms of FOS that are widely found in nature (Fig. 6.3). By definition, oligosaccharides have a DP lower than 9. Inulin is commonly referred to as a FOS, despite the fact it contains polymers with a DP of up to 60 (Gibson *et al.*, 1994).

Although FOS are present in a wide variety of foods that are part of our usual diet, such as asparagus, banana, oats, garlic, artichoke and chicory, the amount contained in food is quite low. The individual average daily intake from these sources ranges from 2 to 10 g in Europe, and from 1 to 4 g in the United States (van Loo *et al.*, 1995). FOS can be produced on a commercial scale, and exhibit a variety of technological properties.

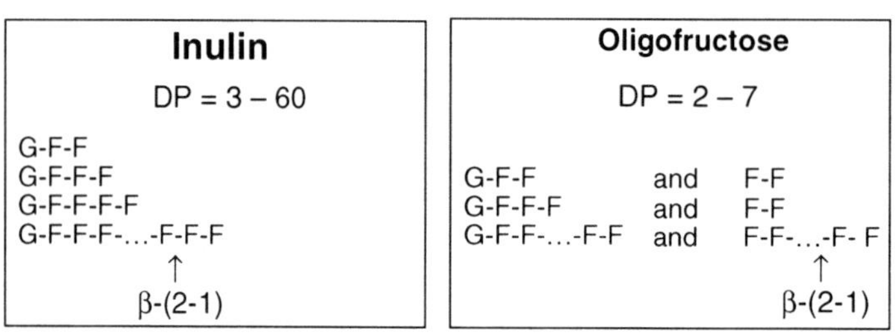


**Fig. 6.3** Chemical structure of inulin and oligofructose.

## 6.5.2 *Technological production of FOS*

The majority of fructans produced commercially are of the inulin type. Production involves the extraction of the naturally occurring inulin (70 g 100 $g^{-1}$ of the dry substance) from chicory roots (*Cichorium intybus* L.), followed by refining and then drying. Common forms of FOS are synthetically derived short-chain fructo-oligosaccharides (scFOS; neosugar) and oligofructose (partially hydrolysed inulin), which are produced using two different manufacturing techniques. Oligofructose can be produced via a partial enzyme hydrolysis of inulin extracted from chicory roots using a specific endo-inulinase, followed by spray-drying. This is the method of choice for manufacturers such as Orafti in Belgium (Fig. 6.4).

The second technique used by manufacturers such as Meiji Seika (Japan) and Beghin-Meiji Industries (France) produces scFOS on a commercial scale by the transfructosylation of sucrose using a beta-fructosyltransferase enzyme derived from micro-organisms or plants. This ingredient is a combination of 1-kestose (1-kestotriose), nystose (1,1-kestotetrasoe) and $1^{F}$-β-fructofuranosylnystose (1,1,1-kestopentaose) oligomers (Bornet, 1994). This process can be divided into two different types: the batch system using soluble enzymes; and the continuous process using immobilised enzymes or whole cells (e.g. *Aspergillus niger*) entrapped in a calcium alginate gel (Meiji Seika Co. Japan) (Fig. 6.5).

The commercial forms of inulin and oligofructose differ by the average DP and digestible carbohydrate (fructose, glucose and saccharose) content.

## 6.5.3 *Characteristics of FOS*

Inulin contains polymers with an average DP of 10, and exhibits a bland neutral taste, which allows it to be combined easily with other ingredients without modifying any delicate flavours. Standard inulin is slightly sweet (10% sweetness compared with sugar), whereas high-performance inulin (DP ~25) does not have a sweet taste. Inulin is moderately soluble in water (10 g 100 $mL^{-1}$ maximum at room temperature), and has a low viscosity (De Leenheer, 1996). It has a lipid-like texture when mixed with water, and may be used as a fat replacer (Roberfroid *et al.*, 1998). High-performance inulin functions effectively to form a gel network when mixed with water or other aqueous liquids, and this can be incorporated

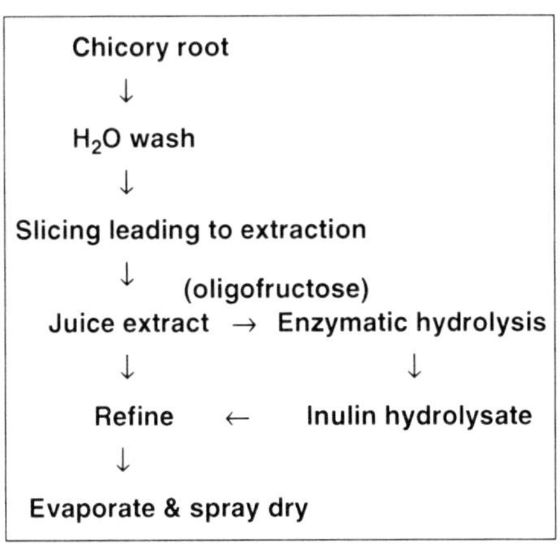


**Fig. 6.4** Inulin and oligofructose manufacturing process. Adapted from Tungland (2003).

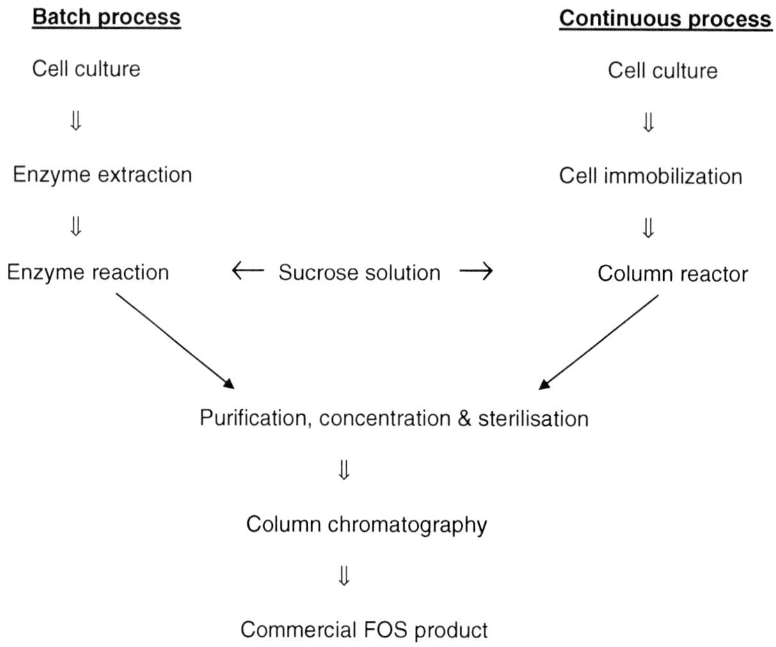


**Fig. 6.5** Production outline of fructo-oligosaccharide (FOS) via transfructosylation of sucrose. Adapted from Tungland (2003).

into foods to replace fat (up to 100 g 100 $g^{-1}$). Inulin has also been shown to replace stabilisers in food products, improving the stability, for example, of foams and emulsions in ice-creams and aerated desserts (Franck & Coussement, 1997). scFOS is 0.4 to 0.6 times as sweet as sucrose, and has an average DP of 3.7. Oligofructose has an average DP of 4.5 and is 0.3 times as sweet as sucrose (Hidaka *et al.*, 1991; Spiegel *et al.*, 1994).

In terms of sweetening profiles, the oligofructoses provide functional qualities similar to sugars or glucose syrups (Gibson & Roberfroid, 1995; Roberfroid *et al.*, 1998). Oligofructose is much more soluble than inulin (80% at room temperature), and exhibits a good microbial stability during food processing. Oligofructose is considered to be an important ingredient to add bulk with fewer calories, and enhances nutritional benefits without compromising on taste and mouth-feel. In Asia, it is also used increasingly in infant formulas and baby foods (Franck & Coussement, 2001). The solubility of FOS also allows fibre incorporation in watery systems such as drinks and dairy products.

Evidence to date suggests that, despite the different processing methods used in the production of the two types of FOS, the same physiological effects are predicted due to their structural similarity (Roberfroid, 1997).

Inulin and oligofructose are considered to be important fibre ingredients, and often improve taste and texture in food products (Franck & Coussement, 1997). They are also considered to be low-calorie ingredients, as the calorific value is limited to that contributed through SCFA generated in the fermentation and partial resorption in the body. Studies relating to energy values indicate that inulin and oligofructose each have a calorific value of ~1.5 kcal $g^{-1}$ (Roberfroid *et al.*, 1993).

### 6.5.4 Health effects of FOS

Similar to GOS, oligofructose and inulin are resistant to digestive enzymes and, therefore, do not undergo hydrolysis due to their specific chemical structure. Consequently, these dietary components reach the colon virtually unaltered. Anaerobic bacteria hydrolyse and ferment these oligomers into bacterial mass, SCFA (acetate, propionate, L-lactate and butyrate) and gases (hydrogen, carbon dioxide and methane) (Delzenne & Roberfroid, 1994). Inulin is active in the more distal regions of the colon because its fermentation occurs more slowly than that of oligofructose (Roberfroid *et al.*, 1998).

The fermentation of oligofructose has several physiological consequences, and has been shown in numerous human studies to exert a preferential stimulatory effect on numbers of health-promoting bacteria such as *Bifidobacterium* spp., whilst maintaining populations of potential pathogens (*Clostridium*, *Escherichia coli*) at relatively low levels. Research by Gibson *et al.* (1995) illustrated that small changes in the diet can produce substantial changes in the intestinal microflora. Subjects ingested controlled diets containing 15 g $d^{-1}$ oligofructose ($n = 8$) or inulin ($n = 4$) for 15 days. Relative to an ingestion of 15 g $d^{-1}$ of sucrose, the faecal bifidobacteria count increased. Subjects consuming inulin demonstrated a decrease in Gram-positive cocci, whilst the levels of clostridia, bacteroides and fusobacteria decreased in subjects consuming oligofructose. Both inulin and oligofructose promoted the growth of bifidobacteria in absolute terms, and in proportion to the number of total anaerobes. Breath hydrogen, faecal wet and dry matter, nitrogen and energy excretion were also increased in both subject groups. FOS demonstrates differing bifidogenic effects at different conditions. *In vitro* studies indicate that FOS and inulin produce an optimum bifidogenic effect at pH 6.8 and 1 g 100 $mL^{-1}$ carbohydrate, equivalent to ~4 g $d^{-1}$ (Palframan *et al.*, 2002).

Extensive research has been carried out into the potential health benefits ascribed to bifidobacteria, which include: (a) competitively inhibiting the growth of harmful micro-organisms such as *Salmonella* spp., clostridia, *S. aureus* and *E. coli* (amongst others); (b) stimulating components of the immune system, lowering blood cholesterol and triglycerides, reducing liver toxins, and reducing food intolerances and allergies; and (c) aiding the absorption of minerals (e.g. calcium) and the synthesis of B vitamins (Tungland, 2003).

Besides their bifidogenic effect, FOS (like GOS) have additional nutritional influence on digestive physiological parameters such as colonic pH and stool bulking, which justifies their classification as dietary fibres (Roberfroid, 1997).

The results of recent investigations indicate that FOS may play a role in the immunoregulatory system. Manhart *et al.* (2003) fed healthy and endotoxic mice a diet supplemented with 10 g $d^{-1}$ FOS or a control diet over a period of 16 days. Populations of B lymphocytes, T lymphocytes, CD4(+) cells and CD8(+) cells were each determined using flow cytometry. The results indicated that FOS supplementation had an immunostimulatory effect on Peyer's patches under healthy and endotoxic conditions. This suggests that FOS administration may affect not only the large intestine, but also the main inductive part of the mucosal immune system in the small intestine. Kleesen *et al.* (2003) fed rats with differing intestinal flora a diet supplemented with oligofructose-long chain inulin (OC-LcIN). The OC-LcIN diet led to increased villus height, deeper crypts, an increased number of goblet cells and increased thickness of the colonic epithelial mucus layer in bacteria-associated rats. It was concluded that stabilisation of the gut mucosal barrier by changes in the mucosal architure via bifidobacteria stimulation by fructans might play an important role in the future treatment and prevention of gastrointestinal disorders.

In addition, recent investigations have been made into the effect of an oligofructose-enriched inulin diet combined with *Lactobacillus rhamnosus* and *Bifidobacterium animalis* subsp. *lactis* on the intestinal immunity of rats with colon cancer. The results indicated that overall synbiotic supplementation in carcinogen-treated rats primarily modulated immune functions in the Peyer's patches, coinciding with a reduced number of colon tumours (Roller *et al.*, 2004).

Both total and triacylglycerols have been implicated as lipid risk factors for coronary heart disease. Studies using animal models have indicated that moderate amounts of inulin and oligofructose may reduce blood cholesterol levels. Rat studies have repeatedly shown that an inulin consumption of 9 to 20 g $d^{-1}$ induces a significant reduction (~40%) in serum and liver triglycerides, and also in total and low-density lipoprotein (LDL) cholesterol concentrations, without affecting high-density lipoprotein (HDL) cholesterol levels. Fiordaliso *et al.* (1995) fed male normo-lipidaemic rats either a control diet or a diet containing 10 g 100 $g^{-1}$ oligofructose. A decrease was observed after one week in the levels of cholesterol, phospholipids and plasma triacylglycerols; this was associated with a decrease in the activity of de-novo lipogenic enzymes, thus reducing plasma very low-density lipoproteins (VLDL).

Further research into the effects on carbohydrate and lipid metabolism after high fructose intakes showed that oligofructose plays a protective role against oxidative stress in rat models (Busserolles *et al.*, 2003). However, relatively few humans studies have been conducted in this respect. Current data suggest that in hyperlipidaemic subjects, any effects that occurred might result primarily in a reduction in cholesterol levels, whereas in normal lipidaemic subjects the effects on serum triglycerides would be the dominant feature (Pereira & Gibson, 2002).

Unlike some other types of fibre (containing phytic or uronic acids), inulin and oligofructose may improve the bioavailability of minerals such as calcium, magnesium and iron. The use of different models has repeatedly shown an increase in the absorption of calcium (and magnesium) at the level of the large intestine, as well as increased calcium uptake into the bone tissue, resulting in improved mineral density (Franck, 2000). Delzenne *et al.* (1995) conducted a study in rats in which the diet was supplemented with either inulin or oligofructose (10 g 100 $g^{-1}$ diet). The results showed a highly significant increase in the absorption of calcium, magnesium and iron. A similar increase in calcium absorption was also observed with rats fed a diet supplemented with oligofructose at 5 g 100 $g^{-1}$ diet (Brommage *et al.*, 1993). Oligofructose (2.5 and 5 g 100 $g^{-1}$ diet) has also been shown to increase calcium and magnesium absorption in ovariectomised rats, and to prevent bone loss caused by oestrogen deficiency (Taguchi *et al.*, 1995). Scholz-Ahrens *et al.* (1998) demonstrated a dose-related effect on the increase of both calcium absorption and bone mineralisation in femur and lumbar vertebrae following oligofructose ingestion (2.5–5.0 and 10 g 100 $g^{-1}$ diet), thereby confirming the increased uptake of calcium into the bone tissue. Oligofructose was also shown to be very effective at preventing ovariectomy-induced loss of trabecular bone (tibia) when dietary calcium levels were high (Scholz-Ahrens *et al.*, 2001).

Studies on human subjects have shown more varied results. Ellegard *et al.* (1997) administered ileostomised volunteers with 15 g $d^{-1}$ of inulin or oligofructose. Mineral excretion (calcium, magnesium, iron, zinc) from the small intestine was not altered and, therefore, absorption in this part of the gut was not affected. It was suggested that any effect that the fructans might have on mineral absorption must occur in the colon. van den Heuvel *et al.* (1998) reported that the consumption of 15 g $d^{-1}$ of GOS, inulin or FOS did not lead to any improvement in calcium or iron absorption in young male subjects. However, measurements were based on the content of calcium isotopes in urine after a 24-h period of collection, which may have been too short to detect the effect of the fructans. The same group demonstrated a significant increase in calcium absorption with 15 g $d^{-1}$ of oligofructose in male adolescents during a 9-day period, indicating that in this group oligofructose may help to maximise peak bone mass (van den Heuvel *et al.*, 2000). More recently, Griffin *et al.* (2002) assessed the effects of 8 g $d^{-1}$ of either oligofructose or a mixture of inulin and oligofructose on calcium absorption in girls near menarche, by using a balanced, randomised crossover design study. Calcium absorption was significantly higher in the group receiving the inulin and oligofructose mixture than in the placebo group, though this difference was not statistically significant. It is important to note that not all fructans are equally efficient in stimulating calcium absorption, and further research is required in this area.

Nevertheless, studies evaluating the potential toxic effects of FOS have not revealed any significant adverse effects. Side effects noted following the ingestion of large quantities include flatulence and soft stools. Excessive consumption may indeed cause intestinal discomfort such as bloating, flatus or colic, but these effects are comparable to those observed with other dietary fibres. In practical use, the level of prebiotic consumption (approx. 2–3 g per serving) is below the amount at which intestinal discomfort and/or laxative effects occur (Salminen *et al.*, 2000).

## 6.6 Conclusions

From a physicochemical perspective, the properties of GOS and FOS suggest that these materials are suitable for use in a wide variety of foods, and their production is likely to increase in the near future (Playne & Crittenden, 1996; Sako *et al.*, 1999). Preliminary research on the physiological benefits of consuming GOS is promising, and in particular the bifidogenic and mineral-enhancing effects. NDOs such as GOS and FOS are increasingly being incorporated into high value-added foods such as enteral feeds and infant formula. In the future, it is foreseeable that a second generation of NDOs capable of enhancing the microflora at the species rather than at the genus level will be developed (Rastall & Gibson, 2002). To date, however, the number of studies on the consumption of existing NDO in humans is limited, and the results are often inconsistent. Thus, further studies – particularly of longer-term duration, and using molecular-based tools to determine changes in the gut flora and other appropriate biomarkers – are urgently warranted in human subjects to delineate fully the beneficial potential of these versatile NDOs.

## References

Alles, M.S., Hartemink, R. Meyboom, S., Harryvan, J.L., van Laere, K.M., Nagengast, F.M. & Hautvast, J.G. (1999) Effect of transgalactooligosaccharides on the composition of the human intestinal microflora and on putative risk markers for colon cancer. *American Journal of Clinical Nutrition*, **69**, 980–991.

Anonymous (1999) *Oligomate 55 – Bifidus Growth-Promoting Substance*. Technical Bulletin, Yakult Pharmaceutical Industries Co. Ltd., Tokyo.

AOAC (2003) Official Methods of Analysis of AOAC International. Vol. 2, 17th edn., current through 2nd revision, pp. 88–91, AOAC, Arlington.

Arai, S., Morinaga, Y., Yoshikawa, T., Ichiishi, E., Kiso, Y., Yamazaki, M., Morotomi, M., Shimizu, M., Kuwata, T. & Kaminogawa, S. (2002) Recent Trends in Functional Food Science and the Industry in Japan. *Bioscience Biotechnology & Biochemistry*, **66**, 2017–2029.

Boehm, G., Lidestri, M., Casetta, P., Jelinek, J., Negretti, F., Stahl, B. & Marini, A. (2002) Supplementation of a bovine milk formula with an oligosaccharide mixture increases counts of faecal bifidobacteria in preterm infants. *Archives of Digestive Child Fetal Neonatal Education*, **86**, 178–181.

Bornet, F.R.J. (1994) Undigestible sugars in food products. *American Journal of Clinical Nutrition*, **59**(Suppl.), 763S–769S.

Bouhnik, Y., Flourie, B., D'Agay-Abensour, L., Pochart, P., Gramet, G., Durand, M. & Rambaud, J.-C. (1997) Administration of transgalacto-oligosaccharides increases fecal bifidobacteria and modifies colonic fermentation metabolism in healthy humans. *Journal of Nutrition*, **127**, 444–448.

Brommage, R., Binacua, C., Antille, S. & Carrie, A.L. (1993) Intestinal calcium absorption in rats is stimulated by dietary lactulose and other resistant sugars. *Journal of Nutrition*, **123**, 2186–2194.

Busserolles, J., Gueux, E., Rock, E., Demigne, C., Mazur, A. & Rayssiguier, Y. (2003) Oligofructose protects against the hypertriglyceridemic and pro-oxidative effects of a high fructose diet in rats. *Journal of Nutrition*, **133**, 1903–1908.

Chaturvedi, P., Warren, C.D., Ruiz-Palacios, G.M., Pickering, L.K. & Newburg, D.S. (1997) Milk oligosaccharide profiles by reversed-phase HPLC of their perbenzoylated derivatives. *Analytical Biochemistry*, **251**, 89–97.

Chonan, O. & Watanuki, M. (1995) Effect of galactooligosaccharides on calcium absorption in rats. *Journal of Nutritional Science and Vitaminology*, **41**, 95–104.

Chonan, O. & Watanuki, M (1996) The effect of 6'-galactooligosaccharides on bone mineralization of rats adapted to different levels of dietary calcium. *International Journal of Vitamin and Nutritional Research*, **66**, 244–249.

Chonan, O., Matsumoto, K. & Watanuki, M. (1995) Effect of galactooligosaccharides on calcium absorption and preventing bone loss in ovariectomized rats. *Bioscience Biotechnology and Biochemistry*, **59**, 236–239.

Chonan, O., Takahashi, R., Yasui, H. & Watanuki, M. (1996) Effects of β-1–4 linked galactooligosaccharides on use of magnesium and calcification of the kidney and heart in rats fed excess dietary phosphorus and calcium. *Bioscience Biotechnology and Biochemistry*, **60**, 1735–1737.

Cummings, J., Antoine, J.-M., Azpiroz, F., Bourdet-Sicard, R., Brandtzaeg, P., Calder, P., Gibson, G., Guarner, F., Isolauri, E., Pannemans, D., Shortt, C., Tuijtelaars, S. & Watzl, B. (2004) PASSCLAIM: Gut health and immunity. *European Journal of Nutrition*, **43**, 118–174.

Deguchi, Y., Makino, A., Iwabuchi, A., Watanuki, M., & Yamashita, T. (1993) Selection of ammonia-assimilating bifidobacteria and their effect on ammonia levels in rat cecal contents and blood. *Microbial Ecology in Health and Disease*, **6**, 85–94.

Deguchi, Y., Matsumoto, K., Ito, T. & Watanuki, M. (1997) Effects of β-1–4 galactooligosaccharides administration on defecation of healthy volunteers with constipation tendency. *Japanese Journal of Nutrition*, **55**, 13–22.

De Leenheer, L. (1996) Production and use of inulin: industrial reality with a promising future, *Carbohydrates as Organic Raw Materials III*, (eds. H. Van Bekkum, H. Roper & A.G.J.), pp. 67–92, VCH Publishers, New York.

Delzenne, N.M. & Roberfroid, M.R. (1994) Physiological effects of non-digestible oligosaccharides. *Lebensmittel-Wissenschaft und Technologie*, **27**, 1–6.

Delzenne, N., Aertssens, J., Verplaetse, N., Roccaro, M. & Roberfroid, M. (1995) Effect of fermentable fructo-oligosaccharides on energy and nutrients absorption in the rat. *Life Sciences*, **57**, 1579–1587.

van Dokkum, W., Wezendonk, B., Strikumar, T.S. & van den Heuvel, E.G.H.M. (1999) Effect of non digestible oligosaccharides on large-bowel functions, blood lipid concentrations and glucose absorption in young healthy male subjects. *European Journal of Clinical Nutrition*, **53**, 1–7.

Durand, M., Cordelet, C., Hannequart, G., Beaumatin, P. & Givet, J.P. (1992) *In vitro* fermentation of a galacto-oligosaccharide by human bacteria in continuous culture. *Proceedings of Nutritional Society*, **51**, 6A.

Ebrahim, G. J. (1997) Breast milk oligosaccharides point the way to new therapeutic strategies. *Journal of Tropical Pediatrics*, **43**, 2–3.

Ellegard, L., Andersson, H. & Bosaeus, I. (1997) Inulin and oligofructose do not influence the absorption of cholesterol, or the excretion of cholesterol, Ca, Mg, Zn, Fe or bile acids but increases energy excretion in ileostomy subjects. *European Journal of Clinical Nutrition*, **51**, 1–5.

Fiordaliso, M., Kok, N., Desager, J.P., Goethaks, F., Deboyser, D., Roberfroid, M. & Delzenne, N. (1995) Dietary oligofructose lowers triglycerides, phospholipids and cholesterol in serum and very low density lipoproteins of rats. *Lipids*, **30**(2), 163–167.

Franck, A. (2000) Prebiotics and calcium absorption, *Functional Foods 2000*, Conference Proceedings, (ed. F. Angus & C. Miller), pp. 108–113, Leatherhead Publishing, Surrey.

Franck, A. & Coussement, P. (1997) Multi-functional inulin. *Food Ingredients and Analysis International*, **October**, 8–10.

Franck, A. & Coussement, P. (2001) Prebiotics, *Guide to Functional Food Ingredients*, (ed. J. Young), pp. 1–19, Leatherhead Publishing, Surrey.

Gibson, G.R., Willis, C.L. & Van Loo, J. (1994) Non digestible oligosaccharides and bifidobacteria – Implications for Health. *International Sugar Journal*, **96**, 381–387.

Gibson, G.R. & Roberfroid, M.B. (1995) Dietary modulation of the human colonic microbiota: introducing the concept of prebiotics. *Journal of Nutrition*, **125**, 1401–1412.

Gibson, G.R., Beatty, E.R., Wang, X. & Cummings, J.H. (1995) Selective stimulation of bifidobacteria in the human colon by oligofructose and inulin. *Gastroenterology*, **108**, 975–982.

Gill, C. & Rowland, I. (2002) Diet and cancer: assessing the risk. *British Journal of Nutrition*, **88**, 73–87.

Griffin, I.J., Dvila, P.M. & Abrams, S.A. (2002) Non-digestible oligosaccharides and calcium absorption in girls with adequate calcium intakes. *British Journal of Nutrition*, **87**, S187–S191.

Hamosh, M. (1997) Bioactive components in milk and development of the neonate: does their absence make a difference? *American Society of Nutritional Sciences*, **127**, 971S–974S.

van den Heuvel, E.G.H.M., Schaafsma, G., Murs, T. & van Dokkum, W. (1998) Nondigestible oligosaccharides do not interfere with calcium and nonheme-iron absorption in young healthy men. *American Journal of Clinical Nutrition*, **67**, 445–451.

van den Heuvel, E.G.H.M., Schoterman, M. & Muis, T. (2000) Transgalactooligosaccharides stimulate calcium absorption in postmenopausal women. *Journal of Nutrition*, **130**, 2938–2942.

Hidaka, H., Hirayama, M. & Yamada, K. (1991) Fructooligosaccharides – enzymatic preparation and biofunctions. *Journal for Carbohydrate Chemistry*, **10**, 509–522.

Holzapfel, W., Haberer P., Snel, H., Schillinger, U. & Huis in't Veld, J. (1998) Overview of the gut microflora and probiotics. *Journal of Food Microbiology*, **41**, 85–101.

Ishikawa, F., Takayama, H., Matsumoto, K., Ito, M., Chonan, O., Deguchi, Y., Kikuchi-Hayakawa, H. & Watanuki, M. (1995) Effects of β-1–4 linked galactooligosaccharides on human fecal microflora. *Bifidus*, **9**, 5–18.

Ito, M., Deguchi, Y., Miyamori, A., Matsumoto, K., Kikuchi, H., Matsumoto, K., Kobayashi, Y., Yajima, T. & Kan, T. (1990) Effects of administration of galactooligosaccharides on the human fecal microflora, stool weight and abdominal sensation. *Microbial Ecology in Health and Disease*, **3**, 285–292.

Ito, M., Deguchi, Y., Matsumoto, K., Kimura, M., Onodera, N. & Yajima, T. (1993) Influence of galactooligosaccharides on the human fecal microflora. *Journal of Nutritional Science and Vitaminology*, **39**, 635–640.

Japanscan (2002) Legal affairs, **19**(12) February, 32 (www.japanscan.co.uk).

Kanamori, Y., Hashizume, K., Kitano, Y., Tanaka, Y., Morotomi, M., Yuki, N. & Tanaka, R (2003) Anaerobic dominant flora was reconstructed by synbiotics in an infant with MRSA enteritis. *Pediatrics International*, **45**, 359–362.

Kikuchi, H., Andrieux, C., Riottot, M., Bensaada, M., Popot, F., Beaumatin, P. & Szylit, O. (1996) Effect of two levels of transgalactosylated oligosaccharide intake in rats associated with human fecal microflora on bacterial glycolytic activity, end-products of fermentation and bacterial steroid transformation. *Journal of Applied Bacteriology*, **80**, 439–446.

Kikuchi, H., Kimura, M. & Watanuki, M. (1997) Adaptation of rate of organic acid production of hindgut bacteria to chronic intake of galactooligosaccharide in the rat. *Journal of Nutritional Science and Vitaminology*, **43**, 357–368.

Kleesen, B., Hartmann, L. & Blaut, M. (2003) Fructans in the diet cause alterations of mucosal architure, released mucins and mucosa-associated bifidobacteria in gnotobiotic rats. *British Journal of Nutrition*, **89**, 587–606.

Kunz, C. & Rudloff, S. (1996) Structural and functional aspects of oligosaccharides in human milk. *Zeitschrift fur Ernahrungswiss*, **35**, 22–31.

Kunz, C., Rudloff, S., Hintelmann, A., Pohlentz, G. & Egge, H. (1996) High-pH anion exchange chromatography with pulsed amperometric detection and molar response factors of human milk oligosaccharides. *Journal of Chromatography B*, **685,** 211–221.

Kwak, H.S. & Jeon, I.J. (1986) Effect of various conditions on the formation of oligosaccharides in milk treated with β-galactosidase. *Journal of Dairy Science*, **69**, 2785–2790.

van Loo, J., Coussement, P., De Leenheer, L., Hoebregs, H. & Smits, G. (1995) On the presence of inulin and oligofructose as natural ingredients in the Western diet. *Critical Reviews in Food Science and Nutrition*, **35**, 525–552.

Manhart, N., Splittler, A., Bergmeiste.r H., Mittlbock, M. & Roth, E. (2003) Influence of fructooligosaccharides on Peyer's patch lymphocyte numbers in healthy and endotoxemic mice. *Nutrition*, **19**, 657–660.

Matsumoto, K., Kobayashi, Y., Ueyama, S., Watanabe, T., Tanaka, R., Kan, T., Kuroda, A., & Sumihara, Y. (1990) Galactooligosaccharides, *Japanese Technology Reviews*, (ed. T. Nakanuki), pp. 90–106, Gordon and Breach Science Publishers, Tokyo.

Miller, J.B., Bull, S., Miller, J. & McVeagh, P. (1994) The oligosaccharide composition of human milk: temporal and individual variations in monosaccharide components. *Journal of Pediatric Gastroenterology and Nutrition*, **19**, 371–376.

Minami, Y., Yazawa, K., Nakamura, K. & Tamura, Z (1985) Selectively and efficiency of utilization of galactosyl-oligosaccharides by bifidobacteria. *Chemical and Pharmaceutical Bulletin*, **33**, 710–714.

Narimiya, M., Yokoi, K., Tajima, N., Sakai, O., Ikeda, Y. & Takayama, H. (1996) The effect of β-1–4 galactooligosaccharides on fecal flora in non insulin dependent diabetic patients with constipation. *Japanese Journal of Clinical Nutrition*, **18**, 44–50.

Newburg, D.S. (1997) Bioactive components in milk and development of the neonate: does their absence make a difference? *American Society of Nutritional Science*, **127**, 980S–984S.

Ohtsuka, K., Benno, Y., Endo, A., Ueda, H., Ozawa, O., Uchida, T. & Mitsuoka, T. (1989) Effects of 4'-galactosyl-lactose on human intestinal microflora. *Bifidus* **2**, 143–149.

Oku, T. (1996) Available energy of nondigestible and/or nonabsorbable saccharides. *Japanese Journal of Nutrition*, **54**, 143–150.

Palframan, R.J., Gibson, G.R. & Rastall, R.A. (2002) Effect of pH and dose on the growth of gut bacteria on prebiotic carbohydrates in vitro. *Anaerobe*, **8**, 287–292.

Pereira, D.I. & Gibson, G.R. (2002) Effects of consumption of probiotics and prebiotics on serum levels in humans. *Critical Reviews in Biochemical Molecular Biology*, **37**, 259–281.

Playne, M.J. & Crittenden, R. (1996) Commercially available oligosaccharides. *Effects of Whey Proteins on Cheese Characteristics – Oligosaccharides and Probiotic Bacteria*, Document No. 313, pp. 10–20, International Dairy Federation, Brussels.

Quemener, B., Thibault, J. & Coussement, P. (1994) Determination of inulin and oligofructose in food products, and integration in the AOAC method for measurement of total dietary fibre. *Lebensmittel-Wissenschaft und Technologie*, **27**, 125–132.

Rastall, R. & Gibson, G. (2002) Prebiotic oligosaccharides: evaluation of biological activities and potential future developments. *Probiotics and Prebiotics, Where are we Going?* (ed. G. Tannock), pp. 107–149, IBT Global, Barking.

Reddy, G.P. & Bush, C.A. (1991) High performance anion exchange-chromatography of neutral milk oligosaccharides and oligosaccharide alditols derived from mucin glycoproteins. *Analytical Biochemistry*, **198**, 278–284.

Roberfroid, M.B. (1997) Health benefits of non-digestible oligosaccharides. *Advances in Experimental Medicine and Biology*, **427**, 211–219.

Roberfroid, M.B., Gibson, G.R. & Delzenne, N. (1993) The biochemistry of oligofructose, a nondigestible fiber: an approach to calculate its calorific value. *Nutrition Reviews*, **51**, 137–146.

Roberfroid, M.B., Van Loo, J.A.E. & Gibson, G.R. (1998) The bifidogenic nature of chicory inulin and its hydrolysis products. *Journal of Nutrition*, **128**, 11–19.

Roller, M., Femia, A.P., Caderni, G., Rechkemmer, G. & Watzl, B. (2004) Intestinal immunity of rats with colon cancer is modulated by oligofructose-enriched inulin combined with *Lactobacillus rhamnosus* and *Bifidobacterium lactis*. *British Journal of Nutrition*, **92**, 931–938.

Rowland, I. & Tanaka, R. (1993) The effects of transgalactosylated oligosaccharides on gut flora metabolism in rats associated with a human fecal microflora. *Journal of Applied Bacteriology*, **74**, 667–674.

Sako, T. Matsumoto, K. & Tanaka, R. (1999). Recent progress on research and applications of non-digestible galacto-oligosaccharides. *International Dairy Journal*, **9**, 69–80.

Salminen, S., Ouwehand, A. & Marteau, P. (2000) Functional foods and ingredients for gut health, *Functional Foods 2000*, Conference Proceedings, (ed. F. Angus & C. Miller), pp. 134–142, Leatherhead Publishing, Surrey.

Scholz-Ahrens, K., Van Loo, J., & Schrezenmeir, J. (1998) Effect of oligofructose on bone mineralisation in ovariectomized rats is affected by dietary calcium, *International Symposium on Probiotics and Prebiotics*, pp. 11–12, Kiel, Germany.

Scholz-Arhens, K., Schaafsma, G., van den Heuval, E. & Schrezenmeir, J. (2001) Effects of probiotics on mineral metabolism. *American Journal of Clinical Nutrition*, **73**, 459–464.

Shortt C. (2004) Perspectives on foods for specific health uses (FOSHU), Food Science and Technology Bulletin, (http://www.foodsciencecentral.com/library.html#ifis/12686).

de Slegte, J. (2002) Determination of trans-galactooligosaccharides in selected food products by ion-exchange chromatography: collaborative study. *Journal of the AOAC International*, **85**, 417–423.

Spiegel, J.E., Rose, R., Karabell, P., Frankos, V.H. & Schmitt, D.F. (1994) Safety and benefits of Fructo-oligosaccharides as food ingredients. *Food Technology*, **January**, 85–89.

Stahl, B., Thurl, S., Zeng, J., Karas, M., Hillenkamp, F., Steup, M. & Sawatzki, G. (1994) Oligosaccharides from human milk as revealed by matrix-assisted laser desorption/ionisation mass spectrometry. *Analytical Biochemistry*, **223**, 218–226.

Taguchi, A., Ohta, A., Abe, M., Baba, S., Ohtsuki, M., Takizawa, T., Yuda, Y. & Adachi, T. (1995) Effects of fructooligosaccharides on bone and mineral absorption in the rat model with ovariectomized osteoporosis. *Meiji Seika Kenyu Nenpo*, **33**, 37–43.

Tamai, S., Ohtsuka, K., Ozawa, O., & Uchida, T. (1992) Effect of a small amount of galactooligosaccharides on fecal *Bifidobacterium*. *Journal of the Japanese Society of Nutrition and Food Science*, **45**, 456–460.

Tanaka, R., Takayama, H., Morotomi, M., Kuroshima, T., Ueyama, S., Matsumoto, K., Kuroda, A. & Mutai, M. (1983) Effects of administration of TOS and *Bifidobacterium breve* 4006 on the human fecal flora. *Bifidobacteria and Microflora* **2**, 17–24.

Teuri, U., Korpela, R., Saxelin, M., Montonen, L. & Salminen, S. (1998) Increased fecal frequency and gastrointestinal symptoms following ingestion of galactooligosaccharide containing yoghurt. *Journal of Nutrition Science & Vitaminology*, **44**, 465–471.

Thurl, S., Offermanns, J., Muller-Werner, B., Sawatzki, G. (1991) Determination of neutral oligosaccharide fractions from human milk by gel permeation chromatography. *Journal of Chromatography*, **568**, 291–300.

Tungland, B.C. (2003) Fructooligosaccharides and other fructans: structures and occurrence, production, regulatory aspects, food applications and nutritional health significance. *Oligosaccharides in Food and Agriculture*, Symposium Series, pp. 135–152, American Chemical Society, Washington.

Watanuki, M., Wada, Y. & Matsumoto, K. (1996) Digestibility and physiological heat of combustions of β-1–4 and β-1–6 galactooligosaccharides in vitro. *Annual Reports of the Yakult Central Institute for Microbiological Research*, **16**, 1–12.

Yamashita, K. & Kobata, A. (1974) Oligosaccharides of human milk. V. Isolation and characterization of a new trisaccharide, 6'-galactosyllactose. *Archives of Biochemistry and Biophysics*, **161**, 164–170.

Yanahira, S., Kobayashi, T., Suguri, T., Nakakoshi, M., Miura, S., Ishikawa, H. & Nakajima, I. (1995) Formation of oligosaccharides from lactose by *Bacillus circulans* β-galctosidase. *Biosciences Biotechnology Biochemistry*, **59**, 1021–1026.

Yazawa, K., & Tamura, Z. (1982) Search for sugar sources for selective increase of bifidobacteria. *Bifidobacteria Microflora*, **1**, 39–44.

# 7 Health Claims Associated with Probiotics

G.A. O'May and G.T. Macfarlane

## 7.1 Introduction

The importance of the colonic microbiota in human physiology and metabolism is being increasingly recognised (Macfarlane & Cummings, 2002). This is due, in part, to recent advances in molecular techniques for studying the composition and activities of the normal microflora, such as, for example, 16S rRNA gene sequencing (Hold *et al.*, 2002), *in situ* hybridisation (Hopkins & Macfarlane, 2003), real-time PCR (Fite *et al.*, 2004) and denaturing gradient gel electrophoresis (DGGE) (Tannock, 2004). These developments have also led to more systematic schemes for the taxonomy of potentially probiotic micro-organisms, such as lactobacilli (Reid *et al.*, 2002).

Probiotics have, by international consensus, been defined as 'live micro-organisms which, when administered in adequate amounts, confer a health benefit on the host' (Reid *et al.*, 2003a; see also Chapter 3). Micro-organisms that have been used as probiotics in humans include yeasts, particularly *Saccharomyces boulardii* (Gaon *et al.*, 2003; Herek *et al.*, 2004); bacilli (Pinchuk *et al.*, 2001; Jadamus *et al.*, 2002); *Escherichia coli* (Tromm *et al.*, 2004); enterococci (Lund *et al.*, 2000); bifidobacteria (Gill *et al.*, 2001; Chouraqui *et al.*, 2004; Mimura *et al.*, 2004); and lactic acid bacteria [lactobacilli (Koebnick *et al.*, 2003; Xiao *et al.*, 2003; Schultz *et al.*, 2004), streptococci and bifidobacteria (Saavedra *et al.*, 1994, 2004), and lactococci (Madsen, 2001)]. For reasons of history and convenience, lactic acid bacteria are the most commonly used probiotics (Reuter, 2001).

Micro-organisms must possess a number of characteristics before they can be regarded as probiotics. These are: (a) a lack of toxicity and pathogenicity; (b) the ability to survive the body's innate defences in the upper gastrointestinal (GI) tract, including surviving exposure to gastric acid and bile salts; (c) an ability to colonise the intestinal mucosa; and (d) the capacity to exert a health improvement on the host. The latter can include protection against pathogens through competitive exclusion (Fujiwara *et al.*, 2001; van der Wielen *et al.*, 2002), contribution to host nutrition (e.g. butyrate production; Kanauchi *et al.*, 1999), possession of anti-cancer effects (Rafter, 2003), as well as the ability to stimulate host immunity within the GI tract (Roller *et al.*, 2004) and in the peripheral circulation (Cross, 2004). Whilst it is unlikely that any one probiotic strain will manifest all of these properties, it is evident from the aforementioned consensus definition that for a strain to be described as a probiotic, some beneficial effect on human health must have been documented. This may be complicated by the fact that, while a particular organism may be safe in one group of patients, under different circumstances the same strain could cause problems, for example, in immunocompromised individuals.

Probiotics are increasingly being made available over the counter in health food shops and, more recently, in supermarkets. However, a recent study has found that many of these products often do not contain the micro-organisms they are supposed to (Weese, 2002). Of 13 products tested, only two were correctly identified on the packaging; however, the remainder contained bacteria with no proven probiotic activity, and some even had pathogenic potential. For this reason, only studies undertaken where the authors might reasonably be expected to have confirmed the identity of the probiotic strain(s) will be included in this chapter.

Evidence relating to the effects of probiotics on human health can be obtained from *in vitro* or *in vivo* studies. The former can be useful in advancing our understanding of the mechanisms whereby probiotic micro-organisms exert their effects. However, because of the vastly increased complexity of living systems and inter-individual variation, *in vitro* experiments cannot provide definitive evidence on the validity of probiotic phenomena. Only *in vivo* investigations in humans can verify probiotic efficacy, and many different types of study exist, ranging from pilots with a handful of patients to double-blinded, randomised placebo-controlled clinical trials involving large numbers of individuals. Information obtained from this type of study is considered to be the most reliable.

## 7.2 Probiotic use in GI tract conditions

### 7.2.1 *Inflammatory bowel disease*

The two major forms of idiopathic inflammatory bowel disease (IBD) are Crohn's disease (CD) and ulcerative colitis (UC). Although their aetiologies are unknown, the pathogenesis of both conditions is thought to involve species belonging to the normal gut microflora. This has resulted in several attempts to treat both UC and CD with probiotics (Table 7.1).

#### *Crohn's disease*

An important factor which differentiates CD from UC is that the disease can involve the entire GI tract, rather than just the colon. It is also characterised by inflammatory processes occurring deeper in the tissues. Typical treatments involve the use of elemental diets, anti-inflammatory drugs, monoclonal antibodies against tumour necrosis factor-alpha (TNF-$\alpha$), antibiotics and, if necessary, surgery. *Lactobacillus rhamnosus* GG was used in a randomised placebo-controlled study aimed at reducing the rate and/or severity of recurrence after surgery (Prantera *et al.*, 2002). The study lasted for 12 months and 38 patients were involved. Those taking the probiotic received $1.2 \times 10^{10}$ colony forming units (cfu) orally per day, while the controls received a placebo containing maltodextrins and sorbitol. Recurrence rates measured clinically or endoscopically were not significantly different between the two groups. Guslandi *et al.* (2000) used the yeast *S. boulardii* in an attempt to prevent relapse in CD. A total of 32 patients was randomised to receive either mesasalazine (1 g three times daily), or mesasalazine (1 g twice daily) plus a *S. boulardii* preparation (1 g $d^{-1}$). After 6 months, clinical relapses occurred in 37.5% of those receiving mesasalazine and in 6.25% of patients receiving mesasalazine and the probiotic.

In a pilot study using the non-pathogenic bacterium *E. coli* strain Nissle 1917 to attempt maintenance of remission in colonic CD (Malchow, 1997), 28 patients were randomised to receive either probiotic ($n = 16$) or placebo ($n = 12$). All patients received 60 mg prednisolone per day until symptoms improved; the dosage was then gradually reduced. Those in the probiotic group were given an increasing dose of $10^9$ cfu $d^{-1}$ and $5 \times 10^9$ cfu $d^{-1}$ for 24 days, and then two capsules per day each containing $2.5 \times 10^{10}$ cfu for the duration of the trial (i.e. one year). No difference in the rate of remission was observed; however, patients

**Table 7.1** Summary of probiotic trials in inflammatory bowel disease (IBD) and pouchitis.

| Study description | Duration | Results | Reference |
|---|---|---|---|
| *Lb. rhamnosus* GG or placebo; 38 patients with active CD[1]; $1.2 \times 10^{10}$ cfu $d^{-1}$ orally | 12 months | No SD[2] between probiotic and placebo | Prantera *et al.* (2002) |
| *S. boulardii*; 32 patients with active CD; 1 g daily with mesalazine, or without | 6 months | Relapses in 6.25% of probiotic group and 37.5% of mesalazine only | Guslandi *et al.* (2000) |
| *E. coli* strain Nissle 1917 or placebo; 28 patients with active CD; $10^9$–$2.5 \times 10^{10}$ cfu $d^{-1}$ | 12 months | No SD in rate of remission; patients in probiotic group entered remission earlier | Malchow (1997) |
| *E. coli* strain Nissle 1917 or mesalazine; 116 patients with active UC[3]; $2.5 \times 10^{10}$ cfu $d^{-1}$ | 12 weeks or until remission | Remission in 75% of mesalazine and 68% of probiotic group | Rembacken *et al.* (1999) |
| VSL#3[4]; 20 UC patients in remission; $10^{12}$ cfu $d^{-1}$ | 12 months | Remission in 75% of patients after 1 year | Venturi *et al.* (1999) |
| *S. boulardii* and mesalazine; 24 patients with active UC; 250 mg daily | 4 weeks | 68% achieved a decrease in CAI[5] from 9 to 5 | Guslandi *et al.* (2003) |
| *B. breve* and *B. bifidum* Yakult-fermented milk; 21 UC patients in remission; 100 mL daily | 12 months | 90% and 27.3% of control and probiotic groups relapsed | Ishikawa *et al.* (2003) |
| *B. longum* and prebiotic or placebo; 18 patients with active UC; $4 \times 10^{11}$ cfu $d^{-1}$ | 4 weeks | Significant improvements in CAI, bowel function and decrease in inflammatory markers | Furrie *et al.* (2004) |
| VSL#3 or placebo; 40 pouchitis patients in remission; $3 \times 10^{12}$ cfu $d^{-1}$ | 6 months | 15% and 100% in the probiotic and placebo groups relapsed ($P < 0.001$). | Gionchetti *et al.* (2000) |
| VSL#3 or placebo; 40 patients post-colectomy; $3 \times 10^{12}$ cfu $d^{-1}$ | 9 months | 15% and 100% of the probiotic and placebo groups developed pouchitis | Gionchetti *et al.* (2003) |
| VSL#3 or placebo; 36 patients with recurrent pouchitis | 12 months | 6% and 85% of the probiotic and placebo groups maintained remission | Mimura *et al.* (2004) |
| Cultura™; 10 patients with mild/moderate pouchitis; $5 \times 10^{10}$ cfu $d^{-1}$ | 4 weeks | No SD in disease activity in the probiotic group | Laake *et al.* (2003) |

[1] CD = Crohn's disease.
[2] SD = Significant difference.
[3] UC = ulcerative colitis.
[4] VSL#3 = probiotic preparation – see text.
[5] CAI = clinical activity index.

receiving the probiotic entered remission sooner than those with the placebo, although this difference was not significant. Only 30% of patients receiving probiotic had experienced a relapse within one year compared to 70% of those in the placebo group. While marked, this effect was not significant, probably due to the small numbers involved. The authors observed that all of the patients receiving probiotic were able to stop prednisolone treatment, but a number of placebos were unable to do so.

Evidence concerning the efficacy of probiotics in the treatment of CD is limited. Further studies are required to determine whether any of the currently available probiotic strains have therapeutic value. More specifically, large-scale double-blinded and placebo-controlled trials on a variety of putatively probiotic strains would generate additional useful information.

### *Ulcerative colitis*

Ulcerative colitis is an acute and chronic form of IBD that only affects the large bowel. This relapsing disease invariably initiates in the rectum and descending colon and may spread proximally along the gut. Much of the evidence for involvement of the microbiota in UC comes from animal models. Genetically susceptible mice only acquire characteristic lesions when populated with a normal commensal microflora (Sadlack *et al.*, 1993), while germ-free animals do not manifest an inflammatory response (Taurog *et al.*, 1994). Studies conducted in humans have suggested that mucosal bacterial populations in UC patients may be shifted towards a more pro-inflammatory profile (Poxton *et al.*, 1997; Macfarlane *et al.*, 2004), and the disease shares some histological features with infective colitis (Tedesco *et al.*, 1983). Attempts to use antibiotics in UC have met with little success (Dickinson *et al.*, 1985; Mantzaris *et al.*, 1994; Turunen *et al.*, 1998) and this, together with the fact that the microflora are known to be involved in its initiation and maintenance, makes the use of probiotics an attractive proposition.

*E. coli* lipopolysaccharide (LPS) is known to induce lymphocyte cytotoxicity against colonic epithelial cells (Shorter *et al.*, 1970), and UC patients are often colonised by haemolytic or necrotoxic *E. coli* strains (Cooke, 1968). For these reasons, Rembacken *et al.* (1999) compared the ability of *E. coli* strain Nissle 1917 with mesasalazine to prevent relapse. A total of 116 patients was randomised to receive two capsules of *E. coli* per day, containing $2.5 \times 10^{10}$ cfu per capsule (57 patients), or 800 mg mesasalazine three times per day (59 patients). All patients received oral gentamicin for 1 week, together with standard medical therapies. Patients were monitored for a maximum of 12 weeks, or until remission. Patients not in remission after 12 weeks were eliminated from the next phase of the trial. During the second phase, the dosage of *E. coli* was reduced to twice daily and mesasalazine reduced to one 400-mg capsule three times per day. Patients were monitored until relapse or until 12 months had passed. Results suggested that administration of *E. coli* strain Nissle 1917 had an effect similar to that of treatment with mesasalazine. Remission was achieved in 75% of patients receiving the drug, and 68% of the probiotic group. During the second phase of the trial, 73% and 67% of the mesasalazine and probiotic groups, respectively, relapsed. The authors concluded that *E. coli* strain Nissle 1917 was a viable alternative to mesasalazine in the treatment of UC. Further studies have also found no significant differences between *E. coli* strain Nissle 1917 and mesasalazine in terms or relapse rates or remission time (Kruis *et al.*, 1997, 2001). Thus, this bacterium seems to have therapeutic potential in UC patients, particularly in cases where side effects of mesasalazine (Das *et al.*, 1973) occur and an alternative is needed. However, there is clearly scope for improved therapies as seen by the high relapse rate experienced by UC patients.

Other probiotic micro-organisms have been employed in UC trials. The multi-species probiotic preparation VSL#3, which contains four lactobacilli, three bifidobacteria and *Streptococcus thermophilus*, was used in an uncontrolled pilot study in patients with UC (Venturi *et al.*, 1999). Twenty patients were treated with $5 \times 10^{11}$ cfu $g^{-1}$ VSL#3 (3 g twice daily for 12 months). After this time, 75% of patients were still in remission, while faecal excretion of lactobacilli, bifidobacteria and *S. thermophilus* increased significantly during treatment. Additionally, a reduction in stool pH from an average of 7.6 to 6.8 was recorded after 90 days of treatment, providing evidence that the metabolic activity of the microflora had been altered by feeding the probiotic. A similar beneficial effect was observed when Fedorak *et al.* (2003) used the same preparation, in conjunction with conventional anti-inflammatory therapies, to induce remission in patients with active UC. Another small pilot study used *S. boulardii* to maintain remission in UC patients experiencing a clinical relapse (Guslandi *et al.*, 2003). Twenty-four patients received 250 mg of *S. boulardii* and mesasalazine alone for 4 weeks; no information concerning the numbers of viable cells administered was provided. After this time, 68% of patients receiving the probiotic had a successful clinical outcome, as measured by the reduction in Rachmilewitz's activity index to less than 5 from a value greater than 9 at the beginning of the trial. These results were confirmed using sigmoidoscopy. The authors concluded that while *S. boulardii* showed promise in their trial, further large-scale trials were needed.

Bifidobacteria have also been used in clinical trials in UC patients. Ishikawa *et al.* (2003) used milk fermented by *Bifidobacterium breve* strain Yakult and *Bifidobacterium bifidum* strain Yakult in a small trial (21 patients, one year) aimed at reducing the rate of relapse. These authors found that probiotic consumption reduced the rate of relapse from nine out of 10 in the control group, to three out of 11 in those treated with probiotic, although no differences between the groups were observed at colonoscopy. Similarly, no difference in total numbers of bifidobacteria was detected during the trial; however, the probiotic strains were recovered from faeces at a concentration of about $10^6$ cfu $g^{-1}$. A strain of *Bifidobacterium longum*, together with a prebiotic, was used by Furrie and colleagues to induce remission in patients with active UC (Furrie *et al.*, 2004). The rationale in using the probiotic in conjunction with a prebiotic (a synbiotic) is that as well as encouraging the growth of the probiotic strain, the prebiotic may enhance the growth of other micro-organisms with potential health benefits that are already present in the large bowel (see Chapter 6). In a previous investigation, this group had found that high numbers of bifidobacteria were present on the rectal mucosae of healthy people and that their numbers were greatly reduced in UC patients (Macfarlane *et al.*, 2004). Because UC initiates in the rectum, it was hypothesised that an alteration in the microflora at this point was a contributory factor in disease initiation. A strain of *B. longum* isolated originally from the rectal mucosa of a healthy individual was selected as the probiotic on the basis of empirical evidence of its suitability in terms of acid, bile salts, resistance to freeze-drying and oxygen tolerance (Kennedy, 2002). Eighteen patients with active UC were randomised to the synbiotic ($n = 9$) and control ($n = 9$) groups. Patients in the synbiotic group each received a capsule containing $2 \times 10^{11}$ cfu $g^{-1}$ *B. longum* and a sachet containing 6 g fructo-oligosaccharide/inulin mix twice daily for 4 weeks. Patients in the control group received an identical capsule containing starch and a sachet containing 6 g powdered maltodextrose. The protocol involved patients taking their capsules and sachets after breakfast and evening meal to minimise the effect of gastric acid on probiotic survival. Each patient was assessed by clinical activity index (CAI) and sigmoidoscopy score. Additionally, each patient kept a bowel habit diary. Rectal biopsies were assessed histologically and immunologically, while C-reactive protein levels in venous blood were measured. A marked reduction in expression of various markers of inflammation was observed between the pre-feeding and post-feeding samples of the synbiotic group. Also, levels of the pro-inflammatory cytokines interleukin (IL)-1 alpha and TNF-$\alpha$ in the mucosal tissue of patients receiving the synbiotic returned to levels measured in healthy individuals. Moreover, the amount of *Bifidobacterium*-specific mRNA in rectal mucosal biopsies increased 42-fold in the synbiotic group, compared to 4.6-fold in the placebo group. Clinical outcome improved in the synbiotic group, particularly with respect to bowel habit diaries and CAI scores. A smaller change was observed in sigmoidoscopy scores, although histological assessment of rectal biopsies showed marked reductions in tissue damage and mucosal regeneration in the synbiotic group. These results demonstrated that treatment with the bifidobacteria synbiotic resulted in marked improvement in the patient's condition.

The efficacy of probiotic therapy in UC is beginning to be recognised as the results of clinical studies are published. At present, there is some evidence that *E. coli* strain Nissle 1917, *S. boulardii*, VSL#3 and various strains of bifidobacteria are able to exert a beneficial effect in UC patients. However, each of the studies described above was performed in a relatively small number of patients, and some studies were neither randomised nor blinded. For these reasons, conclusions regarding the efficacy of probiotic therapy in UC must be regarded as preliminary until larger-scale clinical trials are carried out with these micro-organisms.

### 7.2.2 *Pouchitis*

Pouchitis is a chronic inflammation of the ileal pouch formed after ileal pouch-anal anastomosis, following colectomy in UC patients. A number of studies have investigated the efficacy of probiotics in treating or preventing pouchitis (see Table 7.1). The most convincing evidence of the usefulness of probiotics in treating pouchitis has been obtained from studies conducted by Gionchetti *et al.* (2000), who used VSL#3 to maintain remission in pouchitis. Forty patients were randomised to receive either 6 g VSL#3 per day (equal to $3 \times 10^{12}$ cfu $d^{-1}$) or a maize starch placebo, for 6 months. A significantly greater proportion of subjects receiving the placebo relapsed (100% versus 15%; $P < 0.001$). Patients receiving the probiotic also had significantly increased faecal concentrations of lactobacilli, bifidobacteria and *S. thermophilus* ($P < 0.01$), showing that the probiotic was surviving in the lower gut.

Another study conducted by the same group (Gionchetti *et al.*, 2003) investigated the ability of VSL#3 to prevent the occurrence of pouchitis. Forty patients were randomised to receive either 6 g VSL#3 per day, or a maize starch placebo. The trial continued for 9 months, during which time faecal bacteriology, endoscopic appearance and histology were assessed bi-monthly. Pouchitis occurred in all patients receiving placebo compared to only 15% in the probiotic group. A subsequent investigation evaluated VSL#3 as a treatment for patients with recurrent pouchitis, in whom remission was induced by 4 weeks' treatment with metronidazole and ciprofloxacin (Mimura *et al.*, 2004). Patients were again randomised (20 to VSL#3; 16 to placebo) and the trial was run for one year or until relapse. A significantly higher proportion of patients taking VSL#3 remained in remission (6% versus 85%) for the duration of the study. The IBD quotient (IBDQ), a score used to quantify disease activity, remained high in the probiotic group, but deteriorated significantly in patients receiving placebo.

Other investigations have found that treatment with *Lb. rhamnosus* GG after colectomy also reduced significantly the risk of pouchitis (Gosselink *et al.*, 2004). A further study investigated the effect of Cultura™, a commercially available product containing *Lactobacillus acidophilus* and *Bifidobacterium lactis*, on ileal pouch inflammation and mucosal perfusion in a group of patients with mild to moderate pouchitis (Laake *et al.*, 2003). This product had previously been shown to reduce stool frequency in such individuals (Laake *et al.*, 1999). Ten patients received $5 \times 10^{10}$ cfu $d^{-1}$ of each micro-organism for 4 weeks. Inflammation of the pouch mucosa was assessed before and after treatment by endoscopy and histology, while numbers of lactobacilli and bifidobacteria in patient faeces were assessed using traditional culturing methodologies. Endoscopic observations indicated that there was a reduction in pouchitis disease activity index (PDAI) from a median of 3 (range 0–5) to 1.5 (range 0–5), following probiotic therapy. However, histological assessments in eight of the 10 patients showed a non-significant change, from a median of 1.02 (range 1–1.25) before treatment to 1.05 (range 1–1.22) after treatment. Numbers of lactobacilli increased from a median of $2.8 \times 10^7$ to $7.5 \times 10^7$ cfu $mL^{-1}$ in the stools of eight patients during treatment, although this was not significant. Similarly, numbers of bifidobacteria increased from a median of $2.5 \times 10^7$ to $1.0 \times 10^8$ cfu $mL^{-1}$. Again, this increase was not significant. However, those patients who did have higher numbers of faecal lactobacilli and bifidobacteria also showed the greatest mucosal healing during the study. This suggests both that the increase in bacterial numbers was due to consumption of the probiotic and that the probiotic was reducing mucosal inflammation.

From the foregoing, a number of probiotic preparations seem to be effective in preventing the onset of pouchitis in some individuals. The available data suggest that VSL#3 is able not only to prevent the onset of pouchitis, but also to maintain remission in those patients already suffering from this condition. The studies using VSL#3 involved higher numbers of patients than most other probiotic trials in IBD. Taken together, the results of these investigations indicate that probiotics potentially have an important role in the treatment of inflammatory bowel conditions.

### 7.2.3 *Irritable bowel syndrome*

Irritable bowel syndrome (IBS) is used to describe a heterogeneous group of GI symptoms including diarrhoea, constipation, bloating and abdominal pain. The precise cause of IBS is unknown, but altered GI motility, heightened intestinal sensitivity, psychosocial factors and an abnormal intestinal microflora have all been linked with the disease. The role of the intestinal microflora in IBS is now receiving much attention due to the recognition of a subset of IBS that follows GI infection, known as PI-IBS (McKendrick & Read, 1994), and the fact that disturbance of the GI microflora by oral antibiotics can lead to development of IBS in some individuals (Alun-Jones *et al.*, 1984; Mendall & Kumar, 1998). Lower numbers of enterobacteria, bifidobacteria and lactobacilli have been reported in IBS patients (Balsari *et al.*, 1982), while the dominant anaerobic bacteria in the large bowel of these individuals are shifted from bacteroides and bifidobacteria to clostridia (Bradley *et al.*, 1987). Cummings & Macfarlane (1991) further suggested that this shift in the intestinal microflora causes abnormal colonic fermentation, and that it is this which causes the symptoms seen in IBS. IBS patients have higher levels of intestinal gas than healthy people (Koide *et al.*,

2000), and an exclusion diet led to decreased gas production in these individuals, which correlated with a reduction in symptoms (King *et al.*, 1998). However, given the diversity of symptoms associated with IBS it is unlikely that a single mechanism is responsible for all of them. Because of the mounting evidence that the intestinal microflora is involved in the pathogenesis of at least one subset of IBS cases, a number of probiotic trials have taken place with varying degrees of success (Table 7.2).

*Lactobacillus plantarum* strain 299V has been used in a number of IBS studies. One investigation examined the effect of ingesting $6.25 \times 10^9$ cfu *Lb. plantarum* 299V daily over 4 weeks. Clinical symptoms and colonic fermentation products were monitored during this time. No significant reduction in total hydrogen production, or any clinical improvement, was observed over the course of the trial (Sen *et al.*, 2002). A rose-hip drink was used as a novel way of delivering either $2 \times 10^{10}$ cfu $d^{-1}$ *Lb. plantarum* DSM 9843 or a placebo of identical taste and smell to a cohort of 60 IBS patients (Nobaeck *et al.*, 2000). Treatment was continued for 4 weeks, after which a high proportion (>40%) of patients in the study group reported a >50% reduction in flatulence, compared with only 18% in the placebo group. However, the assessments of flatus in this study were subjective, while there was no significant difference between the groups in terms of stomach bloating. The effect of the form of administration of *Lb. plantarum* 299V was investigated by Niedzielin *et al.* (1998). The probiotic was given either on its own, in combination with trimebutin or merbeverine, or finally in a pasteurised solution. Results suggested that administration of live probiotic was more effective than either the inactivated organisms or the formulations containing either of the drugs (Niedzielin *et al.*, 1998).

*Enterococcus faecium* PR88 was used to treat 28 patients with severe diarrhoea for 12 weeks in a non-placebo-controlled study. Each patient received $10^{10}$ cfu enterococci per day. The probiotic was recovered from stools throughout the study at a concentration of about $10^8$ cfu $g^{-1}$, and a concomitant decrease in faecal excretion of *Enterococcus faecalis* was detected. However, two weeks after the probiotic feeding was stopped *E. faecium* disappeared from the faeces of all test subjects, demonstrating that it was unable to establish in

**Table 7.2** Summary of probiotic trials in irritable bowel syndrome (IBS).

| Study description | Duration | Results | Reference |
|---|---|---|---|
| *Lb. plantarum* 299V or placebo; 12 patients with untreated IBS; $6.25 \times 10^9$ cfu $d^{-1}$ | 4 weeks | No SD[1] in clinical symptoms or hydrogen production | Sen *et al.* (2002) |
| *Lb. plantarum* DSM 9843 or placebo; 60 patients; $2 \times 10^{10}$ cfu $d^{-1}$ in rose-hip drink | 4 weeks | 40% vs. 18% in probiotic and placebo group reported >50% reduction in flatus | Nobaeck *et al.* (2000) |
| *E. faecium* PR88; 28 patients; $10^{10}$ cfu $d^{-1}$ | 12 weeks | IBS symptoms reduced in 67.9% of patients | Hunter *et al.* (1996) |
| *Lb. rhamnosus* GG; 19 patients; $10^{10}$ cfu $d^{-1}$ | 8 weeks | No SD in symptoms; diarrhoea improved in some patients | O'Sullivan & O'Morain (2000) |
| VSL#3; 25 patients; $9 \times 10^{11}$ cfu $d^{-1}$ | 8 weeks | Reduced abdominal bloating; no SD in other symptoms | Kim *et al.* (2003) |

[1] SD = significant difference.

the colonic ecosystem. IBS symptoms were reduced in 19 patients (68%), and there was also a significant decrease in faecal weight, although no detectable alterations in the faecal microflora occurred (Hunter *et al.*, 1996).

*Lb. rhamnosus* GG was used in a double-blind, randomised, crossover clinical trial aimed at reducing the severity of symptoms in IBS (O'Sullivan & O'Morain, 2000). All patients received a single-blind 2-week placebo run-in followed by 8 weeks of receiving either placebo or probiotic. This was followed by a 2-week washout period and a further 8 weeks of either placebo or probiotic. The probiotic was given at a concentration of $10^{10}$ cfu $d^{-1}$, in an enterocoated capsule which aimed at allowing the micro-organism to better survive passage through the upper gut. Nineteen patients completed the trial. No significant differences in either abdominal pain, faecal urgency or bloating were recorded. However, the authors noted that a subgroup of patients in whom diarrhoea had been the predominant IBS symptom did show some improvement with the probiotic.

Given the success achieved in treating pouchitis and CD with VSL#3, its use in studies on IBS is unsurprising. Kim *et al.* (2003) fed $4.5 \times 10^{11}$ cfu $d^{-1}$, or placebo, twice daily for 8 weeks to 25 patients with IBS. GI transit times, bowel function scores and severity of IBS symptoms were monitored. The results, however, were inconclusive, in that no significant differences in GI transit times, bowel function scores or global symptoms of IBS were observed. However, when each symptom was individually analysed, the authors found that abdominal bloating was significantly reduced in the probiotic group, although other symptoms were not. Thus VSL#3 seems to be useful only in reducing abdominal bloating – an effect which is unrelated to any effect on GI transit time or bowel function scores. The lack of effectiveness stands in marked contrast to the results obtained with VSL#3 in IBD patients (Gionchetti *et al.*, 2000, 2003; Mimura *et al.*, 2004). Understanding the reasons for this apparent discrepancy would advance our understanding of the mechanisms underlying probiotic action.

In summary, evidence concerning the effectiveness of probiotics as therapeutic agents in IBS is not strong. Only a few trials involving relatively low numbers of participants have been undertaken, while much of the data generated in these studies have been inconclusive. As with CD, larger and more controlled clinical studies need to be undertaken to determine conclusively whether probiotics are likely to be useful in treating IBS.

### 7.2.4 *Antibiotic-associated diarrhoea and* Clostridium difficile

It has long been recognised that oral antibiotics can disturb the GI microflora, allowing pathogenic micro-organisms to colonise the gut. This occurs because of the breakdown of colonisation resistance, a process in which the presence of the normal microflora serves as a barrier to invasion by allochthonous species. The precise mechanisms are unclear but are thought to include low pH, secretion of antimicrobial compounds and physical occupation of adhesion sites by the normal microbiota. In many patients, one result of this process is the condition of antibiotic-associated diarrhoea (AAD). The incidence of AAD varies markedly from institution to institution, but can occur in up to 40% of those receiving antibiotics (McFarland, 1998), incurring extra costs in treatment and extended hospital stays. A proportion of these patients may become colonised by *C. difficile* and go on to develop more serious conditions such as pseudomembranous colitis. The fact that dysfunction of the normal microbiota is central to the pathogenesis of AAD suggests that probiotic therapy might be an appropriate prophylaxis. In this respect, a number of trials have taken place in both AAD and *C. difficile*-associated disease (CDAD) (Table 7.3).

**Table 7.3** Summary of probiotic trials in antibiotic-associated diarrhoea (AAD) and *C. difficile*-associated diarrhoea (CDAD)

| Study description | Duration | Results | Reference |
|---|---|---|---|
| *Lb. rhamnosus* GG or placebo; 267 patients; $2 \times 10^{10}$ cfu $d^{-1}$ | 2 weeks | Diarrhoea in 29.9% and 29.3% of placebo and probiotic groups ($P = 0.93$) | Thomas *et al.* (2001) |
| *Lb. rhamnosus* GG or placebo; 16 patients on erythromycin; 125 mL twice daily | 1 week | Significantly reduced diarrhoea, bloating, flatus and erythromycin side effects | Siitonen *et al.* (1990) |
| *Lb. rhamnosus* GG or placebo; 119 patients with RTI[2]; $4 \times 10^{10}$ cfu $d^{-1}$ | 3 months | Diarrhoea in 5% and 16% of the probiotic and placebo groups | Arvola *et al.* (1999) |
| *Lb. rhamnosus* GG or placebo; 174 patients; $2 \times 10^{10}$ cfu $d^{-1}$ | Time[1] | Diarrhoea in 8% and 26% of the probiotic and placebo groups; duration reduced ($P = 0.05$) | Vanderhoof *et al.* (1999) |
| *Lb. rhamnosus* GG with or without antibiotics; 120 patients with *H. pylori* | 2 weeks | Significant reduction in diarrhoea | Armuzzi *et al.* (2001) |
| Lactinex™ or placebo; 79 patients on ampicillin | 5 days | Diarrhoea in 8.3% and 21% of the probiotic and placebo groups ($P = 0.21$) | Gotz *et al.* (1979) |
| Lactinex™ or placebo; paediatric patients receiving ampicillin | 10 days | No SD[3] between probiotic and placebo | Tankanow *et al.* (1990) |
| *L. sporogenes* and FOS[4] or placebo; 120 paediatric patients; | 10 days | Diarrhoea in 38% and 71% of the probiotic and placebo groups ($P = 0.002$) | La Rosa *et al.* (2003) |
| *C. butyricum* MIYAIRI or antibiotics; 110 paediatric patients; $4 \times 10^{7}$ cfu $d^{-1}$ | 6 days | Diarrhoea in 5% and 59% of the probiotic and placebo groups | Seki *et al.* (2003) |
| *E. faecium* SF68 (Bioflorin™) or placebo; 123 patients | Time[1] | Diarrhoea in 8% and 27.2% of the probiotic and placebo groups | Wunderlich *et al.* (1989) |
| *B. Longum*; 10 patients on erythromycin | 2 × 3 days; 3 weeks gap | | Colombel *et al.* (1987) |
| *S. boulardii* or placebo; 198 patients; 200 mg $d^{-1}$ | Unknown | Diarrhoea in 4% and 17% of the probiotic and placebo groups ($P < 0.01$) | Adam *et al.* (1977) |
| *S. boulardii* or placebo; 180 patients; 200 mg $d^{-1}$ | 23 months | Diarrhoea in 9.5% and 22% of the probiotic and placebo groups ($P < 0.038$) | Surawicz *et al.* (1989) |
| *S. boulardii* or placebo, 193 patients; 1000 mg $d^{-1}$ | Antibiotic therapy +3 days | Diarrhoea in 7.2% and 14.6% of the probiotic and placebo groups ($P < 0.02$) | McFarland *et al.* (1995) |

(*Continued.*)

**Table 7.3** *(Continued.)*

| Study description | Duration | Results | Reference |
|---|---|---|---|
| *S. boulardii* or placebo; 128 patients on EN[5]; 2000 mg $d^{-1}$ | 21 days or cessation of EN | Mean percentage of days with diarrhoea per day EN reduced from 18.9 to 14.2% by *S. boulardii* ($P = 0.069$) | Bleichner *et al.* (1997) |
| *S. boulardii* or placebo; 69 elderly patients; 226 g $d^{-1}$ | Duration of antibiotic therapy | No SD in incidence or severity of diarrhoea | Lewis *et al.* (1998) |
| *Lb. rhamnosus* GG; 11 patients with recurring colitis | 7–10 days | 72% successful treatment | Gorbach *et al.* (1987) |
| *Lb. rhamnosus* GG; 9 patients with recurring colitis | Unknown | 66.6% successful treatment | Bennet *et al.* (1990) |
| *Lb. rhamnosus* GG; 4 paediatric patients with recurring colitis | Unknown | Colitis resolved in 50% of patients | Biller *et al.* (1995) |
| *Lb. rhamnosus* GG or placebo; 16 patients | Unknown | 30% and 0% of placebo and probiotic groups, respectively, developed recurrent *C. difficile* infection | Pochapin *et al.* (1998) |
| *Lb. acidophilus* and *B. bifidum* with antibiotics or antibiotics alone; 138 patients; $4 \times 10^{10}$ cfu $d^{-1}$ | 20 days | 13% and 16% of placebo and probiotic groups, respectively, were *C. difficile*-positive, of those with diarrhoea, 7.3% and 2.9% were toxin-positive | Plummer *et al.* (2004) |
| *Lb. plantarum* 299V or placebo; 20 patients with recurrent *C. difficile* disease | Unknown | 66.7% and 36.4% of placebo and probiotic groups, respectively, developed recurrent *C. difficile* infection | Wullt *et al.* (2003) |

[1] Not stated, but assumed to be for duration of antibiotic therapy.
[2] RTI = Respiratory tract infection.
[3] SD = Significant difference.
[4] FOS = Fructo-oligosaccharides.
[5] EN = Enteral nutrition.

## 7.2.5 *Antibiotic-associated diarrhoea*

A variety of probiotic micro-organisms, mainly lactobacilli, have been used in trials relating to AAD. A number have involved *Lb. rhamnosus* GG. In one large study, 267 hospitalised patients taking antibiotics were randomised to receive either $2 \times 10^{10}$ cfu $d^{-1}$ ($n = 133$) or placebo ($n = 134$) for 14 days (Thomas *et al.*, 2001). Patients were followed up for 21 days, during which time they recorded stool frequency and consistency. Diarrhoea developed in 40 of 134 (29.9%) individuals taking the placebo and in 39 of 133 (29.3%) of those receiving the probiotic ($P = 0.93$). The probiotic was therefore not effective in reducing the incidence of diarrhoea in these individuals. These results contrast with those reported by Siitonen *et al.* (1990), who investigated the effectiveness of *Lb. rhamnosus* GG in preventing diarrhoea in adult patients receiving erythromycin. Sixteen individuals were randomised to receive

125 mL twice daily of either *Lb. rhamnosus* GG-containing yoghurt or pasteurised non-probiotic-containing yoghurt as a placebo. Patients taking the probiotic had significantly less diarrhoea than the placebos, while other side effects of erythromycin, such as bloating and flatulence, were also reduced in the probiotic group. A microbiological analysis of the faeces demonstrated the presence of *Lb. rhamnosus* GG in patients who received the probiotic. In another investigation, Arvola *et al.* (1999) reported that *Lb. rhamnosus* GG reduced significantly the incidence of AAD in children suffering from respiratory infections. A total of 119 patients was randomised to receive either $2 \times 10^{10}$ cfu of probiotic in capsules, twice daily, or placebo capsules that were indistinguishable from the probiotic. When cases of diarrhoea occurred, stool bacteriology was assessed for the presence of various common pathogens including *C. difficile*. The incidence of diarrhoea was 5% in the probiotic group vs. 16% in the placebo group, but no differences were observed between the two groups with respect to the severity of diarrhoea. *Clostridium difficile* was detected in two cases of diarrhoea. Similar results were obtained by Vanderhoof *et al.* (1999), who entered 188 children in a double-blind, randomised study in which *Lb. rhamnosus* GG was given at a dose of *ca.* $2 \times 10^{10}$ cfu $d^{-1}$. Twenty-five children taking the placebo and seven receiving the probiotic had diarrhoea, a significant reduction (Vanderhoof *et al.*, 1999). In another controlled trial, 120 patients with asymptomatic *Helicobacter pylori* colonisation received *Lb. rhamnosus* GG in conjunction with traditional anti-*Helicobacter* therapies (Armuzzi *et al.*, 2001). Results showed a significant reduction in diarrhoea together with decreases in side effects of treatment such as bloating and disturbance of taste.

Lactinex™ is a commercial probiotic preparation containing *Lb. acidophilus* and *Lactobacillus delbrueckii* subsp. *bulgaricus*. In a double-blind, placebo-controlled trial, 98 adult hospital inpatients were randomised to receive either Lactinex™ or placebo four times per day for the first 5 days of ampicillin therapy. Seventy-nine patients completed the study. Diarrhoea occurred in nine placebo subjects (21%) and in three (8.3%) individuals receiving the probiotic, although the difference was not statistically significant. However, when patients with diarrhoea unrelated to ampicillin were excluded, the incidence of diarrhoea in the probiotic group (0%) was found to be significantly lower than that in the placebo group (14%). Another study found similar results in paediatric patients receiving amoxicillin (Tankanow *et al.*, 1990). The synbiotic approach has also been used in a trial aimed at preventing the onset of AAD, in which *Lactobacillus sporogenes* was used in conjunction with fructo-oligosaccharides (La Rosa *et al.*, 2003). The study was double-blind in design and involved 120 paediatric patients in multiple centres. Among the 98 patients who completed the study, a significant reduction in both incidence and length of diarrhoeal episodes was observed in those receiving probiotic.

A group in Japan have used a novel probiotic micro-organism to treat AAD in children (Seki *et al.*, 2003). *Clostridium butyricum* is a butyrate-producing, Gram-positive strict anaerobe which is already used in anti-diarrhoeal medicines in Japan. A cohort of 110 paediatric patients was randomised into three groups: one group ($n = 25$) received only antibiotics; a second group ($n = 38$) was given antibiotics and *C. butyricum*, from the midpoint of the antibiotic course; and a third group ($n = 45$) was given antibiotics and *C. butyricum* concomitantly for the whole treatment period. The two groups given the probiotic had significantly lower incidences of diarrhoea, while children who received concurrent antibiotic and probiotic treatment also had higher faecal anaerobe counts, particularly *Bifidobacterium* spp. Thus, the protective effect in this study may not have been due directly to the probiotic. When *E. faecium* CF68 (Bioflorin™) was used in a trial aimed at reducing the incidence of

AAD (Wunderlich *et al.*, 1989) there was a small protective effect, reducing the incidence from 27% in the placebo group to 9% in the probiotic group.

Bifidobacteria have been used as probiotics in a small number of AAD studies, which is somewhat surprising in view of their ubiquity in other clinical trials. However, two studies have reported positive effects in the use of: (a) *B. longum* in conjunction with *Lb. acidophilus* in 10 patients receiving clindamycin (Orrhage *et al.*, 1994); and (b) *B. longum* alone in 10 patients undergoing erythromycin treatment (Colombel *et al.*, 1987). Despite this, due to the small sample sizes in these trials, the significance of the apparent probiotic effects is unclear.

*S. boulardii* has been used in a number of investigations on AAD. One study examined the incidence of AAD in 198 patients, randomised to receive either 200 mg of the yeast or placebo per day (Adam *et al.*, 1977). The results suggested that there was a probiotic effect, in that 4% of patients taking the yeast developed diarrhoea compared to 17% of the placebos. A similar result was reported by Surawicz and colleagues when they assessed the efficacy of *S. boulardii* in treating AAD (Surawicz *et al.*, 1989), while other studies showed that 7% of patients taking β-lactam antibiotics and given 500 mg *S. boulardii* twice daily experienced diarrhoeal episodes as compared to 15% of the placebo group ($P = 0.02$) (McFarland *et al.*, 1995). This yeast has also been shown to reduce significantly the incidence of diarrhoea in critically ill tube-fed patients (Bleichner *et al.*, 1997) in a multicentre, double-blind randomised and placebo-controlled trial. Here, 128 patients were randomised to receive either 500 mg *S. boulardii* four times daily, or placebo. Patients receiving the probiotic had a mean of 14.2% days with diarrhoea, compared to 18.9% in the control group. Conversely, no beneficial effects were associated with *S. boulardii* when it was used in the treatment of 69 elderly hospitalised patients with AAD (Lewis *et al.*, 1998). Thus, with few exceptions, the majority of studies with *S. boulardii* in AAD have shown positive effects.

The overall picture of probiotic usefulness in AAD is mixed, as is the case with most other conditions in which probiotics have been used. Clinical data suggest that *S. boulardii* is the most effective probiotic, with the caveat that this may not be the case in elderly people. Other investigations with organisms such as *E. faecium* and bifidobacteria have indicated that there may be some benefits associated with their use, but in most cases the results were obtained from pilot trials; consequently, further research is needed to determine their efficacy in AAD.

### **7.2.6** Clostridium difficile*-associated disease (CDAD)*

*C. difficile* is an opportunistic pathogen which produces two toxins, enterotoxin A and cytotoxin B, and often causes diarrhoea in vulnerable patients. A more serious manifestation of *C. difficile* disease is pseudomembranous colitis, which occurs in some individuals. Colonisation with this micro-organism usually follows a course of antibiotics. Antibiotic treatment of CDAD, typically with metronidazole and vancomycin, is usually effective but the condition often recurs. For this reason – and also because degeneration of the normal gut microbiota is implicated in pathogenesis – probiotics have been used in a number of trials aimed at preventing the occurrence or recurrence of CDAD (see Table 7.3).

*Lb. rhamnosus* GG has been employed in a number of investigations in CDAD. Three separate studies have demonstrated positive effects: Gorbach *et al.* (1987) noted successful treatment of 72% (eight of 11) of patients with recurring *C. difficile* colitis. Similarly, five of nine (67%) of people receiving *Lb. rhamnosus* GG were cured of recurrent colitis

(Bennet *et al.*, 1990). Another small trial in the United States also found that treatment of recurring *C. difficile* colitis with *Lb. rhamnosus* GG was partially effective in that the condition was resolved in 50% (two of four) paediatric patients (Biller *et al.*, 1995). The preliminary results of a randomised, placebo-controlled study published 4 years ago also showed promise (Pochapin, 2000). The author reported that administration of *Lb. rhamnosus* GG reduced the 3-week recurrence rate, and effected an improvement in symptoms such as abdominal cramping and diarrhoea. However, to our knowledge the full results of this trial have yet to be published. Nonetheless, an earlier trial by the same author suggested a benefit for *Lb. rhamnosus* GG in the treatment of CDAD because 30% (three of 10) of the placebo group compared to 0% (zero of six) of those receiving probiotic developed recurrent CDAD (Pochapin *et al.*, 1998).

A double-blind, placebo-controlled study with *Lb. acidophilus* and *B. bifidum* was conducted by Plummer *et al.* (2004). Patients ($n = 69$) received either one capsule per day (containing $2 \times 10^{10}$ cfu $g^{-1}$) together with antibiotics, or antibiotics alone. Post-therapy, nine patients receiving the placebo and 11 taking the probiotic were *C. difficile*-positive, of which seven and five respectively, were toxin-positive. Furthermore, six patients receiving the placebo against only two taking the probiotic suffered diarrhoeal symptoms during the course of the study. The authors observed that consumption of the probiotic was not affecting *C. difficile* colonisation, but instead may have been restricting the organism's ability to cause disease by keeping patients in an asymptomatic carrier state, perhaps because of neutralisation of the toxins (Linevsky & Kelly, 1997) through increased IgA production (Gorbach, 2000). While these results appear promising, the number of patients involved was insufficient to reach statistical significance.

Another pilot double-blind, placebo-controlled study using *Lb. plantarum* 299V, in combination with antibiotics, found that four of 11 patients receiving probiotic (36%) and six of nine on placebo (67%) experienced a recurrence of clinical symptoms (Wullt *et al.*, 2003). Again, while these data appear to show promise, a larger-scale trial is needed to ascertain whether this bacterium has significant probiotic effects in CDAD.

### 7.2.7 *Traveller's diarrhoea*

Diarrhoea which mainly affects tourists travelling to some Third World countries is known as traveller's diarrhoea (TD). The incidence varies from 20 to 50%, depending on both the country of origin and the destination (Gismondo *et al.*, 1999). While not intrinsically life-threatening, the condition can cause considerable inconvenience and disruption to holidays. Various aetiological agents have been identified, but the most common is enterotoxigenic *E. coli* (ETEC), which accounts for 40–70% of cases (Merson *et al.*, 1976). Prophylactic antibiotic therapy has proven effective in preventing occurrence of the disease (DuPont *et al.*, 1986); however, there is increasing reluctance to use antibiotics in this way because of the spread of resistance. Probiotics are therefore seen as an alternative method of prophylaxis and a number of trials have been carried out in this respect (Table 7.4).

A number of clinical trials linked to TD using a variety of putative probiotics have yielded highly variable results. Lactinex™, a mixture of *Lb. acidophilus* and *Lb. delbrueckii* subsp. *bulgaricus* ($2 \times 10^8$ cfu $d^{-1}$) or placebo was given to 48 volunteers who were then challenged orally with $10^8$–$10^{10}$ cfu ETEC $g^{-1}$ (Clements *et al.*, 1981). No significant differences in volume, consistency or frequency of stools were noted, suggesting that Lactinex™ would not be effective in preventing ETEC-associated TD. This may have been due to the intake

**Table 7.4** Summary of probiotic trials in traveller's diarrhoea.

| Study description | Duration | Results | Reference |
|---|---|---|---|
| Lactinex™ or placebo, 48 individuals challenged with $10^8$–$10^{10}$ ETEC[1] | N/A | No SD[2] in stool volume, consistency or frequency | Clements *et al.* (1981) |
| *Lb. acidophilus*, *B. bifidum*, *Lb. delbrueckii* subsp. *bulgaricus*, *S. thermophilus* or placebo, 98 tourists; $10^{10}$ cfu $d^{-1}$ | Duration of stay | Diarrhoea in 34.6% and 59.1% of probiotic and placebo groups respectively ($P = 0.02$) | Black *et al.* (1989) |
| *Lb. rhamnosus* GG or placebo, 756 tourists to Turkey; $2 \times 10^9$ cfu $d^{-1}$ | Duration of stay | 41.0% and 46.5% of probiotic and placebo groups respectively developed diarrhoea ($P = 0.065$) | Oksanen *et al.* (1990) |
| *Lb. acidophilus* or placebo, 202 tourists; $2 \times 10^8$–$2 \times 10^9$ cfu $d^{-1}$ | Duration of stay | 25.7% and 23.8% of probiotic and placebo groups respectively developed diarrhoea | Katelaris *et al.* (1995) |
| *Lb. acidophilus* or placebo, 1231 tourists; $2 \times 10^8$–$2 \times 10^9$ cfu $d^{-1}$ | Duration of stay | 53.2% and 47.0% of probiotic and placebo group respectively developed diarrhoea | Kollaritsch *et al.* (1989) |
| *S. boulardii* or placebo, 3000 tourists; 250 mg or 100 mg $d^{-1}$ | Duration of stay | 39.1%, 31.4% and 28.7% in the placebo, 250-mg probiotic, and 1000-mg probiotic groups respectively ($P < 0.05$) | Kollaritsch *et al.* (1993) |

[1] ETEC = enterotoxigenic *E. coli*.
[2] SD = significant difference.

of a relatively low number of probiotic micro-organisms and very high numbers of the pathogen. The results of another prospective, double-blind study support this hypothesis. Travellers to Egypt ($n = 98$) received either a placebo or $10^{10}$ cfu $d^{-1}$ of a mixture comprising *Lb. acidophilus*, *B. bifidum*, *Lb. delbrueckii* subsp. *bulgaricus* and *S. thermophilus* prior to departure and during their stay (Black *et al.*, 1989). Diarrhoea occurred in 17 individuals who received the probiotic compared to 29 of the placebo group. Similarly, in a larger scale double-blind study, travellers to Turkey ($n = 756$) received either $2 \times 10^9$ cfu $d^{-1}$ *Lb. rhamnosus* GG or placebo (Oksanen *et al.*, 1990). The results showed no significant differences in the frequency of diarrhoea in the probiotic (41%) and placebo groups (47%). Again, fewer than $10^{10}$ cfu of probiotic per day was given to the volunteers, and no statistically significant effects were observed.

These observations are supported by results from two trials using *Lb. acidophilus*. Capsules containing between $2 \times 10^8$ and $2 \times 10^9$ cfu or a placebo were given to 202 individuals (Katelaris *et al.*, 1995). No significant differences were found in the proportions of individuals in each group who had diarrhoea (probiotic 26%, placebo 24%). The same probiotic was used in another study involving 1231 volunteers (Kollaritsch *et al.*, 1989) and a similar result was obtained in that there were no inter-group differences in the relative proportions of people who contracted TD (probiotic 53%, placebo 47%). These results should not be taken to mean that probiotics in general cannot be used as prophylactics in TD, because the bifidobacterial and lactic acid bacteria strains used in the above studies may have been inherently ineffectual as probiotics. Also, insufficient numbers of the probiotic may have been used. Moreover, many cases of TD do not have a microbial aetiology and are caused by other factors such as stress due to long-distance travelling or changes in the normal diet and lifestyle.

*S. boulardii* has been used prophylactically in TD. Either 250 mg or 1000 mg of yeast or an equivalent placebo was given to 3000 Austrian tourists travelling to various destinations (Kollaritsch *et al.*, 1993). The incidence of TD was 39% in the placebo group, 31% in those receiving 250 mg of yeast, and 29% in the group receiving 1000 mg *S. boulardii* ($P < 0.05$ between the 1000-mg and placebo groups). Furthermore, a regional bias in probiotic efficiency was noted, where *S. boulardii* was most effective in preventing TD in tourists visiting North Africa and Turkey, although why this occurred is unknown.

### 7.2.8 *Infant diarrhoea*

Diarrhoeal disease in infants, particularly those in the developing world, is a serious problem. Indeed, it has been estimated that a child dies from such diseases every 15 seconds (Reid *et al.*, 2003a). In the developed world, infantile diarrhoea is most commonly caused by rotaviruses, which are responsible for up to 72% of infant gastroenteritis (Rivest *et al.*, 2004). A number of randomised, placebo-controlled, double-blind clinical trials have been carried out involving large numbers of children (Table 7.5). These have produced evidence for lactobacillus probiosis preventing acute diarrhoea. The mechanism for this is unclear; however, it is thought that immune enhancement (Kaila *et al.*, 1992), competitive blocking of rotavirus binding sites (Bernet *et al.*, 1994), direct killing of virus by probiotic microorganisms (Cadieux *et al.*, 2002) and lactobacilli-induced suppression of gut secretory responses (Reid *et al.*, 2003c) could be involved.

The best evidence for the usefulness of probiotics in the prevention of infantile diarrhoea comes from studies with *Lb. rhamnosus* GG. In one placebo-controlled, randomised and double-blind trial, at least $10^{10}$ cfu per day of probiotic or placebo was given to 291 infants, aged between 1 month and 3 years (Guandalini *et al.*, 2000). All participants in the study also received standard oral rehydration therapy. The probiotic was found to reduce the duration of diarrhoea to 58.3 ± 27.6 h, compared to 71.9 ± 27.6 h in the placebo group ($P = 0.03$). Furthermore, *Lb. rhamnosus* GG reduced the incidence of long-term diarrhoea (those episodes lasting more than 7 days) from 10.7% in the placebo group to 2.7% in children receiving the probiotic ($P = 0.01$). These data are supported by results obtained in another randomised study using 100 infants (Guarino *et al.*, 1997), where the mean duration of diarrhoeal symptoms was reduced from 6 days in the placebo group to 3 days in those receiving the probiotic. All of the patients had also received oral rehydration therapy. A similar effect was observed in a double-blind trial when Szajewska *et al.* (2001) investigated the use of *Lb. rhamnosus* GG in children aged between 1 month and 3 years who had been hospitalised for reasons other than diarrhoea. Eighty-one children were randomised to receive either $6 \times 10^9$ cfu *Lb. rhamnosus* GG twice daily ($n = 45$) or placebo ($n = 36$), until they were discharged from hospital. The results indicated that consumption of *Lb. rhamnosus* GG reduced the risk of infants developing diarrhoea to 6.7% in those receiving the probiotic compared to 33.3% in the placebos. In another trial, 71 infants aged between 4 and 45 months suffering from diarrhoea were randomised to receive either $10^{10}$ cfu *Lb. rhamnosus* GG in a fermented milk product (125 mL), $10^{10}$ cfu of probiotic in lyophilised powder form or a pasteurised yoghurt placebo twice daily for 5 days (Isolauri *et al.*, 1991). The mean duration of diarrhoea was

**Table 7.5** Summary of probiotic trials in infant diarrhoea.

| Study description | Duration | Results | Reference |
|---|---|---|---|
| *Lb. rhamnosus* GG or placebo; 291 hospitalised infants 1 month – 3 years with acute diarrhoea; $>10^{10}$ cfu $d^{-1}$ | Until diarrhoea stopped | Duration of diarrhoea 58.3 ± 28.7 h and 71.9 ± 27.6 h in probiotic and placebo groups ($P = 0.03$) | Guandalini *et al.* (2000) |
| *Lb. rhamnosus* GG or placebo; 100 hospitalised infants with acute diarrhoea; $3 \times 10^{9}$ cfu $d^{-1}$ | Until diarrhoea stopped (max. 5 days) | Duration of diarrhoea 6 and 3 days in probiotic and placebo group | Guarino *et al.* (1997) |
| *Lb. rhamnosus* GG or placebo; 81 hospitalised infants 1 month – 3 years; $6 \times 10^{9}$ cfu $d^{-1}$ | Until discharged | 6.7% and 33.3% of probiotic and placebo groups developed diarrhoea | Szajewska *et al.* (2001) |
| *Lb. rhamnosus* GG or placebo; 71 infants 4–45 months; $10^{10}$ cfu $d^{-1}$ lyophilised or yoghurt | 5 days | Duration of diarrhoea 1.4 ± 0.8 and 2.4 ± 1.1 days in probiotic and placebo groups, respectively ($P < 0.001$) | Isolauri *et al.* (1991) |
| *Lb. rhamnosus* GG, *L. casei* or *S. thermophilus* and *Lb. delbrueckii* subsp. *bulgaricus*; 49 infants 6–35 months with acute diarrhoea; $10^{8}$–$10^{9}$ cfu $d^{-1}$ | 5 days | Duration of diarrhoea 1.8 ± 0.8 d, 2.8 ± 1.2 d and 2.6 ± 1.4 d in the *Lb. rhamnosus* GG, *Lb. casei* and *S. thermophilus* groups, respectively ($P = 0.04$) | Majamaa *et al.* (1995) |
| *Lb. rhamnosus* and *Lb. reuteri* or placebo; 69 hospitalised infants with acute diarrhoea; $10^{10}$ cfu $d^{-1}$ | 5 days | Incidence of diarrhoea 10% and 33% in probiotic and placebo groups respectively ($P = 0.03$). No significant difference in duration of diarrhoea ($P = 0.07$) | Rosenfeldt *et al.* (2002) |
| *Lb. rhamnosus* and *Lb. reuteri* or placebo; 30 non-hospitalised infants with acute diarrhoea; $10^{10}$ cfu $d^{-1}$ | 5 days | Significant reduction in duration of diarrhoea from 116 h in the placebo group to 76 h in the probiotic group ($P = 0.05$) | Rosenfeldt *et al.* (2002) |
| *B. bifidum* and *S. thermophilus* or placebo; 55 infants 5–24 months; ca. $10^{8}$ cfu $d^{-1}$ | 17 months | Incidence of diarrhoea 7% and 31% in probiotic and placebo groups respectively ($P = 0.035$) | Saavedra *et al.* (1994) |
| *Lb. reuteri* or placebo; 40 infants 6–36 months 75% rotavirus positive; $10^{10}$ cfu $d^{-1}$ | 5 days | Duration of diarrhoea 1.7 ± 1.6 and 2.9 ± 2.3 days in probiotic and placebo groups respectively ($P = 0.07$) | Shornikova *et al.* (1997) |
| *S. boulardii*, *Lb. casei* and *Lb. acidophilus* or placebo; 89 infants 6–24 months; $10^{12}$–$10^{14}$ cfu $d^{-1}$ | 5 days | Both formulations caused significant decreases in stool frequency ($P < 0.001$), duration of diarrhoea ($P < 0.005$) and vomiting ($P < 0.002$) | (Gaon *et al.* (2003) |

significantly shorter in those receiving either form of the probiotic at 1.4 ± 0.8 days vs. 2.4 ± 1.1 days in infants taking the placebo ($P < 0.001$).

A number of probiotic lactobacilli and *S. thermophilus* were used in the treatment of acute rotavirus diarrhoea (Majamaa *et al.*, 1995). Following initial rehydration therapy, 49 infants aged between 6 and 35 months received either *Lb. rhamnosus* GG or a combination of *S. thermophilus* and *Lb. delbrueckii* subsp. *bulgaricus* twice daily for 5 days. *Lb. rhamnosus* GG treatment resulted in a significantly greater reduction in the duration of diarrhoea (1.8 ± 0.8 days) compared to the other lactic acid bacteria (2.6 ± 1.4 days with *S. thermophilus* and *Lb. delbrueckii* subsp. *bulgaricus*). Moreover, *Lb. rhamnosus* GG therapy correlated with an increase in rotavirus-specific serum IgA and rotavirus-specific antibody-secreting cells, but not total immunoglobulin-secreting cells, during convalescence. Thus, there is good evidence from a number of studies that *Lb. rhamnosus* GG is effective in reducing acute diarrhoea in infants, possibly due to an increase in the rotavirus-specific immune response. This enhanced immune function may protect against further rotavirus infection. Rosenfeldt *et al.* (2002a) examined the efficacy of a combination of two probiotic lactobacilli, *Lb. rhamnosus* 19070-2 and *Lactobacillus reuteri* DSM12246, in 69 children with acute diarrhoea in a randomised, placebo-controlled study. Patients received either $10^{10}$ cfu of the probiotic combination, or a placebo twice daily for 5 days. The probiotic combination had a significant effect on the incidence of diarrhoea, reducing it from 33% to 10% ($P = 0.03$), but there was no significant effect on the duration of diarrhoea ($P = 0.07$). However, in patients with a duration of diarrhoea of greater than 60 h before being treated, the post-treatment duration of diarrhoea was reduced from 130 h to 80 h ($P = 0.003$).

Another trial by the same investigators used an identical probiotic combination and regimen, this time in 30 non-hospitalised infants (Rosenfeldt *et al.*, 2002b). The outcome was similar, with the exception that the duration of diarrhoea was reduced significantly in this cohort (76 h and 116 h; $P = 0.05$). Taken together, these results indicate that a lactobacillus–probiotic combination was effective in the treatment of infant diarrhoea, although the number of patients in both studies was relatively small. A larger scale clinical trial would provide more definitive evidence for the efficacy of this promising probiotic combination.

Studies with other probiotic bacteria in infant diarrhoea have also shown promise. When *B. bifidum* and *S. thermophilus* were given to 55 infants aged between 5 and 24 months in a double-blind, placebo-controlled trial (Saavedra *et al.*, 1994), the probiotics reduced significantly the incidence of diarrhoea (placebos 31%, probiotics 7%). Additionally, 3% of infants in the probiotic group, compared to 39% of the placebos shed rotavirus during the study. In another investigation, 40 children aged between 6 and 36 months, 75% of whom were rotavirus-positive, were randomised to receive $10^{10}$ cfu *Lb. reuteri* or placebo twice daily for 5 days (Shornikova *et al.*, 1997). The mean duration of diarrhoea was 1.7 ± 1.6 days in the probiotic group and 2.9 ± 2.3 days in those receiving the placebo ($P = 0.07$). However, on the second day of treatment only 26% of the probiotic group compared to 81% of those on placebo had watery diarrhoea. *S. boulardii*, a combination of *Lactobacillus casei* and *Lb. acidophilus* CERELA ($10^{10}$–$10^{12}$ cfu twice daily) or placebo were given to 89 children in a randomised, double-blind, placebo-controlled trial (Gaon *et al.*, 2003). Both probiotic formulations caused a significant reduction in stool frequency ($P < 0.001$), duration of diarrhoea ($P < 0.005$) and reduced vomiting ($P < 0.002$) compared to the placebos. However, not all probiotic treatments have been successful: an Indonesian study employed a *B. bifidum* strain within a high-lactose formula in a randomised controlled clinical trial using 100 infants

(Sudarmo *et al.*, 2003). After 3 days of receiving either placebo or probiotic, no significant difference was detected between the two groups.

In summary, the picture emerging from studies involving the use of probiotics in acute infant diarrhoea is that lactobacilli are particularly effective, especially *Lb. rhamnosus* GG (Isolauri *et al.*, 1991; Majamaa *et al.*, 1995; Guarino *et al.*, 1997; Guandalini *et al.*, 2000; Szajewska *et al.*, 2001). Although other lactobacilli may also be useful, the evidence for these organisms is less convincing, principally because fewer studies have been conducted and they have used smaller numbers of patients. Indeed, *Lb. rhamnosus* GG was found to be more effective in a direct comparison with other lactobacilli (Majamaa *et al.*, 1995). *S. boulardii* was also found to be effective when given in high numbers (Gaon *et al.*, 2003), although *B. bifidum* was not (Sudarmo *et al.*, 2003). Overall, acute diarrhoea in infants is a condition in which there is strong evidence from a number of clinical trials to support the efficacy of probiotics.

## 7.3 Probiotic use in extra-gastrointestinal conditions

While the majority of probiotic trials have focussed on diseases related to the GI tract, a number of studies have investigated the effects that they have on conditions associated with other parts of the body and their immunostimulatory effects (Pessi *et al.*, 2000; Cross *et al.*, 2001; Cross, 2004; Mastrandrea *et al.*, 2004). Most of these investigations have in the main addressed three conditions: atopic dermatitis, food allergies and bacterial vaginosis.

### 7.3.1 *Atopic dermatitis*

Atopic disease is the most common of the chronic skin conditions generally referred to as eczema. It is defined as an excess of inflammation in the skin and mucous membranes of individuals of any age, although it most commonly affects infants and young adults and tends to improve post-adolescence. The condition often runs in families, although in approximately 20% of cases this is not the case. The skin of sufferers becomes irritated and the resultant scratching leads to redness, swelling, cracking of the skin and eventually crusting and scaling. While not life-threatening, atopic dermatitis is uncomfortable and can be unsightly, causing psychological distress to sufferers. The incidence varies, but up to 20% of children have been said to suffer from the condition.

Much of the evidence for the effectiveness of probiotics in the treatment of allergies comes from the work of Isolauri *et al.* (2000) (see Table 7.6), who showed the efficacy of *Lb. rhamnosus* GG in alleviating the symptoms of atopic dermatitis (see also Majamaa & Isolauri, 1997). Twenty-seven infants received either hydrolysed whey formula with ($n = 13$) or without ($n = 14$) $5 \times 10^8$ cfu probiotic per day. In a second group of 10 breast-fed infants, who also suffered from cows' milk allergy, nursing mothers received $2 \times 10^{10}$ cfu $d^{-1}$ of the same probiotic, or placebo. After 1 month, those individuals receiving probiotic were found to have a significantly lower disease activity index ($P = 0.008$). This also fell significantly after 1 month in the second group of infants ($P = 0.007$). Concentrations of faecal alpha$_1$-antitrypsin and TNF-$\alpha$ fell in the probiotic group but not in those infants whose mothers received the placebo. Thus, *Lb. rhamnosus* GG appears able to reduce atopic disease, not only in infants receiving the probiotic directly, but also where the mother was given the bacterium. Some studies have suggested that an increase in IL-10 in the serum of infants

**Table 7.6** Summary of probiotic trials in atopic dermatitis and bacterial vaginosis.

| Study description | Duration | Results | Reference |
|---|---|---|---|
| *Lb. rhamnosus* GG or placebo; 27 infants; $5 \times 10^8$ cfu d$^{-1}$ | 1 month | Significant decrease in disease activity index ($P = 0.008$) | Majamaa and Isolauri (1997) |
| *Lb. rhamnosus* GG and *B. lactis* Bb-12; 27 infants mean age 4.6 months | 2 months | Reduced serum soluble CD4 and eosinophil protein X[2] | Isolauri *et al.* (2000) |
| *Lb. rhamnosus* GG or placebo; 132 mothers and infants, $10^{10}$ cfu d$^{-1}$ | 2–4 weeks pre-natally, 4 months post-natally | 23% and 46% of probiotic and placebo groups respectively developed atopic dermatitis ($P = 0.008$) | Kalliomäki *et al.* (2001) |
| As above[1] | As above | 26% and 46% of probiotic and placebo groups respectively developed atopic dermatitis | Kalliomäki *et al.* (2003) |
| *Lb. rhamnosus* GR-1 and *Lb. fermentum* RC-14; 10 patients; $>10^9$ cfu d$^{-1}$ orally | 14 days | All patients reported improved well-being, probiotic recovered from most | Reid *et al.* (2001b) |
| *Lb. rhamnosus* GR-1 and *Lb. fermentum* RC-14; 64 patients; $>10^9$ cfu d$^{-1}$ orally | 60 days | 13% and 37% of placebo and probiotic group respectively showed a normal microflora ($P = 0.02$) | Reid *et al.* (2003b) |
| *Lb. rhamnosus* GR-1 and *Lb. fermentum* B-54; $10^9$ cfu d$^{-1}$ vaginally | Unknown | Recurrence rate similar to antibiotic therapy | Reid *et al.* (1995) |

[1] Four years' follow-up study.
[2] Serum sCD4 fell from 5601 to 1404 pg mL$^{-1}$ ($P < 0.05$), and eosinophil protein X in the urine fell from ca. 80 to 50 µg mL$^{-1}$; both of these observations were after 2 months' intervention.

receiving *Lb. rhamnosus* GG may be responsible for the probiotic effect (Pessi *et al.*, 2000). IL-10 is an immunomodulatory cytokine that down-regulates pro-inflammatory IL-2, IL-6, IL-12, TNF-$\alpha$ and interferon-gamma (IFN-$\gamma$). In a randomised, double-blind and placebo-controlled study, Isolauri *et al.* (2000) used *Lb. rhamnosus* GG and *B. lactis* Bb-12 to treat 27 babies, of average age 4.6 months, with atopic dermatitis. Results again indicated probiotic effects with *Lb. rhamnosus* GG and also with *B. lactis* Bb-12 after 2 months of treatment. A lowering of soluble CD4 in serum and eosinophil protein X in urine was also detected, suggesting down-regulation of the inflammatory response by the probiotics.

Further evidence has suggested that probiotics not only reduce symptoms of atopic disease, but that they can also prevent such conditions from occurring when given prenatally to mothers (Kalliomäki *et al.*, 2001). When *Lb. rhamnosus* GG ($10^{10}$ cfu d$^{-1}$) was given to mothers with a family history of atopic dermatitis and post-natally to their infants for 4 months, significant differences between the probiotic and placebo groups in the frequency of atopic disease were noted. Moreover, this prophylaxis was still evident at a 4-year follow-up (Kalliomäki *et al.*, 2003). Similarly, other studies have shown that *Lb. rhamnosus* GG, when taken by mothers for the last 4 weeks of pregnancy and during breast feeding (3 months), can increase the anti-inflammatory capacity of breast milk as measured by transforming growth factor-beta (TGF-β) levels (Rautava *et al.*, 2002).

### 7.3.2 *Bacterial vaginosis*

The vaginal microflora of healthy human females typically comprises about 50 bacterial species (Redondo-Lopez *et al.*, 1990), the composition of which can be affected by a number of factors, including the body's reproductive stage, treatment with antibiotics and the use of spermicidal agents. This microbiota is mainly comprised of lactobacilli, which are known to protect the host from colonisation by pathogens. Bacterial vaginosis (BV) is defined as the presence of an abnormal microflora in the lower female genital tract (Reid, 2001), and may be present subclinically in up to 70% of healthy females (Burton & Reid, 2002). A diagnosis of BV is made when three of the following four criteria are met: (a) release of an amine (fishy odour) after addition of potassium hydroxide; (b) a pH of lower than 4.5 in the vagina; (c) clue-cells in the vagina; and (d) a homogeneous milky discharge from the lower genital tract. With respect to criterion (c), a score is used to evaluate the degree of microflora abnormality. A score of 0–3 denotes a normal flora dominated by lactobacilli, 4–6 indicates colonisation by small Gram-negative rods such as *Bacteroides* spp., and a score of 7–10 denotes BV and a flora dominated by known pathogens such as *E. coli* (Nugent *et al.*, 1991). Most of these pathogenic organisms originate from the faecal microflora. In health, epithelial cells lining the vagina are covered with a biofilm in which lactobacilli predominate (Domingue *et al.*, 1991), and under most circumstances pathogens are unable to colonise the biofilm. However, breakdown of the biofilm is known to occur in BV, although the ecological and physiological mechanisms involved are unclear.

Probiotics are increasingly being regarded to be of therapeutic benefit in BV (see Table 7.6), partly because of the large numbers of lactobacilli in the healthy vaginal microflora and also because their loss is implicated in disease pathogenesis. A number of early studies investigated the effectiveness of oral and vaginal yoghurt application. Some reported positive results, but study numbers were small and most were not randomised or placebo-controlled (de Vrese & Schrezenmeis, 2002). However, open and placebo-controlled trials using lyophilised *Lb. acidophilus* applied locally (Hallen *et al.*, 1992; Parent *et al.*, 1996) or *Lb. acidophilus* yoghurt given orally (Hilton *et al.*, 1992) have generated similar results.

Other studies have investigated the therapeutic benefits of *Lb. rhamnosus* GR-1 and *Lactobacillus fermentum* RC-14 given orally. *In vitro* experiments suggested that these bacteria possessed several putative properties, such as adhesion to epithelial cells and production of antimicrobial substances, that could be of clinical value (Reid, 2000). More specifically, *Lb. rhamnosus* GR-1 exhibited antimicrobial activity against Gram-negative bacteria and was resistant to spermicide, while *Lb. fermentum* RC-14 appeared to inhibit Gram-positive cocci and to produce hydrogen peroxide (Reid & Bruce, 2003). A pilot trial with these organisms gave promising results (Reid *et al.*, 2001a) and, in a subsequent larger study, 64 patients were randomised to receive either probiotic or placebo orally for 60 days. The results suggested that there was a probiotic effect: 37% of the probiotics compared to 13% of the placebos ($P = 0.02$) showed restoration of a normal microflora dominated by lactobacilli from an asymptomatic BV flora (Reid *et al.*, 2003b). Furthermore, lactobacilli were more prevalent and occurred in higher numbers in the probiotic group at day 28. The authors also reported significant reductions in enterobacteria and yeasts. Further work has suggested that a dose of at least $10^8$ cfu $d^{-1}$ of *Lb. rhamnosus* GR-1 and *Lb. fermentum* RC-14 is needed for a therapeutic effect (Reid *et al.*, 2001b).

Relative to the number of probiotic studies conducted on BV, little effort has focussed on other genitourinary infections. However, Stamey's studies during the 1970s showed that vaginal colonisation is a factor in urinary tract infection (UTI) (Stamey, 1973). It is therefore likely that any probiotic therapy capable of reducing or preventing BV will also have a beneficial effect on UTI. Indeed, vaginal administration of two capsules per week containing $10^9$ cfu *Lb. rhamnosus* GR-1 and *Lb. fermentum* B-54 was found to reduce the UTI recurrence rate to that expected with antibiotic therapy (Reid *et al.*, 1995).

## 7.4 Conclusion

Probiotic research has expanded rapidly over the past few years. There are many reasons for this, including heightened scientific and public awareness, the need for alternatives or replacements for antibiotics due to the increasing spread of resistance, and the realisation that probiotics have the potential temporarily to colonise the GI tract and other parts of the body and to exert their effects over the long term without requiring continuous medical intervention. This is convenient for both the patient and physician, and also reduces treatment costs.

In order to establish probiotics as legitimate alternatives to more traditional clinical interventions, it is essential that good experimental evidence be available to explain probiotic mechanisms. In most cases, these are unknown, and it is clear that the experimental evidence often varies markedly in probiotic trials. The most consistent evidence for the beneficial effects of probiotics is probably in the treatment of infantile diarrhoea, where a number of trials involving large patient cohorts have demonstrated the usefulness of *Lb. rhamnosus* GG (Isolauri *et al.*, 1991; Majamaa *et al.*, 1995; Guarino *et al.*, 1997; Guandalini *et al.*, 2000; Szajewska *et al.*, 2001). Collectively, the trials described previously involved almost 600 patients, and all five showed significant benefits with the probiotic.

For most of the other conditions in which probiotic therapy has been investigated, the evidence is mixed. Most of the studies that have been conducted are pilot scale, involving fewer than 50 patients, and larger-scale clinical trails involving much larger patient cohorts are needed. At present, only clinical outcomes are measured in most trials but future studies should also examine the microbiological and immunological effects of probiotics in more detail and, where possible, investigate the activities of the probiotic micro-organism at the site of interest. This is increasingly possible with molecular techniques such as DGGE and real-time PCR which allow the detection of – and, in the latter case, quantification of – specific strains in a complex microbiota.

Despite recent advances, the study of probiotics is still an emerging field that has yet to realise its full potential. With larger-scale and more frequent studies it is likely that probiotics will become major weapons in the treatment of disease.

## References

Adam, J., Barret, C. & Barret-Bellet, A. (1977) Essai's cliniques contrôlés en double insu de l'ultra-levure lyophilseé. Étude multicentrique par 25 medecins de 388 cas. *Gazette Medicale de France*, **84**, 2072–2081.

Alun-Jones, V., Wilson, A., Hunter, J. & Robinson, R. (1984) The aetiological role of antibiotic prophylaxis with hysterectomy in irritable bowel syndrome. *Journal of Obstetrics and Gynaecology*, **5**(Suppl 1), S22–S23.

Armuzzi, A., Cremonini, F., Ojetti, V., Bartolozzi, F., Canducci, F., Candelli, M., Santarelli, L., Cammarota, G., De Lorenzo, A., Pola, P., Gasbarrini, G. & Gasbarrini, A. (2001) Effect of *Lactobacillus* GG supplementation on antibiotic-associated gastrointestinal side effects during *Helicobacter pylori* eradication therapy: a pilot study. *Digestion*, **63**, 1–7.

Arvola, T., Laiho, K., Torkkeli, S., Mykkanen, H., Salminen, S., Maunula, L. & Isolauri, E. (1999) Prophylactic *Lactobacillus* GG reduces antibiotic-associated diarrhea in children with respiratory infections: a randomized study. *Pediatrics*, **104**(5), 64–67.

Balsari, A., Ceccarilli, A., Dubini, F., Fesce, E. & Poli, G. (1982) The fecal microbial population in the irritable bowel syndrome. *Microbiologica*, **5**, 185–194.

Bennet, R., Laughon, B. & Lindsay, J. (1990) *Lactobacillus* GG treatment of *Clostridium difficile* infection in nursing home patients, Poster, 3rd International Conference on Nosocomial Infection, Atlanta, Georgia.

Bernet, M.F., Brassart, D., Neeser, J.R. & Servin, A.L. (1994) *Lactobacillus acidophilus* LA 1 binds to cultured human intestinal cell lines and inhibits cell attachment and cell invasion by enterovirulent bacteria. *Gut*, **35**, 483–489.

Biller, J.A., Katz, A.J., Flores, A.F., Buie, T.M. & Gorbach, S.L. (1995) Treatment of recurrent *Clostridium difficile* colitis with *Lactobacillus* GG. *Journal of Pediatric Gastroenterology and Nutrition*, **21**, 224–226.

Black, F., Andersen, P., Ørskov, J., Ørskov, F., Gaarslev, K. & Laulund, S. (1989) Prophylactic efficacy of lactobacilli on traveller's diarrhea. *Travel Medicine*, **8**, 1750–1753.

Bleichner, G., Blehaut, H., Mentec, H. & Moyse, D. (1997) *Saccharomyces boulardii* prevents diarrhea in critically ill tube-fed patients. A multicenter, randomized, double-blind placebo-controlled trial. *Intensive Care Medicine*, **23**, 517–523.

Bradley, H., Wyatt, G., Bayliss, C. & Hunter, J. (1987) Instability in the faecal flora of a patient suffering from food-related irritable bowel syndrome. *Journal of Medical Microbiology*, **23**, 29–32.

Burton, J.P. & Reid, G. (2002) Evaluation of the bacterial vaginal flora of 20 postmenopausal women by direct (Nugent score) and molecular (polymerase chain reaction and denaturing gradient gel electrophoresis) techniques. *Journal of Infectious Disease*, **186**, 1770–1780.

Cadieux, P., Burton, J., Gardiner, G., Braunstein, I., Bruce, A.W., Kang, C.Y. & Reid, G. (2002) *Lactobacillus* strains and vaginal ecology. *Journal of the American Medical Association*, **287**, 1940–1941.

Chouraqui, J.P., van Egroo, L.D. & Fichot, M.C. (2004) Acidified milk formula supplemented with *Bifidobacterium lactis*: impact on infant diarrhea in residential care settings. *Journal of Pediatric Gastroenterology and Nutrition*, **38**, 288–292.

Clements, M.L., Levine, M.M., Black, R.E., Robins-Browne, R.M., Cisneros, L.A., Drusano, G.L., Lanata, C.F. & Saah, A.J. (1981) *Lactobacillus* prophylaxis for diarrhea due to enterotoxigenic *Escherichia coli*. *Antimicrobial Agents and Chemotherapy*, **20**, 104–108.

Colombel, J.F., Cortot, A., Neut, C. & Romond, C. (1987) Yoghurt with *Bifidobacterium longum* reduces erythromycin-induced gastrointestinal effects. *Lancet*, **2**(8549), 43.

Cooke, E.M. (1968) Properties of strains of *Escherichia coli* isolated from the faeces of patients with ulcerative colitis, patients with acute diarrhoea and normal persons. *Journal of Pathology and Bacteriology*, **95**, 101–113.

Cross, M.L. (2004) Immune-signalling by orally-delivered probiotic bacteria: effects on common mucosal immunoresponses and protection at distal mucosal sites. *International Journal of Immunopathology and Pharmacology*, **17**, 127–134.

Cross, M.L., Stevenson, L.M. & Gill, H.S. (2001) Anti-allergy properties of fermented foods: an important immunoregulatory mechanism of lactic acid bacteria? *International Immunopharmacology*, **1**, 891–901.

Cummings, J.H. & Macfarlane, G.T. (1991) The control and consequences of bacterial fermentation in the human colon. *Journal of Applied Bacteriology*, **70**, 443–459.

Das, K., Eastwood, M., McManus, J. & Sircus, W. (1973) Adverse reaction during salicylazosulphapyridine therapy and the relation with drug metabolism and acetylator phenotype. *New England Journal of Medicine*, **289**, 491–495.

Dickinson, R.J., O'Connor, H.J., Pinder, I., Hamilton, I., Johnston, D. & Axon, A.T. (1985) Double blind controlled trial of oral vancomycin as adjunctive treatment in acute exacerbations of idiopathic colitis. *Gut*, **26**, 1380–1384.

Domingue, P.A., Sadhu, K., Costerton, J.W., Bartlett, K. & Chow, A.W. (1991) The human vagina: normal flora considered as an in situ tissue-associated, adherent biofilm. *Genitourinary Medicine*, **67**, 226–231.

DuPont, H.L., Ericsson, C.D., Johnson, P.C. & Cabada, F.J. (1986) Antimicrobial agents in the prevention of travellers' diarrhea. *Reviews in Infectious Diseases*, **8**(Suppl 2), S167–S171.
Fedorak, N., Gionchetti, P., Campieri, M., De Simone, C., Madsen, K.L., Isaacs, K. & Sartor, R. (2003) VSL#3 probiotic mixture induces remission in patients with active ulcerative colitis. *Gastroenterology*, **124**, A377.
Fite, A., Macfarlane, G.T., Cummings, J.H., Hopkins, M.J., Kong, S.C., Furrie, E. & Macfarlane, S. (2004) Identification and quantitation of mucosal and faecal desulfovibrios using real time polymerase chain reaction. *Gut*, **53**, 523–529.
Fujiwara, S., Seto, Y., Kimura, A. & Hashiba, H. (2001) Establishment of orally-administered *Lactobacillus gasseri* SBT2055SR in the gastrointestinal tract of humans and its influence on intestinal microflora and metabolism. *Journal of Applied Microbiology*, **90**, 343–352.
Furrie, E., Macfarlane, S., Kennedy, A., Cummings, J., Walsh, S., O'Neil, D. & Macfarlane, G.T. (2004) Synbiotic therapy (*Bifidobacterium longum*/Synergy I™) initiates resolution of inflammation in patients with active ulcerative colitis. Gut, **54**, 242–249.
Gaon, D., Garcia, H., Winter, L., Rodriguez, N., Quintas, R., Gonzalez, S.N. & Oliver, G. (2003) Effect of *Lactobacillus* strains and *Saccharomyces boulardii* on persistent diarrhea in children. *Medicina (B Aires)*, **63**, 293–298.
Gill, H.S., Rutherfurd, K.J., Cross, M.L. & Gopal, P.K. (2001) Enhancement of immunity in the elderly by dietary supplementation with the probiotic *Bifidobacterium lactis* HN019. *American Journal of Clinical Nutrition*, **74**, 833–839.
Gionchetti, P., Rizzello, F., Helwig, U., Venturi, A., Lammers, K. M., Brigidi, P., Vitali, B., Poggioli, G., Miglioli, M. & Campieri, M. (2003) Prophylaxis of pouchitis onset with probiotic therapy: a double-blind, placebo-controlled trial. *Gastroenterology*, **124**, 1202–1209.
Gionchetti, P., Rizzello, F., Venturi, A., Brigidi, P., Matteuzzi, D., Bazzocchi, G., Poggioli, G., Miglioli, M. & Campieri, M. (2000) Oral bacteriotherapy as maintenance treatment in patients with chronic pouchitis: a double-blind, placebo-controlled trial. *Gastroenterology*, **119**, 305–309.
Gismondo, M.R., Drago, L. & Lombardi, A. (1999) Review of probiotics available to modify gastrointestinal flora. *International Journal of Antimicrobial Agents*, **12**, 287–292.
Gorbach, S.L. (2000) Probiotics and gastrointestinal health. *American Journal of Gastroenterology*, **95**(Suppl 1), S2–S4.
Gorbach, S.L., Chang, T.W. & Goldin, B. (1987) Successful treatment of relapsing *Clostridium difficile* colitis with *Lactobacillus* GG. *Lancet*, **2**(8574), 1519.
Gosselink, M.P., Schouten, W.R., Van Lieshout, L.M., Hop, W.C., Laman, J.D. & Ruseler-Van Embden, J.G. (2004) Delay of the first onset of pouchitis by oral intake of the probiotic strain *Lactobacillus rhamnosus* GG. *Diseases of the Colon and Rectum*, **47**(6), 876–884.
Gotz, V., Romankiewicz, J.A., Moss, J. & Murray, H.W. (1979) Prophylaxis against ampicillin-associated diarrhea with a *Lactobacillus* preparation. *American Journal of Hospital Pharmacy*, **36**, 754–757.
Guandalini, S., Pensabene, L., Zikri, M.A., Dias, J.A., Casali, L.G., Hoekstra, H., Kolacek, S., Massar, K., Micetic-Turk, D., Papadopoulou, A., de Sousa, J.S., Sandhu, B., Szajewska, H. & Weizman, Z. (2000) *Lactobacillus* GG administered in oral rehydration solution to children with acute diarrhea: a multicenter European trial. *Journal of Pediatric Gastroenterology and Nutrition*, **30**, 54–60.
Guarino, A., Canani, R.B., Spagnuolo, M.I., Albano, F. & Di Benedetto, L. (1997) Oral bacterial therapy reduces the duration of symptoms and of viral excretion in children with mild diarrhea. *Journal of Pediatric Gastroenterology and Nutrition*, **25**, 516–519.
Guslandi, M., Giollo, P. & Testoni, P.A. (2003) A pilot trial of *Saccharomyces boulardii* in ulcerative colitis. *European Journal of Gastroenterology and Hepatology*, **15**, 697–698.
Guslandi, M., Mezzi, G., Sorghi, M. & Testoni, P.A. (2000) *Saccharomyces boulardii* in maintenance treatment of Crohn's disease. *Digestive Diseases and Science*, **45**, 1462–1464.
Hallen, A., Jarstrand, C. & Pahlson, C. (1992) Treatment of bacterial vaginosis with lactobacilli. *Sexually Transmitted Diseases*, **19**, 146–148.
Herek, O., Kara, I.G. & Kaleli, I. (2004) Effects of antibiotics and *Saccharomyces boulardii* on bacterial translocation in burn injury. *Surgery Today*, **34**, 256–260.
Hilton, E., Isenberg, H., Alperstein, P., France, K. & Borenstein, M. (1992) Ingestion of yoghurt containing *Lactobacillus acidophilus* as prophylaxis for *Candida* vaginitis. *Annals of Internal Medicine*, **116**, 353–357.
Hold, G., Pryde, S., Russell, V., Furrie, E. & Flint, H. (2002) Assessment of diversity in human colonic samples by 16S rDNA sequence analysis. *FEMS Microbiology Ecology*, **39**, 33–39.

Hopkins, M.J. & Macfarlane, G.T. (2003) Nondigestible oligosaccharides enhance bacterial colonization resistance against *Clostridium difficile* in vitro. *Applied and Environmental Microbiology*, **69**, 1920–1927.

Hunter, J., Lee, A., King, T., Barratt, M., Linggood, M. & Blades, J. (1996) Enterococcus faecium strain PR88 – an effective probiotic. *Gut*, **38**(Suppl 1), A62.

Ishikawa, H., Akedo, I., Umesaki, Y., Tanaka, R., Imaoka, A. & Otani, T. (2003) Randomized controlled trial of the effect of bifidobacteria-fermented milk on ulcerative colitis. *Journal of the American College of Nutrition*, **22**, 56–63.

Isolauri, E., Arvola, T., Sutas, Y., Moilanen, E. & Salminen, E. (2000) Probiotics in the management of atopic eczema. *Clinical and Experimental Allergy*, **30**, 1604–1610.

Isolauri, E., Juntunen, M., Rautanen, P., Sillanaukee, P. & Koivula, T. (1991) A human *Lactobacillus* strain (*Lactobacillus casei* sp. strain GG) promotes recovery from acute diarrhea in children. *Pediatrics*, **88**, 90–97.

Jadamus, A., Vahjen, W., Schafer, K. & Simon, O. (2002) Influence of the probiotic strain *Bacillus cereus* var. *toyoi* on the development of enterobacterial growth and on selected parameters of bacterial metabolism in digesta samples of piglets. *Journal of Animal Physiology and Animal Nutrition (Berlin)*, **86**(1–2), 42–54.

Kaila, M., Isolauri, E., Soppi, E., Virtanen, E., Laine, S. & Arvilommi, H. (1992) Enhancement of the circulating antibody secreting cell response in human diarrhea by a human *Lactobacillus* strain. *Pediatrics Research*, **32**, 141–144.

Kalliomäki, M., Salminen, S., Arvilommi, H., Kero, P., Koskinen, P. & Isolauri, E. (2001) Probiotics in primary prevention of atopic disease: a randomised placebo-controlled trial. *Lancet*, **357**(9262), 1076–1079.

Kalliomäki, M., Salminen, S., Poussa, T., Arvilommi, H. & Isolauri, E. (2003) Probiotics and prevention of atopic disease: 4-year follow-up of a randomised placebo-controlled trial. *Lancet*, **361**(9372), 1869–1871.

Kanauchi, O., Fujiyama, Y., Mitsuyama, K., Araki, Y., Ishii, T., Nakamura, T., Hitomi, Y., Agata, K., Saiki, T., Andoh, A., Toyonaga, A. & Bamba, T. (1999) Increased growth of *Bifidobacterium* and *Eubacterium* by germinated barley foodstuff, accompanied by enhanced butyrate production in healthy volunteers. *International Journal of Molecular Medicine*, **3**, 175–179.

Katelaris, P.H., Salam, I. & Farthing, M. J. (1995) Lactobacilli to prevent traveller's diarrhea? *New England Journal of Medicine*, **333**(20), 1360–1361.

Kennedy, A. (2002) *Mucosal bacterial populations in ulcerative colitis and the use of synbiotics in disease therapy*, PhD Thesis, University of Dundee, Dundee, Scotland.

Kim, H.J., Camilleri, M., McKinzie, S., Lempke, M.B., Burton, D.D., Thomforde, G.M. & Zinsmeister, A.R. (2003) A randomized controlled trial of a probiotic, VSL#3, on gut transit and symptoms in diarrhoea-predominant irritable bowel syndrome. *Alimentary Pharmacology and Therapeutics*, **17**, 895–904.

King, T.S., Elia, M. & Hunter, J.O. (1998) Abnormal colonic fermentation in irritable bowel syndrome. *Lancet*, **352**(9135), 1187–1189.

Koebnick, C., Wagner, I., Leitzmann, P., Stern, U. & Zunft, H.J. (2003) Probiotic beverage containing *Lactobacillus casei* Shirota improves gastrointestinal symptoms in patients with chronic constipation. *Canadian Journal of Gastroenterology*, **17**, 655–659.

Koide, A., Yamaguchi, T., Odaka, T., Koyama, H., Tsuyuguchi, T., Kitahara, H., Ohto, M. & Saisho, H. (2000) Quantitative analysis of bowel gas using plain abdominal radiograph in patients with irritable bowel syndrome. *American Journal of Gastroenterology*, **95**, 1735–1741.

Kollaritsch, H., Holst, H., Grobara, P. & Wiedermann, G. (1993) Prevention of traveller's diarrhea with *Saccharomyces boulardii*. Results of a placebo controlled double-blind study. *Fortschung Medicine*, **111**, 152–156.

Kollaritsch, H., Kremsner, P., Wiedermann, G. & Schneier, O. (1989) Prevention of traveller's diarrhoea: comparison of different non-antibiotic preparations. *Travel Medicine International*, **1989**, 9–17.

Kruis, W., Schutz, E., Fric, P., Fixa, B., Judmaier, G. & Stolte, M. (1997) Double-blind comparison of an oral *Escherichia coli* preparation and mesasalazine in maintaining remission of ulcerative colitis. *Alimentary Pharmacology and Therapeutics*, **11**, 853–858.

Kruis, W., Kalk, E., Fric, P. & Stolte, M. (2001) Maintenance of remission in ulcerative colitis is equally effective with *Escherichia coli* Nissle 1917 and with standard mesasalazine. *Gastroenterology*, **120**, A127.

La Rosa, M., Bottaro, G., Gulino, N., Gambuzza, F., Di Forti, F., Ini, G. & Tornambe, E. (2003) Prevention of antibiotic-associated diarrhea with *Lactobacillus sporogens* and fructo-oligosaccharides in children. A multicentric double-blind vs. placebo study. *Minerva Pediatrics*, **55**, 447–452.

Laake, K., Bjorneklett, A., Bakka, A., Midtvedt, T., Norin, K. & Eide, T.J. (1999) Influence of fermented milk on clinical state, faecal bacterial count and biochemical characteristics in patients with ileal-pouch-anal-anastomosis. *Microbial Ecology in Health and Disease*, **11**, 211–217.

Laake, K.O., Line, P.D., Aabakken, L., Lotveit, T., Bakka, A., Eide, J., Roseth, A., Grzyb, K., Bjorneklett, A. & Vatn, M.H. (2003) Assessment of mucosal inflammation and circulation in response to probiotics in patients operated with ileal pouch anal anastomosis for ulcerative colitis. *Scandinavian Journal of Gastroenterology*, **38**, 409–414.

Lewis, S.J., Potts, L.F. & Barry, R.E. (1998) The lack of therapeutic effect of *Saccharomyces boulardii* in the prevention of antibiotic-related diarrhoea in elderly patients. *Journal of Infection*, **36**, 171–174.

Linevsky, J. & Kelly, C. (1997). Clostridium difficile colitis in gastrointestinal infections. *Gastrointestinal infections: Diagnosis and Management*, (ed. J. LaMont), pp. 293–325, Marcel Dekker, New York.

Lund, B., Edlund, C., Barkholt, L., Nord, C.E., Tvede, M. & Poulsen, R. L. (2000) Impact on human intestinal microflora of an *Enterococcus faecium* probiotic and vancomycin. *Scandinavian Journal of Infectious Disease*, **32**, 627–632.

Macfarlane, G.T. & Cummings, J. (2002) Probiotics, infection and immunity. *Current Opinion in Infectious Diseases*, **15**, 501–506.

Macfarlane, S., Furrie, E., Cummings, J. & Macfarlane, G. (2004) Chemotaxonomic analysis of bacterial populations colonizing the rectal mucosa in patients with ulcerative colitis. *Clinical Infectious Diseases*, **38**(12), 1690–1699.

Madsen, K.L. (2001) The use of probiotics in gastrointestinal disease. *Canadian Journal of Gastroenterology*, **15**, 817–822.

Majamaa, H. & Isolauri, E. (1997) Probiotics: a novel approach in the management of food allergy. *Journal of Allergy and Clinical Immunology*, **99**, 179–185.

Majamaa, H., Isolauri, E., Saxelin, M. & Vesikari, T. (1995) Lactic acid bacteria in the treatment of acute rotavirus gastroenteritis. *Journal of Pediatric Gastroenterology and Nutrition*, **20**, 333–338.

Malchow, H.A. (1997) Crohn's disease and *Escherichia coli* – a new approach in therapy to maintain remission of colonic Crohn's disease? *Journal of Clinical Gastroenterology*, **25**, 653–658.

Mantzaris, G.J., Hatzis, A., Kontogiannis, P. & Triadaphyllou, G. (1994) Intravenous tobramycin and metronidazole as an adjunct to corticosteroids in acute, severe ulcerative colitis. *American Journal of Gastroenterology*, **89**, 43–46.

Mastrandrea, F., Coradduzza, G., Serio, G., Minardi, A., Manelli, M., Ardito, S. & Muratore, L. (2004) Probiotics reduce the CD34+ hemopoietic precursor cell increased traffic in allergic subjects. *Allergy and Immunology (Paris)*, **36**(4), 118–122.

McFarland, L.V. (1998) Epidemiology, risk factors and treatments for antibiotic-associated diarrhea. *Digestive Diseases*, **16**(5), 292–307.

McFarland, L.V., Surawicz, C.M., Greenberg, R.N., Elmer, G.W., Moyer, K.A., Melcher, S.A., Bowen, K.E. & Cox, J.L. (1995) Prevention of beta-lactam-associated diarrhea by *Saccharomyces boulardii* compared with placebo. *American Journal of Gastroenterology*, **90**, 439–448.

McKendrick, M.W. & Read, N.W. (1994) Irritable bowel syndrome – post salmonella infection. *Journal of Infection*, **29**, 1–3.

Mendall, M. & Kumar, D. (1998) Antibiotic use, childhood affluence and irritable bowel syndrome. *European Journal of Gastroenterology and Hepatology*, **10**, 59–62.

Merson, M., Morris, G., Sack, D., Well, J., Feeley, J., Sack, R., Creech, W., Kapikian, A. & Gangarosa, E. (1976) Traveller's diarrhea in Mexico. *New England Journal of Medicine*, **1976**, 1299–1305.

Mimura, T., Rizzello, F., Helwig, U., Poggioli, G., Schreiber, S., Talbot, I.C., Nicholls, R.J., Gionchetti, P., Campieri, M. & Kamm, M.A. (2004) Once daily high dose probiotic therapy (VSL#3) for maintaining remission in recurrent or refractory pouchitis. *Gut*, **53**, 108–114.

Niedzielin, K., Kordecki, K. & Kosik, R. (1998) New possibility in the treatment of irritable bowel syndrome: probiotics as a modification of the microflora of the colon. *Gastroenterology*, **114**, A402.

Nobaeck, S., Johansson, M., Molin, G., Ahrne, S. & Jeppsson, B. (2000) Bacterial supplementation in the irritable bowel syndrome. A randomised double-blind placebo-controlled crossover study. *Digestive Diseases and Science*, **32**, 302–304.

Nugent, R.P., Krohn, M.A. & Hillier, S.L. (1991) Reliability of diagnosing bacterial vaginosis is improved by a standardized method of gram stain interpretation. *Journal of Clinical Microbiology*, **29**, 297–301.

O'Sullivan, M.A. & O'Morain, C.A. (2000) Bacterial supplementation in the irritable bowel syndrome. A randomised double-blind placebo-controlled crossover study. *Digestive and Liver Diseases*, **32**, 294–301.

Oksanen, P.J., Salminen, S., Saxelin, M., Hamalainen, P., Ihantola-Vormisto, A., Muurasniemi-Isoviita, L., Nikkari, S., Oksanen, T., Porsti, I. & Salminen, E. (1990) Prevention of travellers' diarrhoea by *Lactobacillus* GG. *Annals of Medicine*, **22**, 53–56.

Orrhage, K., Brignar, B. & Nord, C. (1994) Effects of supplements of *Bifidobacterium longum* and *Lactobacillus acidophilus* on the intestinal microbiota during administration of clindamycin. *Microbial Ecology in Health and Disease*, **7**, 17–25.

Parent, D., Bossens, M., Bayot, D., Kirkpatrick, C., Graf, F., Wilkinson, F. & Kaiser, R. (1996) Therapy of bacterial vaginosis using endogenously applied Lactobacilli acidophilus and a low dose of estriol: a placebo-controlled multicentric clinical trial. *Arzneimittel Forschung*, **46**, 68–73.

Pessi, T., Sutas, Y., Hurme, M. & Isolauri, E. (2000) Interleukin-10 generation in atopic children following oral *Lactobacillus rhamnosus* GG. *Clinical and Experimental Allergy*, **30**, 1804–1808.

Pinchuk, I.V., Bressollier, P., Verneuil, B., Fenet, B., Sorokulova, I.B., Megraud, F. & Urdaci, M.C. (2001) In vitro anti-*Helicobacter pylori* activity of the probiotic strain *Bacillus subtilis* 3 is due to secretion of antibiotics. *Antimicrobial Agents and Chemotherapy*, **45**, 3156–3161.

Plummer, S., Weaver, M.A., Harris, J.C., Dee, P. & Hunter, J. (2004) *Clostridium difficile* pilot study: effects of probiotic supplementation on the incidence of *C. difficile* diarrhoea. *International Microbiology*, **7**, 59–62.

Pochapin, M. (2000) The effect of probiotics on *Clostridium difficile* diarrhea. *American Journal of Gastroenterology*, **95**(Suppl 1), S11–S13.

Pochapin, M., Oltikar, A., Pringle-Smith, R. & Schreiber, C. (1998) A prospective randomised placebo-controlled trial of *Lactobacillus* GG in combination with standard antibiotics for the treatment of *Clostridium difficile* infection. *American Journal of Gastroenterology*, **93**, 1697 (abstract).

Poxton, I.R., Brown, R., Sawyerr, A. & Ferguson, A. (1997) Mucosa-associated bacterial flora of the human colon. *Journal of Medical Microbiology*, **46**, 85–91.

Prantera, C., Scribano, M.L., Falasco, G., Andreoli, A. & Luzi, C. (2002) Ineffectiveness of probiotics in preventing recurrence after curative resection for Crohn's disease: a randomised controlled trial with *Lactobacillus* GG. *Gut*, **51**, 405–409.

Rafter, J. (2003) Probiotics and colon cancer. *Best Practice and Research in Clinical Gastroenterology*, **17**, 849–859.

Rautava, S., Kalliomaki, M. & Isolauri, E. (2002) Probiotics during pregnancy and breast-feeding might confer immunomodulatory protection against atopic disease in the infant. *Journal of Allergy and Clinical Immunology*, **109**, 119–121.

Redondo-Lopez, V., Cook, R.L. & Sobel, J.D. (1990) Emerging role of lactobacilli in the control and maintenance of the vaginal bacterial microflora. *Reviews in Infectious Diseases*, **12**, 856–872.

Reid, G. (2000) Probiotic therapy and functional foods for prevention of urinary tract infections: state of the art and science. *Current Infectious Disease Reports*, **2**, 518–522.

Reid, G. (2001) Probiotic agents to protect the urogenital tract against infection. *American Journal of Clinical Nutrition*, **73**(Suppl 2), 437S–443S.

Reid, G., Beuerman, D., Heinemann, C. & Bruce, A.W. (2001b) Probiotic lactobacillus dose required to restore and maintain a normal vaginal flora. *FEMS Immunology and Medical Microbiology*, **32**, 37–41.

Reid, G. & Bruce, A.W. (2003) Urogenital infections in women: can probiotics help? *Postgraduate Medical Journal*, **79**, 428–433.

Reid, G., Bruce, A. & Taylor, M. (1995) Instillation of *Lactobacillus* and stimulation of indigenous organisms to prevent recurrence of urinary tract infections. *Microecology Therapy*, **23**, 32–45.

Reid, G., Bruce, A.W., Fraser, N., Heinemann, C., Owen, J. & Henning, B. (2001a) Oral probiotics can resolve urogenital infections. *FEMS Immunology and Medical Microbiology*, **30**, 49–52.

Reid, G., Charbonneau, D., Erb, J., Kochanowski, B., Beuerman, D., Poehner, R. & Bruce, A.W. (2003b) Oral use of *Lactobacillus rhamnosus* GR-1 and *L. fermentum* RC-14 significantly alters vaginal flora: randomized, placebo-controlled trial in 64 healthy women. *FEMS Immunology and Medical Microbiology*, **35**, 131–134.

Reid, G., Gan, B.S., She, Y.M., Ens, W., Weinberger, S. & Howard, J. C. (2002) Rapid identification of probiotic *Lactobacillus* biosurfactant proteins by ProteinChip tandem mass spectrometry tryptic peptide sequencing. *Applied and Environmental Microbiology*, **68**, 977–980.

Reid, G., Jass, J., Sebulsky, M.T. & McCormick, J.K. (2003a) Potential uses of probiotics in clinical practice. *Clinical Microbiology Reviews*, **16**(4), 658–672.

Reid, G., Sanders, M.E., Gaskins, H.R., Gibson, G.R., Mercenier, A., Rastall, R., Roberfroid, M., Rowland, I., Cherbut, C. & Klaenhammer, T.R. (2003c) New scientific paradigms for probiotics and prebiotics. *Journal of Clinical Gastroenterology*, **37**, 105–118.

Rembacken, B.J., Snelling, A.M., Hawkey, P.M., Chalmers, D.M. & Axon, A.T. (1999) Non-pathogenic *Escherichia coli* versus mesasalazine for the treatment of ulcerative colitis: a randomised trial. *Lancet*, **354**(9179), 635–639.

Reuter, G. (2001) The *Lactobacillus* and *Bifidobacterium* microflora of the human intestine: composition and succession. *Current Issues in Intestinal Microbiology*, **2**, 43–53.

Rivest, P., Proulx, M., Lonergan, G., Lebel, M. & Bedard, L. (2004) Hospitalisations for gastroenteritis: the role of rotavirus. *Vaccine*, **7**(22), 2013–2017.

Roller, M., Rechkemmer, G. & Watzl, B. (2004) Prebiotic inulin enriched with oligofructose in combination with the probiotics *Lactobacillus rhamnosus* and *Bifidobacterium lactis* modulates intestinal immune functions in rats. *Journal of Nutrition*, **134**, 153–156.

Rosenfeldt, V., Michaelsen, K.F., Jakobsen, M., Larsen, C.N., Moller, P.L., Pedersen, P., Tvede, M., Weyrehter, H., Valerius, N.H. & Paerregaard, A. (2002a) Effect of probiotic *Lactobacillus* strains in young children hospitalized with acute diarrhea. *Pediatric Infectious Diseases Journal*, **21**, 411–416.

Rosenfeldt, V., Michaelsen, K.F., Jakobsen, M., Larsen, C.N., Moller, P.L., Tvede, M., Weyrehter, H., Valerius, N.H. & Paerregaard, A. (2002b) Effect of probiotic *Lactobacillus* strains on acute diarrhea in a cohort of nonhospitalized children attending day-care centers. *Pediatric Infectious Diseases Journal*, **21**, 417–419.

Saavedra, J.M., Bauman, N.A., Oung, I., Perman, J.A. & Yolken, R.H. (1994) Feeding of *Bifidobacterium bifidum* and *Streptococcus thermophilus* to infants in hospital for prevention of diarrhoea and shedding of rotavirus. *Lancet*, **344**(8929), 1046–1049.

Saavedra, J.M., Abi-Hanna, A., Moore, N. & Yolken, R.H. (2004) Long-term consumption of infant formulas containing live probiotic bacteria: tolerance and safety. *American Journal of Clinical Nutrition*, **79**, 261–267.

Sadlack, B., Merz, H., Schorle, H., Schimpl, A., Feller, A.C. & Horak, I. (1993) Ulcerative colitis-like disease in mice with a disrupted interleukin-2 gene. *Cell*, **75**, 253–261.

Schultz, M., Timmer, A., Herfarth, H.H., Sartor, R.B., Vanderhoof, J.A. & Rath, H.C. (2004) *Lactobacillus* GG in inducing and maintaining remission of Crohn's disease. *BMC Gastroenterology*, **4**, 5–8.

Seki, H., Shiohara, M., Matsumura, T., Miyagawa, N., Tanaka, M., Komiyama, A. & Kurata, S. (2003) Prevention of antibiotic-associated diarrhea in children by *Clostridium butyricum* MIYAIRI. *Pediatrics International*, **45**, 86–90.

Sen, S., Mullan, M.M., Parker, T.J., Woolner, J.T., Tarry, S.A. & Hunter, J.O. (2002) Effect of *Lactobacillus plantarum* 299v on colonic fermentation and symptoms of irritable bowel syndrome. *Digestive Diseases and Science*, **47**, 2615–2620.

Shornikova, A.V., Casas, I.A., Mykkanen, H., Salo, E. & Vesikari, T. (1997) Bacteriotherapy with *Lactobacillus reuteri* in rotavirus gastroenteritis. *Pediatric Infectious Diseases Journal*, **16**, 1103–1107.

Shorter, R.G., Cardoza, M.R., ReMine, S.G., Spencer, R.J. & Huizenga, K.A. (1970) Modification of in vitro cytotoxicity of lymphocytes from patients with chronic ulcerative colitis or granulomatous colitis for allogenic colonic epithelial cells. *Gastroenterology*, **58**, 692–698.

Siitonen, S., Vapaatalo, H., Salminen, S., Gordin, A., Saxelin, M., Wikberg, R. & Kirkkola, A.L. (1990) Effect of *Lactobacillus* GG yoghurt in prevention of antibiotic associated diarrhoea. *Annals of Medicine*, **22**, 57–59.

Stamey, T. (1973) The role of enterobacteria in recurrent urinary infections. *Journal of Urology*, **109**, 467–472.

Sudarmo, S.M., Ranuh, R.G., Rochim, A. & Soeparto, P. (2003) Management of infant diarrhea with high-lactose probiotic-containing formula. *Southeast Asian Journal of Tropical Medicine and Public Health*, **34**, 845–848.

Surawicz, C.M., Elmer, G.W., Speelman, P., McFarland, L.V., Chinn, J. & Van Belle, G. (1989) Prevention of antibiotic-associated diarrhoea by *Saccharomyces boulardii*: A prospective study. *Gastroenterology*, **96**, 981–988.

Szajewska, H., Kotowska, M., Mrukowicz, J. Z., Armanska, M. & Mikolajczyk, W. (2001) Efficacy of *Lactobacillus* GG in prevention of nosocomial diarrhea in infants. *Journal of Pediatrics*, **138**, 361–365.

Tankanow, R.M., Ross, M.B., Ertel, I.J., Dickinson, D.G., McCormick, L.S. & Garfinkel, J.F. (1990) A double-blind, placebo-controlled study of the efficacy of Lactinex in the prophylaxis of amoxicillin-induced diarrhea. *Drug Intelligence and Clinical Pharmacy*, **24**, 382–384.

Tannock, G.W. (2004) Can the gut microflora of infants be modified by giving probiotics to mothers? *Journal of Pediatric Gastroenterology and Nutrition*, **38**(3), 244–246.

Taurog, J.D., Richardson, J.A., Croft, J.T., Simmons, W.A., Zhou, M., Fernandez-Sueiro, J.L., Balish, E. & Hammer, R.E. (1994) The germfree state prevents development of gut and joint inflammatory disease in HLA-B27 transgenic rats. *Journal of Experimental Medicine*, **180**, 2359–2364.

Tedesco, F.J., Hardin, R.D., Harper, R.N. & Edwards, B.H. (1983) Infectious colitis endoscopically simulating inflammatory bowel disease: a prospective evaluation. *Gastrointestinal Endoscopy*, **29**, 195–197.

Thomas, M.R., Litin, S.C., Osmon, D.R., Corr, A.P., Weaver, A.L. & Lohse, C.M. (2001) Lack of effect of *Lactobacillus* GG on antibiotic-associated diarrhea: a randomized, placebo-controlled trial. *Mayo Clinic Proceedings*, **76**, 883–889.

Tromm, A., Niewerth, U., Khoury, M., Baestlein, E., Wilhelms, G., Schulze, J. & Stolte, M. (2004) The probiotic *E. coli* strain Nissle 1917 for the treatment of collagenous colitis: first results of an open-label trial. *Zeitschrift fur Gastroenterologie*, **42**, 365–369.

Turunen, U.M., Farkkila, M.A., Hakala, K., Seppala, K., Sivonen, A., Ogren, M., Vuoristo, M., Valtonen, V.V. & Miettinen, T.A. (1998) Long-term treatment of ulcerative colitis with ciprofloxacin: a prospective, double-blind, placebo-controlled study. *Gastroenterology*, **115**, 1072–1078.

Vanderhoof, J.A., Whitney, D.B., Antonson, D.L., Hanner, T.L., Lupo, J.V. & Young, R.J. (1999) *Lactobacillus* GG in the prevention of antibiotic-associated diarrhea in children. *Journal of Pediatrics*, **135**, 564–568.

Venturi, A., Gionchetti, P., Rizzello, F., Johansson, R., Zucconi, E., Brigidi, P., Matteuzzi, D. & Campieri, M. (1999) Impact on the composition of the faecal flora by a new probiotic preparation: preliminary data on maintenance treatment of patients with ulcerative colitis. *Alimentary Pharmacology and Therapeutics*, **13**, 1103–1108.

de Vrese, M. & Schrezenmeis, J. (2002) Probiotics and non-intestinal infectious conditions. *British Journal of Nutrition*, **88**(Suppl 1), S59–S66.

Weese, J.S. (2002) Microbiologic evaluation of commercial probiotics. *Journal of the American Veterinary Medicine Association*, **220**(6), 794–797.

van der Wielen, P.W., Lipman, L.J., van Knapen, F. & Biesterveld, S. (2002) Competitive exclusion of *Salmonella enterica* serovar *Enteritidis* by *Lactobacillus crispatus* and *Clostridium lactatifermentans* in a sequencing fed-batch culture. *Applied and Environmental Microbiology*, **68**, 555–559.

Wullt, M., Hagslatt, M.L. & Odenholt, I. (2003) *Lactobacillus plantarum* 299v for the treatment of recurrent *Clostridium difficile*-associated diarrhoea: a double-blind, placebo-controlled trial. *Scandinavian Journal of Infectious Disease*, **35**, 365–367.

Wunderlich, P.F., Braun, L., Fumagalli, I., D'Apuzzo, V., Heim, F., Karly, M., Lodi, R., Politta, G., Vonbank, F. & Zeltner, L. (1989) Double-blind report on the efficacy of lactic acid-producing *Enterococcus* SF68 in the prevention of antibiotic-associated diarrhoea and in the treatment of acute diarrhoea. *Journal of International Medical Research*, **17**, 333–338.

Xiao, S.D., Zhang D.Z., Lu, H., Jiang, S.H., Liu, H.Y., Wang, G.S., Xu, G.M., Zhang, Z.B., Lin, G.J. & Wang, G.L. (2003) Multicenter, randomized, controlled trial of heat-killed *Lactobacillus acidophilus* LB in patients with chronic diarrhea. *Advances in Therapeutics*, **20**, 253–260.

# 8 Production of Vitamins, Exopolysaccharides and Bacteriocins by Probiotic Bacteria

E.B. O'Connor, E. Barrett, G. Fitzgerald, C. Hill, C. Stanton and R.P. Ross

## 8.1 Introduction

Probiotics are 'live micro-organisms which, when consumed in adequate amounts, confer a health benefit on the host' (FAO/WHO, 2001). In this respect, for a culture to be termed a probiotic it needs to have associated clinical evidence of health promotion in humans or animals. While this is the case for a limited number of strains, the precise mechanism by which probiotics exert a health effect *in vivo* is often not understood. One aspect that is clear, however, is that some intestinal strains produce certain health-promoting metabolites – including proteins and fatty acids – which are desirable from a nutritional and/or health-promotion perspective. The production of such compounds in fermented products means that many are often produced in the food prior to consumption. However, it should be emphasised that the ingestion of probiotic fermented foods opens up the possibility that these health-promoting metabolites may also be produced *in vivo*. This chapter will detail the production of three types of complex biomolecules by probiotic bacteria, namely vitamins, exopolysaccharides (EPS) and bacteriocins, and discuss their potential for health promotion in humans.

## 8.2 Vitamin production by lactic acid bacteria (LAB) and *Bifidobacterium* spp.

### 8.2.1 *Introduction*

Vitamins play a crucial role in human health, where their molecular functions can vary from being involved in genetic regulation to their antioxidant properties (Olson, 1994). Most vitamins must be obtained from the diet due to the inability of humans to synthesise them (Mathews & van Holde, 1990), although it is well known that some intestinal bacteria do produce certain vitamins. In this respect, it has been suggested that vitamin production is one of a number of functional characteristics associated with probiotic bacteria and gastrointestinal microbiota (Vaughan *et al.*, 1999; Holzapfel & Schillinger, 2002). Furthermore, a large number of LAB and *Bifidobacterium* spp. have been reported to produce vitamins, such as folate, cobalamin, menaquinone (vitamin $K_2$), riboflavin and thiamine. As such, the use of these cultures in food fermentation potentially provides routes not only to enhance the nutritional profile of the food but also to deliver micro-organisms to the gut, where they can synthesise such vitamins *in vivo*.

## 8.2.2 *Folate*

Folate is an essential component of the human diet, and is involved in nucleotide and cofactor biosynthesis in many metabolic reactions. Folate is a generic term used to describe the salts of folic acid (pteroylmonoglutamic acid). Folates are receiving increasing attention due to the link between folate deficiency and neural tube defects in developing embryos during pregnancy (Collins, 1994), and their ability to protect against certain types of cancer (Ames, 1999). Furthermore, a link has been established between decreasing levels of plasma homocysteine, a risk factor in the development of coronary heart disease, and the consumption of folic acid (Morrison *et al.*, 1996). Dietary folate is essential for humans, since it cannot be synthesised by mammalian cells (Sarma & Duttagupta, 1995) and is required at high levels by tissues with high growth rates, such as leucocytes, erythrocytes and the intestinal mucosa (Crittenden *et al.*, 2003). This vitamin is synthesised by linking pteridine, *p*-aminobenzoic acid and glutamic acid, with poly-γ-glutamic acid tails of varying lengths (Fig. 8.1). It then undergoes further enzymatic modifications to generate derivatives, which play essential roles as acceptors and donors of $C_1$ in metabolic pathways (Hamm-Alvarez *et al.*, 1989; Bailey & Gregory, 1999).

It has been reported that milk contains between 20 μg and 50 μg $L^{-1}$ of folate (Hugenholtz *et al.*, 2002b); however, the levels of folate in fermented milk have been shown to be higher, with folate concentrations in excess of 140 μg $L^{-1}$ detected in yoghurt (Smid *et al.*, 2001), with *Streptococcus thermophilus*, and not *Lactobacillus delbrueckii* subsp. *bulgaricus*, the reported producer (Rao *et al.*, 1984). This increased level of folate is due to the metabolic activity of LAB during the fermentation process. Indeed, a large number of LAB and *Bifidobacterium* spp. have been reported to produce folate (Smid *et al.*, 2001; Crittenden *et*

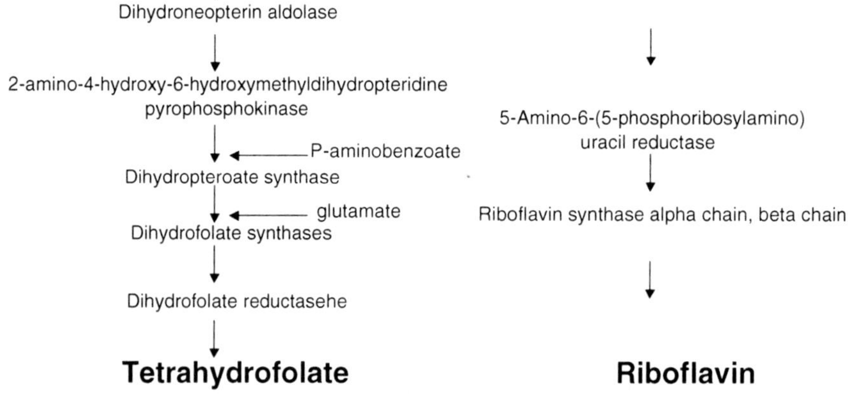


**Fig. 8.1** Enzymatic pathways involved in folate and riboflavin biosynthesis.

*al.*, 2003), including two *Bifidobacterium longum*, *Lactobacillus acidophilus* and *Lb. delbrueckii* subsp. *bulgaricus* strains, which were found to produce folate levels in reconstituted skimmed milk far in excess of the levels produced in synthetic media (M17, MRS) (Lin & Young, 2000). The levels of folate produced after 6 h of incubation varied from 53.9 μg $L^{-1}$ by *Lb. acidophilus* 4356 to 99.2 μg $L^{-1}$ by *B. longum* B6.

Bifidobacteria have been reported to produce folate in a number of studies. For instance, animal trials have reported that rats fed human milk solids had increased levels of both plasma folate and total caecal material folate, which coincided with a seven- and one-fold increase in caecal and colonic *Bifidobacterium* spp. densities, respectively (Krause *et al.*, 1996). In addition, a recent study has reported that a number of other probiotic species of bifidobacteria, which included *Bifidobacterium animalis* subsp. *lactis*, *Bifidobacterium animalis* subsp. *animalis*, *Bifidobacterium infantis* and *Bifidobacterium breve*, produce folate in reconstituted skimmed milk, with values ranging from 20 ng to 45 ng $g^{-1}$ (Crittenden *et al.*, 2003). The use of mixed cultures, which include bifidobacteria, can also provide further increases in the levels of folate. For example, the folate levels in probiotic yoghurt, containing *B. animalis* subsp. *lactis*, were over 33 ng $g^{-1}$, compared with levels under 25 ng $g^{-1}$ in conventional yoghurt, while a mixed culture of *B. animalis* subsp. *animalis* and *S. thermophilus* generated over 70 ng $g^{-1}$ folate (Crittenden *et al.*, 2003).

In contrast to bifidobacteria, it has been reported that strains of *Lactobacillus*, used as both starter and probiotics, generally utilise more folate than they produce. There are however, exceptions with a number of *Lactobacillus* strains able to generate excess folate in the fermentation of dairy products, including *Lactobacillus plantarum*, *Lb. acidophilus* and *Lb. delbrueckii* subsp. *bulgaricus* (Lin *et al.*, 1999; Crittenden *et al.*, 2003; Sybesma *et al.*, 2003a). Since many potential probiotic *Lactobacillus* strains utilise folate, researchers have employed metabolic engineering strategies to develop a folate over-producing *Lactobacillus* spp. A recent study has described the transformation of a folate-utilising probiotic strain, *Lactobacillus gasseri*, into a folate producer through metabolic engineering (Wegkamp *et al.*, 2004). The genetic determinants for folate biosynthesis by *Lactococcus lactis* subsp. *lactis* had been identified (Sybesma *et al.*, 2003a), and were introduced into *Lb. gasseri* on a broad-spectrum host range vector. The resulting strain was capable of producing up to 75 ng $mL^{-1}$ total folate (Wegkamp *et al.*, 2004). This and other examples, using *Lac. lactis* subsp. *lactis* (Sybesma *et al.*, 2003b, 2003c), show the potential which exists to develop genetically modified probiotic strains to over-produce folate, thereby, possibly providing 100% of the required folate intake in the human diet instead of the current 15–20% (Hugenholtz *et al.*, 2002b).

In addition to the aforementioned probiotic strains, two *Leuconostoc* strains – namely, *Leuconostoc lactis* and *Leuconostoc paramesenteroides* – also produce folate (Sybesma *et al.*, 2003b). Even though *Leuconostoc* spp. may not be considered as probiotics, they are, however, used in probiotic products (Goldin, 1998; Holzapfel *et al.*, 1998). Furthermore, there is increasing evidence to support the potential of propionibacteria as probiotic cultures (Zarate *et al.*, 2002a, 2002b, 2004), a genus that has been reported to produce the bioactive fatty acid conjugated linoleic acid (CLA) (Jiang *et al.*, 1998), vitamin $B_{12}$ and folate (Hugenholtz *et al.*, 2002a). It has been reported that the levels of folate produced by different propionibacteria were greater than or equal to the levels produced by the well-known producer *S. thermophilus* (Hugenholtz *et al.*, 2002a). Fourteen strains of propionibacteria were reported to produce folate, with total levels ranging from 9 to 78 ng $mL^{-1}$, with strains possessing varying abilities to excrete the vitamin (Hugenholtz *et al.*, 2002a).

### 8.2.3 Vitamin $B_{12}$ production

Vitamin $B_{12}$, also known as cobalamin, is an important cofactor for the metabolism of amino acids, carbohydrates, fatty acids and nucleic acids (Quesada-Chanto *et al.*, 1994). Although very few food-grade micro-organisms possess the ability to produce this vitamin (Hugenholtz *et al.*, 2002a), there are some members of the genera *Propionibacterium*, *Bifidobacterium* and *Lactobacillus* which have been reported to have this rare activity (Deguchi *et al.*, 1985; Hugenholtz *et al.*, 2002a; Taranto *et al.*, 2003). This vitamin is produced commercially by strains of *Pseudomonas denitrificans*, *Bacillus megaterium* and propionibacteria. Therefore, since dairy propionibacteria are the only food-grade commercial producers of vitamin $B_{12}$, the pathway used for its biosynthesis has been well characterised and is reported to involve at least 25 steps (Hugenholtz *et al.*, 2002a). The fermentation process for the production of cobalamin can be optimised by controlling the aerobic and anaerobic phases of the process to obtain yields on glucose of up to 200 mg vitamin $B_{12}$ $kg^{-1}$ fermentation mesh (Hunik, 2002).

A number of bifidobacteria species have also been reported to produce cobalamin, such as *Bifidobacterium adolesentis* (0.35 ng $mL^{-1}$), *B. bifidum* (0.65 ng $mL^{-1}$), *B. breve* (0.49 ng $mL^{-1}$), *B. infantis* (0.39 ng $mL^{-1}$) and *B. longum* (0.46 ng $mL^{-1}$) (Deguchi *et al.*, 1985). In addition to propionibacteria and bifidobacteria, *Lactobacillus reuteri* CRL1098, a probiotic strain which exhibits a hypocholesterolaemic effect in animal trials (Taranto *et al.*, 2000), has also been shown to produce cobalamin (Taranto *et al.*, 2003). In this case, the sourdough-isolate was capable of generating the cobalamin required by the enzyme glycerol dehydratase, which converts glycerol to 3-hydroxypropionaldehyde, in a cobalamin-free glucose-glycerol co-fermentation, while the RP-HPLC isolated cobalamin displayed a spectrum identical to that of the standard vitamin $B_{12}$ (Taranto *et al.*, 2003). One drawback associated with the use of this strain in the supplementation of foods with cobalamin is that the vitamin is not excreted from the cells. However, the genes encoding the pathways involved in cobalamin biosynthesis were identified, and this may allow metabolic engineering strategies to be exploited to over-express the genes in a strain which would release the vitamin during gastrointestinal transit (Taranto *et al.*, 2003).

### 8.2.4 Vitamin K production

Vitamin K is an important cofactor, involved in the post-translational carboxylation of glutamate residues to produce γ-carboxyglutamic acid (Gla), found in proteins involved in blood clotting, tissue calcification, in atherosclerotic plaque and tissues including bones and kidneys (Price *et al.*, 1976; Olson, 1984; Suttie, 1985). This vitamin occurs in two forms: (a) phylloquinone (vitamin $K_1$), which is present in green plants; and (b) menaquinone (MK) (vitamin $K_2$), which is produced by some intestinal bacteria (Briggs & Calloway, 1979; Conly & Stein, 1992). A range of LAB from a number of genera have been screened for the ability to produce MK; these included strains from the genera *Lactococcus*, *Lactobacillus*, *Enterococcus*, *Bifidobacterium*, *Leuconostoc* and *Streptococcus* (Morishita *et al.*, 1999). Four lactococcal strains and *Lac. lactis* subsp. *lactis* YIT 3001, which produced 648 nmol quinines $g^{-1}$ lyophilised cells, yielded in excess of 250 nmol quinines $g^{-1}$ lyophilised cells. While lactococci are not intestinal microbes *per se*, they are applied in probiotic products (Goldin, 1998; Holzapfel *et al.*, 1998), and can be incorporated into 'multi-species' products, which have been defined as 'containing strains of different probiotic species that belong to one or preferentially more genera, e.g. *Lb. acidophilus*, *B. longum*, *Enterococcus faecium* and *Lac. lactis* subsp. *lactis* (Timmerman *et al.*, 2004).

### 8.2.5 *Riboflavin and thiamine*

It has been reported that bifidobacteria can produce a further two B vitamins, riboflavin and thiamine. Deficiencies in riboflavin (vitamin $B_2$), a co-enzyme involved in numerous redox reactions, can lead to both liver and skin disorders (Lakshini, 1998). Deficiencies in thiamine (vitamin $B_1$), also a co-factor of key metabolic enzymes, can cause changes in brain glucose metabolism (Hakim & Pappius, 1981). Two bifidobacteria strains have been reported to increase the levels of both B vitamins during a 48-h fermentation in soymilk (Hou *et al.*, 2000). In these cases, the riboflavin content increased from 7.36 mg to 8.34 mg 100 $mL^{-1}$ and 8.88 mg 100 $mL^{-1}$ during the fermentations with *B. infantis* CCRC 14633 and *B. longum* B6, respectively, while the levels of thiamine increased from 0.33 mg to 0.38 mg 100 $mL^{-1}$ and 0.37 mg 100 $mL^{-1}$ during the fermentations with *B. infantis* CCRC 14633 and *B. longum* B6, respectively.

As with folate, there is huge potential to develop genetically modified probiotic strains with the ability to over-produce riboflavin. Indeed, strategies have been employed to over-produce riboflavin in association with folate in a riboflavin-utilizing strain of *Lac. lactis* subsp. *lactis* (Sybesma *et al.*, 2004). A spontaneous roseoflavin (riboflavin analogue) -resistant *Lac. lactis* subsp. *lactis* mutant was generated and the resultant riboflavin-producing variant was further transformed with a plasmid harbouring the *folKE* gene, encoding amino-4-hydroxy-6-hydroxymethyldihydropteridine pyrophosphokinase and GTP cyclohydrolase I involved in folate biosynthesis, to generate a folate and riboflavin over-producer. The extracellular level of riboflavin produced by the variant was 1200 ng $mL^{-1}$ greater than the level produced by the wild-type, while the variant produced 100 ng $mL^{-1}$ extracellular folate, with the control strain producing 10 ng $mL^{-1}$ folate (Sybesma *et al.*, 2004). Since some *Lactobacillus* strains possess all four genes required for riboflavin biosynthesis (Burgess *et al.*, 2004), it is possible that similar strategies could be applied to generate a genetically modified riboflavin-over-producing probiotic strain.

## 8.3 Exopolysaccharides (EPS) production by LAB and *Bifidobacterium* spp.

### 8.3.1 *Introduction*

A number of LAB, propionibacteria and bifidobacteria can synthesise EPS, which are excreted from the cell and which may or may not be loosely attached to the cell wall (Laws *et al.*, 2001). In cases where they are actually bound to the cell surface, they are referred to as 'capsular polysaccharides' (Cerning, 1994). EPS can contribute to the improved stability, rheology and texture of fermented dairy products, and may also offer protection to cells against phage attack, desiccation and osmotic stress (Ruas-Madiedo *et al.*, 2002). In addition to the technological characteristics they confer on dairy products, EPS have been reported to exert a number of beneficial health effects, including prebiotic (Gibson & Roberfroid, 1995), immunostimulatory (Hosono *et al.*, 1997), anti-tumoral (Kitazawa *et al.*, 1991), and blood cholesterol-lowering effects (Nakajima *et al.*, 1992). A large number of strains from the genus *Lactobacillus* have been reported to produce EPS (Ricciardi & Clementi,

2000; Ruas-Madiedo *et al.*, 2002), while it is a considerably rarer phenomenon associated with bifidobacteria (Sreekumar & Hosono, 1998). The levels of EPS formed are dependent on growth phase, temperature, pH and carbohydrate source, and are not a direct function of growth (Cerning, 1995). Further information regarding the role of EPS in fermented milks will be detailed in a separate book (*Fermented Milks*) to be published later within this Technical Series prepared on behalf of the Society of Dairy Technology (SDT) in the United Kingdom (Tamime, 2005).

### 8.3.2 *Classification of EPS*

EPS can be categorised into two types: (a) homopolysaccharides, which consist of a single type of monosaccharide; and (b) heteropolysaccharides, which consist of repeating units of polysaccharides and non-carbohydrate units, including phosphate, acetyl and glycerol (Ruas-Madiedo *et al.*, 2002). Homopolysaccharides are synthesised by excreted or anchored enzymes (glycansucrases) following the assembly of monosaccharides outside the cell, and require the substrate sucrose, which provides the energy for elongation (Duboc & Mollet, 2001; Jolly *et al.*, 2002; Ruas-Madiedo *et al.*, 2002). They are further categorised into two groups, fructans and glucans (Monsan *et al.*, 2001). The fructans include levan- and inulin-type homopolysaccharides, while the glucans include dextran, mutan, alteran and β-1,3 glucan (Monsan *et al.*, 2001). On the other hand, the precursors for heteropolysaccharides are synthesised within the cell, and involve isoprenoid glycosyl-carried lipids; they are subsequently transferred across the membrane by glycosyltransferases and assembled extracellularly (De Vuyst *et al.*, 2001; Ruas-Madiedo *et al.*, 2002).

Heteropolysaccharides generally contain D-glucose, D-galactose and L-rhamnose and in some cases *N*-acetylglucosamine, *N*-acetylgalactosamine or glucuronic acid (Ruas-Madiedo *et al.*, 2002). Their synthesis is closely related to carbon metabolism and the production of nucleotide sugars in the cell (Fig. 8.2) (De Vuyst & Degeest, 1999).

### 8.3.3 *Health benefits of EPS*

The majority of studies on EPS tend to focus on their technological aspects rather than their physiological benefits, with many of the investigations focusing on EPS structure, genetics, yield improvement and industrial applications. In this section of the chapter, we will focus on the potential health benefits associated with the consumption of EPS produced by LAB and bifidobacteria, and also outline the potential of EPS-producing bacteria as probiotic cultures and their incorporation into functional foods.

#### *Prebiotic effect of EPS*

One of the suggested health benefits of EPS consumption is its prebiotic effect. A prebiotic is described as 'a nondigestible food ingredient that beneficially affects the host by selectively stimulating the growth and/or activity of one or a limited number of bacteria in the colon and thus improves "host health"' (Gibson & Roberfroid, 1995). Thus, it is essential that prebiotics are not degraded in the stomach and small intestine (Crociani *et al.*, 1994). Indeed, EPS may meet this criterion as they have been reported to withstand '*in vivo*' passage through the gastrointestinal (GI) tract using a model simulating the human digestive process. Rats fed porridge containing 102 mg of EPS produced by *Lactococcus lactis* subsp. *cremoris* NZ4010

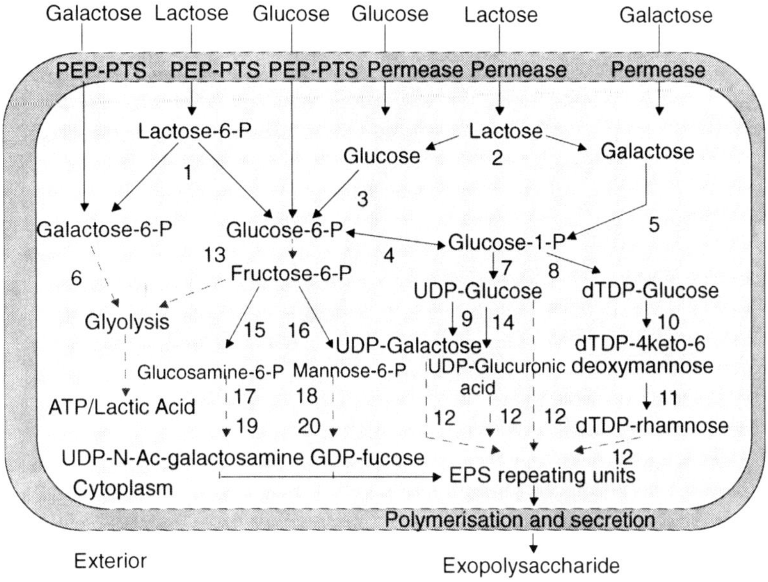


**Fig. 8.2** Schematic diagram of the conversion of lactose, galactose and glucose to nucleotide sugars in lactic acid bacteria (LAB). Note: The numbers refer to the enzymes/pathways involved; 1, Phospho-β-galactosidase; 2, β-galactosidase; 3, Glucokinase; 4, Phosphoglucomutase; 5, Leloir pathway; 6, Tagatose pathway; 7, UDP-glucose pyrophosphorylase; 8, dTDP-glucose pyrophosphorylase; 10, dTDP-glucose 4,6-dehydratase; 11, dTDP-mannose synthetic enzyme system; 12, Glycosyltransferases; 13, Phosphoglucose isomerase; 14, UDP-glucose dehydrogenase; 15, glucosamine-P isomerase; 16, mannose-6P isomerase; 17, *N*-acetylglucosamine-P deacetylase; 18, phosphomannomutase; 19, UDP-*N*-acetylglucosamine pyrophosphorylase; 20, GDP-mannose-4, 6 dehydratase.

resulted in the recovery of 96% of the EPS in their faeces (98 mg EPS per 4.44 g faeces), demonstrating that it was not degraded in the GI tract (Looijesteijn *et al.*, 2001). It was also reported that EPS produced by strains from different genera were biodegraded to varying extents, with EPS produced by *Lactobacillus sakei* 0-1 and *Lactobacillus helveticus* Lh59 being among the most durable (Ruijssenaars *et al.*, 2000). Moreover, it has been reported that the levan-type EPS produced by *Lactobacillus sansfranciscensis* TMW1.392 supported the growth of *Bifidobacterium bifidum*, *B. breve*, *B. infantis* and *B. adolescentis* (Korakli *et al.*, 2002), all of which are strains that are commonly employed for use as probiotic cultures (Goldin, 1998; Holzapfel *et al.*, 1998). During pH-controlled fermentation, all of the *Bifidobacterium* spp. exhibited diauxic growth, with fructose first being metabolised and the EPS being utilised after a second lag phase (Korakli *et al.*, 2002). These results would suggest that there is potential for EPS to be used as prebiotics; however, further '*in vivo*' studies regarding the degradability of EPS in the stomach and small intestine and utilisation by probiotic strains are required to substantiate these reports.

### *Immunostimulatory activity of EPS*

A number of reports have suggested that LAB can induce immunological responses in macrophage and T cells (Marin *et al.*, 1998), and that the EPS which they produce can play a role in eliciting these immunopotentiating activities. The EPS of strains from a number of food-grade genera, including *Bifidobacterium*, *Lactococcus* and *Lactobacillus*, have been reported to be immunostimulatory. For example, the water-soluble polysaccharide fraction of *B. adolescentis* M101-4 cells was reported to increase the [$^3$H]-thymidine uptake of murine splenocytes and Peyer's patches, an indication of cell proliferation, relative to whole cells (Hosono *et al.*, 1997). Furthermore, 75.9 g 100 g$^{-1}$ of the water-soluble fraction of sonicated cells consisted of polysaccharides containing glucose, galactose and *N*-acetylmuramic acid (Hosono *et al.*, 1997). In addition to EPS produced by bifidobacteria, the EPS of *Lac. lactis* subsp. *cremoris* KVS20 stimulated a mitogenic response in B cells (Kitazawa *et al.*, 1992). In this case, a dose of 100 μg mL$^{-1}$ slime products (EPS) enhanced the mitogenic activity of murine spleen cells [Stimulation Index (SI) = 6.29], stimulating B cells but not T cells. Furthermore, neither whole cells nor cell-wall components induced mitogenic activity in the absence of slime products (Kitazawa *et al.*, 1992). Similarly, the EPS of yet another LAB, *Lb. gasseri* LA158, stimulated mitogenic activity in murine spleen cells (SI = 5.97) and Peyer's patches (SI = 4.97), and once again the B cells, but not T cells, were induced by EPS (Takeda *et al.*, 1997).

In addition to the induced stimulation of B cells, administration of extracellular slime products from *Lac. lactis* subsp. *cremoris* SBT 0495, containing glucose, galactose and rhamnose, also enhanced the activity of specific murine antibodies (Nakajima *et al.*, 1995). Doses of 100–500 μg slime product per mouse resulted in optimum *in vivo* DNP-specific antibody production (Nakajima *et al.*, 1995). The immunopotentiating capabilities of EPS have also been reported following the production of the cytokines released by macrophages to communicate with other immune cells (Nishimura-Uemura *et al.*, 2003), interferon (IFN) and interleukin-1 (IL-1) by EPS-induced murine macrophages (Kitazawa *et al.*, 1996). For example, the phosphopolysaccharide produced by *Lac. lactis* subsp. *cremoris* KVS20 stimulated murine spleen Mϕ macrophages, part of the gut-associated lymphoid tissue (GALT) and producer cells of IFN and IL-1 in response to immune stimuli, to produce IFN titres of 27.6 IU mL$^{-1}$ compared to 9.7 IU mL$^{-1}$ for the control. Indeed, it was noted that EPS stimulation of the macrophages, to secrete cytokines, was accompanied by IFN-mRNA expression (Kitazawa *et al.*, 1996). The addition of the same EPS to spleen macrophages caused an increase in IL-1 activity, a phenomenon which was also accompanied by IL-1 mRNA expression after stimulation with the EPS (Kitazawa *et al.*, 1996).

The immunostimulation of cytokine-producing macrophages is not limited to EPS produced by lactococci. For example, *Lactobacillus rhamnosus* RW-9595M has also been reported to stimulate a number of cytokines, tumour necrosis factor (TNF), IL-6, IL-12 and IFN-γ (Chabot *et al.*, 2001). Indeed, the cellular growth (130.8% relative to the control) of murine splenocytes stimulated with a high molecular-weight fraction (F1) of the EPS isolated from *Lb. rhamnosus* RW-9595M was comparable to splenocyte cellular growth (132.7%) stimulated with 0.1 μg mL$^{-1}$ of lipopolysaccharide (LPS), a toxin known to be a strong immunostimulating factor and B-cell mitogen (Takada & Kotani, 1989; Kitazawa *et al.*, 1996; Chabot *et al.*, 2001). In addition to cellular growth, the entire EPS and its fraction F1 induced murine cells to secrete the cytokine TNF and also elicited splenocytes to secrete IL-6 and produce IFN-γ. Many of these studies relate to murine cell lines, but human peripheral blood mononuclear cells (PBMC) were also stimulated to secrete TNF-α by EPS and exhibited IL-6 and IL-12p40 responses to stimulation by EPS isolated from *Lb. rhamnosus* RW-9595M (Chabot *et al.*, 2001).

It is also interesting that the yoghurt strain *Lb. delbrueckii* subsp. *bulgaricus* OLL1073R-1 has been reported to produce an EPS capable of stimulating a mitogenic response in murine B lymphocytes (Kitazawa *et al.*, 1998). More specifically, when the EPS was fractionated into two components by ion-exchange chromatography to yield a neutral polysaccharide (NPS) and an acidic polysaccharide (APS), the APS produced by *Lb. delbrueckii* subsp. *bulgaricus* OLL1073R-1 – termed APS 1073 – enhanced mitogenic activity (Kitazawa *et al.*, 1998). The highest mitogenic activity of spleen cells induced by APS 1073 was obtained at a concentration of 200 $\mu g\ mL^{-1}$ while APS 1073 also stimulated a mitogenic response in murine Peyer's patches (SI = 2). In this study, it was found that the phosphate group of the APS was vital for induction of mitogenic activity (Kitazawa *et al.*, 1998). The importance of the phosphate group was also noted when dextran, produced by *Leuconostoc mesenteroides* subsp. *mesenteroides*, was chemically phosphorylated (P-dextran) to generate a B-cell-specific mitogen (Sato *et al.*, 2004). Macrophage phagocytosis has also been enhanced *in vivo* and *in vitro* by APS 1073, with NPS having little or no effect (Kitazawa *et al.*, 2000). The stimulation of macrophages with APS 1073 led to changes in macrophage morphology and gene expression of cytokines following induction with APS 1073 (Nishimura-Uemura *et al.*, 2003). Interestingly, the morphology of the murine macrophage-like cell line J774.1 changed considerably when stimulated with APS, while mRNA expression (as measured by RT-PCR) showed that APS stimulation of J774.1 resulted in IL-1, IL-6, IL-12p40 and IL-10 production (Nishimura-Uemura *et al.*, 2003). These results clearly show the ability of EPS from LAB and bifidobacteria to elicit immune responses in cells in animal studies. It is important to note that while it has been reported that *Lb. delbrueckii* subsp. *bulgaricus* cannot inhabit the intestine, *Lac. lactis* subsp. *cremoris* ARH74 was seen to adhere to cultured intestinal epithelial cells using a phosphopolysaccharide. Thus, it was suggested that *Lb. delbrueckii* subspp. *bulgaricus* OLL1073R-1 may have the potential to be used as a probiotic culture if adherence to epithelial cells could be demonstrated (Kitazawa *et al.*, 2000).

### *Anti-tumoral activity of EPS*

It has been suggested that yoghurt demonstrates anti-tumour activity (Perdigon *et al.*, 1998), and that one of the mechanisms may involve the EPS produced by the cultures during yoghurt production (Kitazawa *et al.*, 2000). Indeed, the extracts from milk fermented by *Lb. delbrueckii* subsp. *bulgaricus* and *S. thermophilus* were shown to have anti-mutagenic activity (Bodana & Rao, 1990). Furthermore, the macrophage Mϕ, which is activated by ThioGlycolate, was stimulated by the phosphopolysaccharide fraction (APS) of the EPS produced by *Lb. delbrueckii* subsp. *bulgaricus* OLL 1073R-1 (Kitazawa *et al.*, 2000). The cytostatic activity of the macrophage Mϕ was greatly enhanced against S180 sarcoma tumour cells. The APS stimulation resulted in just under 90% cytostaticity, while LPS caused just over 90% cytostaticity by the macrophages (Kitazawa *et al.*, 2000). The APS was also reported to be an effective stimulator of Mϕ cytostaticity against P388 tumour cells, with cytostaticity reaching 60% for both APS and LPS (Kitazawa *et al.*, 2000). Once again, the macrophage stimulation was shown to be dependent on the phosphate group of the APS. In a further study, a polysaccharide isolated from the supernatant of *B. longum* $PS^+$ was reported to exhibit an anti-mutagenic effect against a known mutagen (Sreekumar & Hosono, 1998) whereby, 60 μL of crude polysaccharide solution (1.323 $g\ L^{-1}$) caused 89.4% inhibition of the mutagenicity of Trp-P-1.

*EPS and blood cholesterol-lowering effects*

The EPS from *Lac. lactis* subsp. *cremoris* has been reported to have a cholesterol-lowering effect in rats (Nakajima *et al.*, 1992). In this study, the serum cholesterol concentration of rats fed ropy fermented milk (84.0 mg $d^{-1}$) was lower than that in rats fed non-ropy fermented milk (95.7 mg $d^{-1}$) or milk acidified with lactic acid (102.0 mg $d^{-1}$). In addition, the high-density lipoprotein (HDL) cholesterol/total cholesterol ratio was significantly higher in rats fed ropy fermented milk than in animals fed with the other two products.

### 8.3.4 *Genetic engineering for microbial production of EPS*

At present, there is a huge potential for the development of systems for heterologous expression and over-production of EPS in different starter and probiotic bacteria. Indeed, such strains could be exploited to increase the EPS content of foods, thereby improving their technological and potentially physiological/health characteristics. Strategies have involved the over-expression of genes central to EPS synthesis such as UDP glucose pyrophosphorylase and phosphoglucomutase (Levander *et al.*, 2002). Moreover, the *eps* gene cluster of *Lac. lactis* subsp. *lactis* NIZO B40 has been characterised, and reported to be plasmid encoded; therefore, it may be expressed in other lactococcal strains (van Kranenburg *et al.*, 1999a, 1999b). Indeed, the entire *eps* gene cluster from the plasmid pNZ4000 of *Lac. lactis* subsp. *lactis* NIZO B40 has been homologously expressed in *Lac. lactis* subsp. *lactis* on a high copy number vector, pIL253. The resultant strain produced 343 mg $L^{-1}$ EPS, while a strain harbouring a low copy number plasmid produced 93 mg $L^{-1}$ EPS. Furthermore, the *eps* gene cluster from *S. thermophilus* Sfi6 has been heterologously expressed in *Lac. lactis* subsp. *lactis* MG1363, a non-EPS-producing strain (Stingele *et al.*, 1999; Germond *et al.*, 2001), demonstrating the potential of heterologously expressing different EPS systems in a variety of LAB probiotic strains.

## 8.4 Production of bacteriocins by probiotic cultures

### 8.4.1 *Introduction*

It has been suggested that one of the desirable properties of a probiotic strain is an ability to produce antimicrobial substances such as bacteriocins, which offer the potential to provide a competitive advantage in colonisation and competition in the GI tract. Bacteriocins are generally defined as peptides which are produced by bacteria and kill other related and unrelated micro-organisms. The best known example is nisin, a broad host-range bacteriocin produced by many strains of *Lac. lactis* subsp. *lactis* and considered to be the prototype LAB bacteriocin. Nisin was first described in 1928, when Rogers and Whittier observed inhibitory metabolites to other LAB (Rogers & Whittier, 1928). Nisin has since been studied extensively and, in 1988, was awarded FDA approval for its use as a biopreservative in many processed foods. This major breakthrough, with nisin being the first bacteriocin to gain widespread commercial application, led to considerable interest in bacteriocins produced by generally regarded as safe (GRAS) organisms. As a result, the field has grown dramatically, and this has led to the discovery and detailed characterisation of a great number of bacteriocins from LAB.

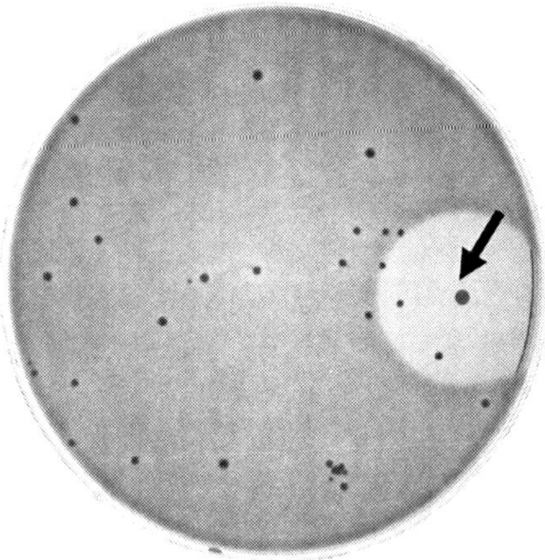

**Fig. 8.3** Production of bacteriocins by intestinal LAB is a relatively common trait. The figure shows a faecal isolate producing a large bacteriocin inhibitory zone. Data with permission of Gillian Gardiner (personal communication).

Concomitant with the discovery of these new bacteriocins, several intestinal and/or probiotic strains have been shown to display these inhibitory peptides (Fig. 8.3). In this respect, most of the probiotic bacteriocins characterised to date are of *Lactobacillus* origin. It is fair to say, though, that the *in vivo* significance of bacteriocin production in relation to strain dominance, competitiveness and pathogen inhibition is not known at this time due to a scarcity of *in vivo* studies. However, looking at their influence on the intestinal gut flora, animal studies have shown reductions in Enterobacteriaceae due to these antimicrobial metabolites (Gardiner *et al.*, 2004), and clinical trials have reported that humans fed *Lactobacillus johnsonii*-fermented milk had significantly reduced densities of *Helicobacter pylori*, a pathogen which is associated with stomach ulcers (Felley *et al.*, 2001). Such potent inhibitors provide a clear opportunity for survival and proliferation of organisms if they can eliminate a competing undesirable organism. Since the most widely used cultures for probiotic applications belong to the genera *Lactobacillus* and *Bifidobacterium*, the antimicrobial activities of species belonging to these genera are the subject of this section.

## 8.4.2 *Production of antimicrobials as a probiotic trait*

There are many examples where bacteriocins have been shown to have a dominant influence on complex microbial populations. A very good example of this is in the oral cavity, where *Streptococcus mutans* producing the bacteriocin mutacin provided the selective advantage to persist, colonise and aggressively displace the indigenous *S. mutans* population (Hillman *et al.*, 1985, 1987; Hillman, 2002). It was found that, after 14 years post-treatment of a single concentrated cell suspension of the recombinant non-acidogenic *S. mutans* mutacin-producing strain, all of the available subjects remained colonised by this strain and that no other *mutans* streptococci were observed. This example clearly illustrates the potential of bacteriocins within complex ecosystems. Furthermore, bacteriocins play a role in predominance over other micro-organisms in food fermentation systems. When the bacteriocin-producing *Lb. plantarum* was used to ferment Spanish-style green olives, the bacteriocin producer outnumbered all other individual naturally occurring *Lactobacillus* populations over the course of the 12-week fermentation (Ruiz-Barba *et al.*, 1994; Ruiz-Barba & Jimenez-Diaz, 1994). In contrast, the derivative bacteriocin-negative strain was not detected after 7 weeks. Similarly, lacticin 3147 is able to control food pathogens in a variety of fermentations, with Ryan *et al.* (1996) showing that lacticin 3147-producing starter cultures could control the developing microflora in ripened fermented products. Ample evidence shows that bacteriocins play a fundamental role in influencing complex microbial populations. A constant search for novel probiotic strains with bacteriocin antagonistic traits such as a selective advantage to persist and colonise, are actively screened for every day. Therefore, mining the gut flora metagenome for novel bacteriocins and antimicrobial compounds presents a rational approach for the selection of probiotics.

### 8.4.3 *Classification of bacteriocins*

Generally, bacteriocins are ribosomally synthesised polypeptides, and are produced by micro-organisms that are immune to their own action. They are normally modified post-translationally to some degree, with the secreted mature peptides usually ranging in size from 20 to 60 amino acids, and possessing bactericidal activity (Joerger & Klaenhammer, 1986). Almost all bacteriocin peptides have a net charge at neutral or slightly acidic pH, and they usually contain stretches of sequence that are hydrophobic and/or amphiphilic (Eijsink *et al.*, 2002). As the list of bacteriocins continues to grow, significant diversity in their structure and activity is evident, and this has meant that their classification continues to be updated. The current classification divides bacteriocins into three main classes (Table 8.1; see also Tamime, 2005).

#### *Class I: Lantibiotics*

Class I bacteriocins, known as the lantibiotics, were initially broadly grouped according to structure, with type A being elongated amphiphilic peptides and type B more compact and globular (Jung, 1991). More recently, Twomey *et al.* (2002) further subdivided lantibiotics into six subgroups based on primary sequence comparisons (see Table 8.1). They contain a set of unique features including complex, post-translationally modified peptides containing the unusual amino acids lanthionine, and β-methyllanthionine, which form characteristic intramolecular (thioether) ring structures (Gross & Morell, 1971; Jack & Sahl, 1995; Sahl *et al.*, 1995; McAuliffe *et al.*, 2000; Twomey *et al.*, 2002). Many lantibiotics also possess other modified residues, such as dehydro amino acids, D-alanine residues (Skaugen *et al.*, 1994; Ryan *et al.*, 1999), N-terminal alpha keto amines (Kellner *et al.*, 1991), and oxidative decarboxylation of C-terminal cysteine to yield a C-terminal S-aminovinylcysteine (Schnell *et al.*, 1988; Bierbaum *et al.*, 1996). Most of the characterised lantibiotics appear to have a common mode of action in that they dissipate the proton motive force in target organisms through the formation of pores in the cytoplasmic membrane (Garcia-Garcera *et al.*, 1993; Montville & Bruno, 1994; Abee *et al.*, 1995; Moll *et al.*, 1996). Interestingly, some of these bacteriocins use a mechanism which first involves binding to the cell wall precursor lipid II and, thus, they employ a dual mechanism which includes inhibition of cell wall synthesis (Brotz *et al.*, 1998a, 1998b). Indeed many lantibiotics, such as mersacidin and lacticin 3147, share a motif which, in the case of mersacidin, has been shown to be involved in the interaction with lipid II (Brotz *et al.*, 1998a). The genetic determinants of several lantibiotics from

**Table 8.1** Outline of classification of bacteriocins for Gram-positive bacteria including examples of *Lactobacillus* and *Bifidobacterium* spp. characterised bacteriocins.

| Class | Subgroup/subclass | Bacteriocin | Micro-organism | Reference(s) |
|---|---|---|---|---|
| Class I: Lantibiotics | Subgroup I - Nisin | Nisin | *Lac. lactis* subsp. *lactis* | Roger and Whiter (1928) |
| | Subgroup II - Lacticin 481 | Plantaricin C | *Lb. plantarum* | Gonzalez *et al.* (1994) |
| | Subgroup III - Mersacidin | Plantaricin plwa | *Lb. plantarum* | Holo *et al.* (2001) |
| | Subgroup IV - LtnA2 | Plantaricin plwb | *Lb. plantarum* | Holo *et al.* (2001) |
| | Subgroup V - Cytolysis group | Cytolysin CylLL & CylLS | *E. faecalis* | Gilmore *et al.* (1994) |
| | Subgroup VI - Lactocin S | Lactocin S | *Lb. sake* | Skaugen *et al.* (1994) |
| Class II: Bacteriocins | Subclass IIa - Pediocin-like | Acidocin A | *Lb. acidophilus* TK9201s | Kanatani *et al.* (1995) |
| | | Acidocin CH5 | *Lb. acidophilus* CH5 | Chumchalova *et al.* (2004) |
| | | Bavaricin A | *Lb. sake* M1401 | Larsen *et al.* (1993) |
| | | Bifidocin 1454 | *B. bifidum* | Yildirim *et al.* (1998) |
| | | Curvacin A | *Lb. curvatus* LTH1174 | Tichaczek *et al.* (1993) |
| | | Bavaricin MN | *Lb. sake* MN | Kaiser and Montville (1996) |
| | | Sakacin A | *Lb. sake* LB706 | Holck *et al.* (1992) |
| | | Sakacin G | *Lb. sake* 2512 | Simon *et al.* (2002) |
| | | Sakacin P | *Lb. sake* LB674 | Tichaczek *et al.* (1992); Huhne *et al.* (1996) |
| | Subclass IIa - Two peptides | ABP-118 | *Lb. salivarius* subsp. *salivarius* UCC118 | Flynn *et al.* (2002) |
| | | Acidocin LF221 | *Lb. gasseri* LF221 | Bogovic-Matijasic *et al.* (1998); Majhenic *et al.* (2004) |
| | | Acidocin JI132 | *Lb. acidophilus* JCM 1132 | Tahara *et al.* (1996) |
| | | Lactacin F | *Lb. johnsonii* VPI 11088 | Fremaux *et al.* (1993); Abee *et al.* (1994) |
| | | Lactobin A | *Lb. amylovorus* LMG P-13139 | Contreras *et al.* (1997) |
| | | Lactocin 705 | *Lb. casei* CRL 705 | Cuozzo *et al.* (2000) |
| | | Plantaricin EF | *Lb. plantarum* C11 | Anderssen *et al.* (1998) |
| | | Plantaricin JK | *Lb. plantarum* C11 | Anderssen *et al.* (1998) |
| | | Plantaricin NC8 | *Lb. plantarum* NC8 | Maldonado *et al.* (2003) |
| | | Plantaricin S | *Lb. plantarum* LPCO10 | Jimenez-Diaz *et al.* (1995) |
| | Subclass IIc - Sec-dependent | Acidocin B | *Lb. acidophilus* M46 | Leer *et al.* (1995) |
| | | Gassericin A | *Lb. gasseri* LA39 | Kawai *et al.* (1994) |
| Class III: Bacteriocins | Class III | Helveticin J | *Lb. helveticus* 481 | Joerger and Klaenhammer (1986) |
| | | Helveticin V-1829 | *Lb. helveticus* 1829 | Vaughan *et al.* (1992) |
| | | Lactacin B | *Lb. acidophilus* N2 | Barefoot and Klaenhammer (1984) |

LAB have now been fully determined, including nisin (Kaletta & Entian, 1989; Kuipers *et al.*, 1993), lacticin 3147 (Dougherty *et al.*, 1998; Ryan *et al.*, 1999), cytolysin (Gilmore *et al.*, 1996), lactocin S (Skaugen *et al.*, 1994), lacticin 481 (van den Hooven *et al.*, 1996b) and plantaricin W (Holo *et al.*, 2001). While high-resolution structural characterisation of some lantibiotics have been elucidated, including nisin (van den Hooven *et al.*, 1996a), mersacidin (Prasch *et al.*, 1997) and lacticin 3147 (Martin *et al.*, 2004), further investigations are essential to unfold the precise mechanisms involved in immunity, post-translational modifications, the mode of action and structure–function relationships.

In the context of this section there are three *Lactobacillus* lantibiotic bacteriocins – plantaricin C (Gonzalez *et al.*, 1994), plantaricin W (Holo *et al.*, 2001) and lactosin S (Mortvedt *et al.*, 1991) – that have been characterised to date. Plantaricin C is a broad-spectrum bacteriocin produced by *Lb. plantarum* isolated from ripening cheese (Gonzalez *et al.*, 1994). Determination of the solution structure of this bacteriocin (Turner *et al.*, 1999) revealed a 27-amino acid peptide with a linear N-terminal end and a globular C-terminus. Structure similarity with lacticin 481 designated it a member of subgroup II lacticin 481 (see Table 8.1) (Twomey *et al.*, 2002). In contrast, plantaricin W is a two-component bacteriocin, with both peptides displaying inherent antimicrobial activity (Holo *et al.*, 2001). This is unusual given that the other two characterised two-component lantibiotics, lacticin 3147 (Ryan *et al.*, 1999) and staphylococcin C55 (Navaratna *et al.*, 1998), have only one peptide which displays activity alone. The two peptides plwα and plwβ belong to separate subgroups, that is, the mersacidin and ltnA2 groups, respectively (see Table 8.1). Another *Lactobacillus* lantibiotic lactocin S, which is produced by *Lb. sake* L45, is not subgrouped with any other lantibiotic group as its primary sequence shares little similarity with any other known lantibiotics. Interestingly, molecular characterisation of this bacteriocin showed that it contained D-alanine, which the authors suggest is derived from serine (Skaugen *et al.*, 1994).

From a probiotic applications point of view, the lantibiotic class are possibly the most interesting group of bacteriocins, especially as their inhibition spectra can vary from medium to very broad. One such example is plantaricin, which displays a medium spectrum of inhibition in contrast to the broad-spectrum, commercially available lantibiotic nisin (Rogers & Whittier, 1928; Marki *et al.*, 1991; Holo *et al.*, 2001).

### *Class II: Bacteriocins*

The class II bacteriocins are known to be small (<10 kDa), non-modified, heat-stable bacteriocins, and include three subclasses as follows (see Table 8.1).

#### *Subclass IIa*

The best known are members of subclass IIa, known as pediocin-like bacteriocins. These are characterised by an N-terminal consensus sequence motif YGNGVXCXXXXCXV, which includes two cysteines that can form a disulphide bridge (Eijsink *et al.*, 1998, 2002). This group has attracted much attention due to its strong anti-listerial activity (Ennahar *et al.*, 2000a, 2000b). Bifidocin B, a member of this subclass, and produced by *B. bifidum* NCFB 1454, is the only bifidobacteria bacteriocin characterised fully to date (Yildirim & Johnson, 1998; Yildirim *et al.*, 1999). Bifidocin B displays activity against a number of food-borne pathogens and food spoilage bacteria, such as *Listeria*, *Enterococcus*, *Bacillus*, *Lactobacillus*, *Leuconostoc* and *Pediococcus* species. Curing experiments have indicated that the genetic determinants of bifidocin B are associated with an 8-kb plasmid.

*Subclass IIb*

These are two-component bacteriocins with both peptides required for full activity (see Table 8.1). Members of this subgroup include lactacin F and plantaricin NC8, each of which requires two peptides for full activity. Lactacin F, produced by *Lb. johnsonii* VPI 11088 (previously classified as *Lb. acidophilus* 11088), has a relatively narrow spectrum of inhibition, and is bactericidal against only closely related bacteria (Abee *et al.*, 1994). Another example is plantaricin NC8; this is produced by *Lb. plantarum* NC8, and is unusually only produced when co-cultured with other Gram-positive bacteria such as *Pediococcus*, *Lactococcus* and *Listeria* species (Maldonado *et al.*, 2004a, 2004b). Recently, ABP-118 – a novel two-component bacteriocin produced by the probiotic bacterium *Lactobacillus salivarius* subsp. *salivarius* UCC118 – was purified and characterised (Flynn *et al.*, 2002). *Lb. salivarius* subsp. *salivarius* UCC118 is a human intestinal isolate chromosomally encoding the genetic determinants for ABP-118. It exhibits a broad spectrum of inhibition, and is capable of inhibiting medically significant pathogens such as *Bacillus*, *Listeria*, *Enterococcus* and *Staphylococcus* species. Although the health-contributing effects of this bacteriocin-producing probiotic are yet to be assessed, given its spectrum it should provide the bacterium with a competitive advantage in the complex microbial environment of the gut.

*Subclass IIc*

This subclass is a diverse set of bacteriocins containing all non-lantibiotics that do not belong to classes IIa or IIb, including sec-dependent secreted bacteriocins. Table 8.1 includes two *Lactobacillus* bacteriocins which belong to this group. One of these is gassericin A, which is produced by the probiotic *Lb. gasseri* strain LA39 and was isolated from human infant faeces (Kawai *et al.*, 1994). This chromosomally encoded bacteriocin has an uncommon cyclic structure, and displays inhibitory activity against a wide variety of food-borne pathogenic bacteria including *Listeria monocytogenes*, *Bacillus cereus* and *Staphylococcus aureus*.

#### *Class III: Bacteriocins*

The class III bacteriocins are large, heat-labile protein bacteriocins of which very few have been described from *Lactobacillus* and bifidobacteria. They are less well-characterised, and examples include helveticin J produced by *Lb. helveticus* 481 (Joerger & Klaenhammer, 1986), helveticin V produced by *Lb. helveticus* 1829 (Vaughan *et al.*, 1992), and lactacin B produced by *Lb. acidophilus* N2 (Barefoot & Klaenhammer, 1984). They all share a narrow inhibitory spectrum, and are antagonistic only to closely related species.

### **8.4.4** *Antimicrobial potential of* Lactobacillus *spp.*

The application of *Lactobacillus* antimicrobial agents has been demonstrated in many food systems, which in many cases indicates the effectiveness of these potent inhibitors to control undesirable bacteria. Lactocin 705, for example, reduced numbers of the food-borne pathogen *L. monocytogenes* from $1.4 \times 10^4$ to $1 \times 10^2$ colony forming units (cfu)$mL^{-1}$ after 24 h in a ground meat slurry (Vignolo *et al.*, 1996; Palacios *et al.*, 1999). In addition, the same food pathogen was inhibited by the bacteriocin producer *Lactobacillus bavaricus* (currently known as *Lb. sake* or *Lactobacillus curvatus*) in three different beef systems at refrigeration temperatures (Winkowski *et al.*, 1993). Acidocin A, which was produced by the starter organism for the production of fermented milk, *Lb. acidophilus* TK901, inhibits food spoilage bacteria and food pathogens including *Propionibacterium* spp. and *Enterococcus faecalis* (Kanatani *et al.*, 1995). Another example is plantaricin C, which was isolated from ripened cheese and is active against a wide range of Gram-positive bacteria, including *Clostridium* spp. (Gonzalez *et al.*, 1994). Furthermore, gassericin A, which is produced by *Lb. gasseri* LA39 and isolated from human infant faeces, shows bactericidal activity against pathogens such as *L. monocytogenes*, *B. cereus* and *S. aureus* (Kawai *et al.*, 1994). This intestinal strain is permitted in fermented products to promote health and inhibit pathogens. Interestingly, the Food Hygiene Law in Japan prohibits the use of bacteriocins such as nisin in food; however, this delivery system for bacteriocins is permitted (Saito, 2004). In addition, lactobacilli strains such as *Lb. johnsonii* La1, *Lb. gasseri* K7 and *Lb. acidophilus* IBB 801 exhibited bacteriocin production when grown in milk supplemented with yeast extract (Avonts *et al.*, 2004; see also Majhenic *et al.*, 2003).

The potential benefits for applications of antimicrobial-producing *Lactobacillus* strains in animal feed systems are becoming apparent (Haddadin *et al.*, 1997). For example, chicken *Lactobacillus* isolates with uncharacterised antimicrobial attributes, *Lactobacillus animalis* (currently known as *Lactobacillus murinus*) and *Lb. fermentum,* show potential as chicken probiotic feed ingredients intended to control salmonellosis and improve poultry sanitation (Gusils *et al.*, 1999). In addition, trials with pigs have shown a reduction in *Enterobacteriaceae* on administration of a bacteriocin-producing *Lb. salivarius* subsp. *salivarius* strain, which also has important probiotic properties, such as bile and acid tolerance (Casey *et al.*, 2004; Gardiner *et al.*, 2004). These bacteriocin-producing strains may offer potential as an alternative to antibiotics, and may be useful as a means of controlling pathogen carriage; they may, therefore, be highly suitable as microbial feed additives.

Several *in vitro* studies have also examined the antimicrobial potential of *Lactobacillus* towards clinically significant pathogens and infections. For example, *Lactobacillus* spent culture supernatants significantly inhibited the proliferation of a protozoan, *Giardia intestinalis,* which causes diarrhoeal disease worldwide (Perez *et al.*, 2001). In addition, the human *Lb. acidophilus* LB strain displays antagonistic activity against both Gram-positive and Gram-negative pathogens, including antagonistic activity against *H. pylori* and *Salmonella typhimurium* infection *in vitro* and *in vivo* (Coconnier *et al.*, 1997, 1998, 2000). Interestingly, certain *Lactobacillus* strains have the ability to interfere with the adherence and growth of uropathogenic bacteria (McGroarty & Reid, 1988). This interaction is believed to be important in the maintenance of a normal urogenital flora and in the prevention of infection in females. For example, vaginal *Lactobacillus* isolates displaying bacteriocin-like substances that are inhibitory towards *E. faecalis*, *E. faecium* and *Neisseria gonorrheae* (Ocana *et al.*, 1999) and *Gardnerella vaginalis* (Aroutcheva *et al.*, 2001; Alpay-Karaoglu, 2003) show potential application in a probiotic to prevent urogenital infections. Combinations of *Lactobacillus* strains have been selected for the preparation of vaginal tablets to treat such infections; the performance of the formulation was optimised *in vitro*, though *in vivo* studies must also be performed. Interestingly, among the characteristics of selected *Lactobacillus* strains was an ability to inhibit the growth of *G. vaginalis* (Mastromarino *et al.*, 2002).

The role of *Lactobacillus* antimicrobial agents as being a desirable property of a probiotic is becoming apparent from the results of clinical studies. For example, *Lb. acidophilus* La1 culture supernatant was shown to have a partial, acid-independent long-term suppressive effect on *H. pylori* in clinical trials (Michetti *et al.*, 1999). More recently, it was reported that humans fed *Lb. johnsonii* La1-fermented milk which demonstrated bactericidal activity (presently not characterised) had significant reductions in the density of *H. pylori* (Felley *et al.*, 2001). Such results suggest that the consumption of *Lb. johnsonii*-acidified milk might down-regulate *H. pylori* infection and gastritis.

### 8.4.5 *Antimicrobial potential of* Bifidobacterium *spp.*

Unlike the situation for *Lactobacillus*, only a limited number of studies have been performed to demonstrate the production of antimicrobial compounds or bacteriocins among bifidobacterial strains. In this respect, bifidobacteria are generally inhibitory to a wide range of micro-organisms due to their intense production of lactic and acetic acids as part of their normal metabolic processes (Eklund, 1983). However, Gibson and Wang (1994) showed that the antagonistic activity of several species of bifidobacteria towards both Gram-positive and Gram-negative pathogens was not solely attributed to the presence of acids. These authors showed that eight strains of bifidobacteria were able to excrete bactericidal or bacteriostatic substances. Bifidocin B, the only bifidobacteria bacteriocin characterised to date, is a class IIa bacteriocin; this material is homologous to pediocin-like bacteriocins, and strongly inhibits food-borne pathogens, supposedly by binding to specific cell wall receptors such as lipoteichoic acid to form pores (Yildirim *et al.*, 1999). *In vitro* studies showed that bifidobacteria inhibited enterohaemorrhagic *Escherichia coli* 0157:H7 (Gagnon *et al.*, 2004) and *L. monocytogenes* due to the production of inhibitory compounds (Toure *et al.*, 2003). Interestingly, it was found that anti-enterotoxigenic *E. coli* activity was more effective in acidic conditions (Gopal *et al.*, 2001). Food trial results on cottage cheese inoculated with inhibitor-producing bifidobacteria strains indicated that overall levels of *Pseudomonas* were significantly reduced (O'Riordan & Fitzgerald, 1998). Following evidence of anti-*S. typhimurium* activity *in vitro*, *in vivo* tests demonstrated a protective effect against *S. typhimurium* lethal infection in mice (Lievin *et al.*, 2000). However, bifidobacterial antimicrobial compounds displayed *in vitro* do not always confer reductions in pathogen numbers *in vivo*. For example, a higher survival during a *Salmonella* challenge of mice pre-treated with *B. longum* Bb46 was reported (Silva *et al.*, 2004), but the effect was attributed to a reduced inflammatory response mediated by the probiotic treatment rather than to population antagonism. Similar findings for *B. animalis* subsp. *lactis* showed a 10-fold increase in the survival rate of *S. typhimurium*-challenged mice that was attributed to enhanced immune function (Shu *et al.*, 2000). *B. breve* YIT4064 enhanced antigen-specific IgA antibodies directed against rotavirus in the mouse mammary gland, thereby protecting pups that received milk against rotavirus challenge (Yasui *et al.*, 1995). Other antimicrobial compounds were identified, including a proteinaceous proposed binding inhibitor, produced by *B. longum* that interfered with *Enterobacteriaceae* adherence to a common bacterial binding GA1 site (Fujiwara *et al.*, 1997). Furthermore, using agar well assay technique, Christopher *et al.* (2004) reported antimicrobial activity against *Shigella dysenteriae*, *E. coli*, *Salmonella typhi* and *Yersinia enterocolitica* by *B. animalis* subsp. *lactis* BB-12 and *B. bifidum* DSM 20456; thus, it was suggested that these strains could be used as dietary adjuncts in fermented milk products. In a recent study, Saleh and El-Sayed (2004) reported the activity of two bacteriocins produced by *B. animalis* subsp. *lactis* Bb-12 (known as biflact Bb-12) and *B.* longum Bb-46 (known as biflong Bb-46). SDS-PAGE profiles indicated a heterogeneous composition for both bacteriocins: Biflact Bb-12 consisted of four bands with molecular weights between 25 and 89 kDa, whilst 13 bands were obtained in biflong Bb-46 with molecular weights between 25 and 127 kDa. Both bacteriocins were sensitive to pepsin and trypsin, but were resistant to $\alpha$-amylase or lipase when tested by agar diffusion assay or by RP-HPLC.

### 8.4.6 *Heterologous expression*

The development of heterologous expression systems for bacteriocins offers a number of advantages over the natural host system. These include the possibility of elevated production levels, controlled bacteriocin gene expression, broadened inhibitory spectrum and increased efficacy within certain ecosystems. Moreover, it opens up the possibility of expressing a number of potent bacteriocin systems, which are not of intestinal origin in probiotic cultures. This may in turn offer a competitive advantage in the gut. Advanced genetic tools have led to an escalating numbers of heterologous expression systems, which are now well-developed for lactobacilli, but less so for bifidobacteria. For example, the lantibiotic, lactocin S, was successfully expressed in heterologous hosts *Lb. sakei* 790 and *Lb. plantarum* NC8 (Skaugen *et al.*, 1999). Likewise, acidocin A – a class II bacteriocin produced by the starter strain *Lb. acidophilus* TK9201 – was successfully expressed in heterologous hosts *Lb. casei* TK9008 and *Lb. acidophilus* TK1-5 (Kanatani *et al.*, 1995). Helveticin J, a narrow-spectrum class III bacteriocin, was the first reported case of heterologous production of a LAB bacteriocin (Joerger & Klaenhammer, 1986). The recombinant plasmid pTRK135 encoding the genes for helveticin J production was also heterologously expressed in the *Lb. acidophilus* host with identical activity spectrum to the wild-type strain.

The powerful anti-listerial activity of pediocin PA-1 makes this bacteriocin highly desirable, particularly to the dairy industry. Pediocin PA-1 has been expressed in many backgrounds including *E. coli*, *Pediococcus*, *Lactococcus* and *Lactobacillus* spp. (Marugg *et al.*, 1992; Chikindas *et al.*, 1995; Venema *et al.*, 1995; Coderre & Somkuti, 1999; Horn *et al.*, 1999, 2004; Martinez *et al.*, 2000). The *Lactobacillus* host, *Lb. sake* Lb790, a bacteriocin-negative meat-associated strain, heterologously expressed the strong anti-listerial activity of pediocin PA-1, thereby enhancing its commercial potential (Axelsson *et al.*, 1998). Many bacteriocin-producing probiotic strains are only active against Gram-positive bacteria; heterologous expression systems could overcome this problem and enable these strains to co-express specific Gram-negative targeting bacteriocins such as colicin. The co-production of two bacteriocins – one targeting Gram-positive bacteria and the second inhibitory to Gram-negative bacteria – has already been achieved in the lactococcal host, *Lac. lactis* subsp. *lactis* FI10054 (Horn *et al.*, 2004).

Heterologous expression has certainly opened up the possibility of taking known probiotics and superimposing them with bacteriocin production machinery. It would be of great interest to follow these 'super strains' in relation to their persistence, colonisation and pathogen interaction *in vivo*.

## 8.5 Overall conclusions

For millennia, milk has been preserved by fermentation using lactic acid bacteria, the primary role of which is to convert lactose to lactic and other organic acids, thus lowering the pH. In addition, many of these food cultures can produce a range of secondary metabolites during fermentation, which may influence not only the quality, safety, flavour and texture of the fermented food, but also its nutritional and health status. Probiotics are described as 'live micro-organisms which, when consumed in adequate amounts, confer a health benefit on the host' (FAO/WHO, 2001). Lactobacilli, bifidobacteria and, to a lesser extent propionibacteria, are being increasingly exploited in probiotic dairy products such as cheese, yoghurt and milk drinks, given the accumulating clinical evidence for their human health-promoting activities. Such probiotic-containing dairy foods are associated with a range of health claims, including the alleviation of symptoms of lactose intolerance, treatment of infectious diarrhoea, cancer suppression and the reduction of blood cholesterol. A number of mechanisms have been proposed to explain these health benefits, including the secretion by probiotic bacteria of beneficial nutrients such as short-chain fatty acids, vitamins, bioactive peptides and fatty acids, bacterial–host signalling molecules and antimicrobial substances (Fooks *et al.*, 1999).

In this chapter, we have presented some examples whereby the bioactivities of LAB – particularly probiotic bacteria – can be exploited and/or accentuated to produce complex biomolecules. For example, the (over)production of vitamins by LAB provides a very attractive approach to improve the nutritional profile of fermented foods, while EPS production is associated with enhanced rheological properties, in addition to the biogenic effects associated with their ingestion. The ability of probiotic LAB to produce antimicrobial substances such as bacteriocins may provide an advantage in terms of survival and proliferation in the gut, and may have a controlling influence on the composition of the gut microflora. Furthermore, the production of these antimicrobials offers a type of self-preservation to foods by protecting them against certain pathogenic and spoilage micro-organisms. It is important to emphasise that, while production of these biomolecules may have important nutritional and safety implications for the food itself, what may be even more important is the production by probiotics of bioactive molecules *in vivo* in the gut, where they may have a greater impact on human health. In addition, such endeavours will be greatly facilitated by genomic-based studies, which will contribute towards our greater understanding of the mechanisms underlying these health-promoting effects.

## Acknowledgements

The preparation of this chapter was supported by the Department of Agriculture and Food under the National Development Plan (2000–2006), and Science Foundation Ireland Centre for Science Engineering and Technology (SFI CSET).

## References

Abee, T., Klaenhammer, T.R. & Letellier, L. (1994) Kinetic studies of the action of lactacin F, a bacteriocin produced by *Lactobacillus johnsonii* that forms poration complexes in the cytoplasmic membrane. *Applied and Environmental Microbiology*, **60**, 1006–1013.

Abee, T., Krockel, L. & Hill, C. (1995) Bacteriocins: modes of action and potentials in food preservation and control of food poisoning. *International Journal of Food Microbiology*, **28**, 169–185.

Alpay-Karaoglu, S., Aydin, F., Kilic, S.S & Kilic, A.O. (2003) Antimicrobial activity and characteristics of bacteriocins produced by vaginal lactobacilli. *Turkish Journal of Medical Science*, **33**, 7–13.

Ames, B.N. (1999) Micronutrient deficiencies cause DNA damage and cancer. *Food Science and Agricultural Chemistry*, **1**, 1–15.

Anderssen, E.L., Diep, D.B., Nes, I.F., Eijsink, V.G.H. & Nissen-Meyer, J. (1998) Antagonistic activity of *Lactobacillus plantarum* C11: two new two-peptide bacteriocins, plantaricins EF and JK and the induction factor plantaricin A. *Applied and Environmental Microbiology*, **64**, 2269–2272.

Aroutcheva, A.A., Simoes, J.A. & Faro, S. (2001) Antimicrobial protein produced by vaginal *Lactobacillus acidophilus* that inhibits *Gardnerella vaginalis*. *Infectious Diseases in Obstetrics and Gynecology*, **9**, 33–39.

Avonts, L., van Uytven, E. & De Vuyst, L. (2004) Cell growth and bacteriocin production of probiotic *Lactobacillus* strains in different media. *International Dairy Journal*, **14**, 947–955.

Axelsson, L., Katla, T., Bjornslett, M., Eijsink, V.G. & Holck, A. (1998) A system for heterologous expression of bacteriocins in *Lactobacillus sake*. *FEMS Microbiology Letters*, **168**, 137–143.

Bailey, L.B. & Gregory III, J.F. (1999) Folate requirements and metabolism. *Journal of Nutrition*, **129**, 779–782.

Barefoot, S.F. & Klaenhammer, T.R. (1984) Purification and characterization of the *Lactobacillus acidophilus* bacteriocin lactacin B. *Antimicrobial Agents and Chemotherapy*, **26**, 328–334.

Bierbaum, G., Gotz, F., Peschel, A., Kupke, T., van de Kamp, M. & Sahl, H.G. (1996) The biosynthesis of the lantibiotics epidermin, gallidermin, Pep5 and epilancin K7. *Antonie van Leeuwenhoek*, **69**, 119–127.

Bodana, A.R. & Rao, D.R. (1990) Antimutagenic activity of milk fermented by *Streptococcus thermophilus* and *Lactobacillus bulgaricus*. *Journal of Dairy Science*, **73**, 3379–3384.

Bogovic-Matijasic, B., Rogelj, I., Nes, I.F. & Holo, H. (1998) Isolation and characterization of two bacteriocins of *Lactobacillus acidophilus* LF221. *Applied Microbiology and Biotechnology*, **49**, 606–612.

Briggs, G. & Calloway, D. (1979) *Bogert's Nutrition and Physical Fitness*, 10th edn, W.B. Saunders Co., Philadelphia.

Brotz, H., Bierbaum, G., Leopold, K., Reynolds, P.E. & Sahl, H.G. (1998a) The lantibiotic mersacidin inhibits peptidoglycan synthesis by targeting lipid II. *Antimicrobial Agents and Chemotherapy*, **42**, 154–160.

Brotz, H., Josten, M., Wiedemann, I., Schneider, U., Gotz, F., Bierbaum, G. & Sahl, H.G. (1998b) Role of lipid-bound peptidoglycan precursors in the formation of pores by nisin, epidermin and other lantibiotics. *Molecular Microbiology*, **30**, 317–327.

Burgess, C., O'Connell-Motherway, M., Sybesma, W., Hugenholtz, J. & van Sinderen, D. (2004) Riboflavin production in *Lactococcus lactis*: Potential for in situ production of vitamin-enriched foods. *Applied and Environmental Microbiology*, **70**, 5769–5777.

Casey, P.G., Casey, G.D., Gardiner, G.E., Tangney, M., Stanton, C., Ross, R.P., Hill, C. & Fitzgerald, G.F. (2004) Isolation and characterization of anti-Salmonella lactic acid bacteria from the porcine gastrointestinal tract. *Letters in Applied Microbiology*, **39**, 431–438.

Cerning, J. (1994) Polysaccharides exocellulaires produits par les bacteries lactiques, *Bacteries Lactiques – Aspects Fondamentaux et Technologiques*, (eds H. de Roissart & F.M. Luquet), pp. 309–329, Lorica, Grenoble.

Cerning, J. (1995) Production of exopolysaccharides by lactic acid bacteria and dairy propionibacteria. *Lait*, **75**, 463–472.

Chabot, S., Yu, H.L., de Leseleuc, L., Cloutier, D., van Clasteren, M.R., Lessard, M., Roy, D., Lacroix, M. & Oth, D. (2001) Exopolysaccharides from *Lactobacillus rhamnosus* RW-9595M stimulate TNF, IL-6 and IL-12 in human and mouse cultured immunocompetent cells, and IFN-γ in mouse splenocytes. *Lait*, **81**, 683–697.

Chikindas ML, Venema K, Ledeboer AM, Venema G & Kok J (1995) Expression of lactococcin A and pediocin PA-1 in heterologous hosts. *Letters in Applied Microbiology*, **21**, 183–189.

Christopher, M.D., Reddy, V.P., Mrudula, N. & Venkateswarulu, K. (2004) Assessment of antimicrobial activity of *Bifidobacterium* ssp. *Cheiron*, **32**, 132–134.

Chumchalova, J., Stiles, J., Josephsen, J. & Plockova, M. (2004) Characterization and purification of acidocin CH5, a bacteriocin produced by *Lactobacillus acidophilus* CH5. *Journal of Applied Microbiology*, **96**, 1082–1089.

Coconnier, M.H., Lievin, V., Bernet-Camard, M.F., Hudault, S. & Servin, A.L. (1997) Antibacterial effect of the adhering human *Lactobacillus acidophilus* strain LB. *Antimicrobial Agents and Chemotherapy*, **41**, 1046–1052.

Coconnier, M.H., Lievin, V., Hemery, E. & Servin, A.L. (1998) Antagonistic activity against *Helicobacter* infection *in vitro* and *in vivo* by the human *Lactobacillus acidophilus* strain LB. *Applied and Environmental Microbiology*, **64**, 4573–4580.

Coconnier, M.H., Lievin, V., Lorrot, M. & Servin, A.L. (2000) Antagonistic activity of *Lactobacillus acidophilus* LB against intracellular *Salmonella enterica* serovar Typhimurium infecting human enterocyte-like Caco-2/TC-7 cells. *Applied and Environmental Microbiology*, **66**, 1152–1157.

Coderre, P.E. & Somkuti, G.A. (1999) Cloning and expression of the pediocin operon in *Streptococcus thermophilus* and other lactic fermentation bacteria. *Current Microbiology*, **39**, 295–301.

Collins, K. (1994) Folic acid supplements provide useful benefits. *Better Nutrition for Today's Living*, **56**(1), 18–19.

Conly, J.M. & Stein, K. (1992) The production of menaquinones (vitamin $K_2$) by intestinal bacteria and their role in maintaining coagulation homeostasis. *Progresses in Food and Nutritional Science*, **16**, 307–343.

Contreras, B.G.L., de Vuyst, L., Devreese, B., Busanyova, K., Raymaeckers, J., Bosman, F., Sablon, E. & Vandamme, E.J. (1997) Isolation, purification and amino acid sequence of lactobin A, one of the two bacteriocins produced by *Lactobacillus amylovorus* LMG P-13139. *Applied and Environmental Microbiology*, **63**, 13–20.

Crittenden, R.G., Martinez, N.R. & Playne, M.J. (2003) Synthesis and utilisation of folate by yoghurt starter cultures and probiotic bacteria. *International Journal of Food Microbiology*, **80**, 217–222.

Crociani, F., Alessandrini, A., Mucci, M.M. & Biavati, B. (1994) Degradation of complex carbohydrates by *Bifidobacterium* spp. *International Journal of Food Microbiology*, **24**, 199–210.

Cuozzo, S.A., Sesma, F., Palacios, J.M., de Ruiz Holgado, A.P. & Raya, R.R. (2000) Identification and nucleotide sequence of genes involved in the synthesis of lactocin 705, a two-peptide bacteriocin from *Lactobacillus casei* CRL 705. *FEMS Microbiology Letters*, **185**, 157–161.

Deguchi, Y., Morishita, T. & Mutai, M. (1985) Comparative studies on synthesis of water-soluble vitamins among human species of bifidobacteria. *Agricultural and Biological Chemistry*, **49**, 13–19.

De Vuyst, L. & Degeest, B. (1999) Heteropolysaccharides from lactic acid bacteria. *FEMS Microbiology Reviews*, **23**, 153–177.

De Vuyst, L., De Vin, F., Vaningelgem, F. & Degeest, B. (2001) Recent developments in the biosynthesis and applications of heteropolysaccharides from lactic acid bacteria. *International Journal of Food Microbiology*, **11**, 687–707.

Dougherty, B.A., Hill, C., Weidman, J.F., Richardson, D.R., Venter, J.C. & Ross, R.P. (1998) Sequence and analysis of the 60 kb conjugative, bacteriocin-producing plasmid pMRC01 from *Lactococcus lactis* DPC3147. *Molecular Microbiology*, **29**, 1029–1038.

Duboc, P. & Mollet, B. (2001) Applications of exopolysaccharides in the dairy industry. *International Dairy Journal*, **11**, 759–768.

Eijsink, V.G., Skeie, M., Middelhoven, P.H., Brurberg, M.B. & Nes, I.F. (1998) Comparative studies of class IIa bacteriocins of lactic acid bacteria. *Applied and Environmental Microbiology*, **64**, 3275–3281.

Eijsink, V.G., Axelsson, L., Diep, D.B., Havarstein, L.S., Holo, H. & Nes, I.F. (2002) Production of class II bacteriocins by lactic acid bacteria; an example of biological warfare and communication. *Antonie van Leeuwenhoek*, **81**, 639–654.

Eklund, T. (1983) The antimicrobial effect of dissociated and undissociated sorbic acid at different pH levels. *Journal of Applied Bacteriology*, **54**, 383–389.

Ennahar, S., Deschamps, N. & Richard, J. (2000a) Natural variation in susceptibility of *Listeria* strains to class IIa bacteriocins. *Current Microbiology*, **41**, 1–4.

Ennahar, S., Sashihara, T., Sonomoto, K. & Ishizaki, A. (2000b) Class IIa bacteriocins: biosynthesis, structure and activity. *FEMS Microbiology Reviews*, **24**, 85–106.

FAO/WHO (2001) *Evaluation of Health and Nutritional Properties of Probiotics in Food Including Powder Milk with Live Lactic Acid Bacteria*, Report of a Joint FAO/WHO Expert Consultation, Food and Agriculture Organization of the United Nations, Rome.

Felley, C.P., Corthesy-Theulaz, I., Rivero, J.L., Sipponen, P., Kaufmann, M., Bauerfeind, P., Wiesel, P.H., Brassart, D., Pfeifer, A., Blum, A.L. & Michetti, P. (2001) Favourable effect of an acidified milk (LC-1) on *Helicobacter pylori* gastritis in man. *European Journal of Gastroenterology and Hepatology*, **13**, 25–29.

Flynn, S., van Sinderen, D., Thornton, G.M., Holo, H., Nes, I.F. & Collins, J.K. (2002) Characterization of the genetic locus responsible for the production of ABP-118, a novel bacteriocin produced by the probiotic bacterium *Lactobacillus salivarius* ssp. *salivarius* UCC118. *Microbiology*, **148**, 973–984.

Fooks, L.J., Fuller, R. & Gibson, G.R. (1999) Prebiotics, probiotics and human gut microbiology. *International Dairy Journal*, **9**, 53–61.

Fremaux, C., Ahn, C. & Klaenhammer, T.R. (1993) Molecular analysis of the lactacin F operon. *Applied and Environmental Microbiology*, **59**, 3906–3915.

Fujiwara, S., Hashiba, H., Hirota, T. & Forstner, J.F. (1997) Proteinaceous factor(s) in culture supernatant fluids of bifidobacteria which prevents the binding of enterotoxigenic *Escherichia coli* to gangliotetraosylceramide. *Applied and Environmental Microbiology*, **63**, 506–512.

Gagnon, M., Kheadr, E.E., Le Blay, G. & Fliss, I. (2004) In vitro inhibition of *Escherichia coli* O157:H7 by bifidobacterial strains of human origin. *International Journal of Food Microbiology*, **92**, 69–78.

Garcia-Garcera, M.J., Elferink M.G., Driessen A.J. & Konings, W.N. (1993) In vitro pore-forming activity of the lantibiotic nisin: role of proton motive force and lipid composition. *European Journal of Biochemistry*, **212**, 417–422.

Gardiner, G.E., Casey, P.G., Casey, G., Lynch, P.B., Lawlor, P.G., Hill, C., Fitzgerald, G.F., Stanton, C. & Ross, R.P. (2004) Relative ability of orally administered *Lactobacillus murinus* to predominate and persist in the porcine gastrointestinal tract. *Applied and Environmental Microbiology*, **70**, 1895–1906.
Germond, J.E., Delley, M., D'Amico, N. & Vincent, S.J. (2001) Heterologous expression and characterization of the exopolysaccharide from *Streptococcus thermophilus* Sfi39. *European Journal of Biochemistry*, **268**, 5149–5156.
Gibson, G.R. & Roberfroid, M. (1995) Dietary modulation of the human colonic microbiota: introducing the concept of prebiotics. *Journal of Nutrition*, **124**, 1401–1412.
Gibson, G.R. & Wang, X. (1994) Regulatory effects of bifidobacteria on the growth of other colonic bacteria. *Journal of Applied Bacteriology*, **77**, 412–420.
Gilmore, M.S., Segarra, R.A., Booth, M.C., Bogie, C.P., Hall, L.R. & Clewell, D.B. (1994) Genetic structure of the *Enterococcus faecalis* plasmid pAD1-encoded cytolytic toxin system and its relationship to lantibiotic determinants. *Journal of Bacteriology*, **176**, 7335–7344.
Gilmore, M.S., Skaugen, M. & Nes, I. (1996) *Enterococcus faecalis* cytolysin and lactocin S of *Lactobacillus sake*. *Antonie van Leeuwenhoek*, **69**, 129–138.
Goldin, B.R. (1998) Health benefits of probiotics. *British Journal of Nutrition*, **80**, S203–S207.
Gonzalez, B., Arca, P., Mayo, B. & Suarez, J.E. (1994) Detection, purification, and partial characterization of plantaricin C, a bacteriocin produced by a *Lactobacillus plantarum* strain of dairy origin. *Applied and Environmental Microbiology*, **60**, 2158–2163.
Gopal, P.K., Prasad, J., Smart, J. & Gill, H.S. (2001) *In vitro* adherence properties of *Lactobacillus rhamnosus* DR20 and *Bifidobacterium lactis* DR10 strains and their antagonistic activity against an enterotoxigenic *Escherichia coli*. *International Journal of Food Microbiology*, **67**, 207–216.
Gross, E. & Morell, J.L. (1971) The structure of nisin. *Journal of American Chemical Society*, **93**, 4634–4635.
Gusils, C., Perez Chaia, A., Gonzalez, S. & Oliver, G. (1999) Lactobacilli isolated from chicken intestines: potential use as probiotics. *Journal of Food Protection*, **62**, 252–256.
Haddadin, M.S.Y., Abdulrahim, S., Odetallah, N.H. & Robinson, R.K. (1997) A proposed protocol for checking the suitability of *Lactobacillus acidophilus* for use during feeding trials with chickens. *Tropical Science*, **74**, 16–20.
Hakim, A.M. & Pappius, H.M. (1981) The effect of thiamine deficiency on local cerebral glucose utilization. *Annals of Neurology*, **9**, 334–339.
Hamm-Alvarez, S., Sancar, A. & Rajagopalan, K.V. (1989) Role of enzyme-bound 5,10-methenyltetrahydropteroylpolyglutamate in catalysis by *Escherichia coli* DNA photolyase. *Journal of Biological Chemistry*, **264**, 9649–9656.
Hillman, J.D. (2002) Genetically modified *Streptococcus mutans* for the prevention of dental caries. *Antonie van Leeuwenhoek*, **82**, 361–366.
Hillman, J.D., Yaphe, B.I. & Johnson, K.P. (1985) Colonization of the human oral cavity by a strain of *Streptococcus mutans*. *Journal of Dental Research*, **64**, 1272–1274.
Hillman, J.D., Dzuback, A.L. & Andrews, S.W. (1987) Colonization of the human oral cavity by a *Streptococcus mutans* mutant producing increased bacteriocin. *Journal of Dental Research*, **66**, 1092–1094.
Holck, A., Axelsson, L., Birkeland, S.E., Aukrust, T. & Blom, H. (1992) Purification and amino acid sequence of sakacin A, a bacteriocin from *Lactobacillus sake* Lb706. *Journal of General Microbiology*, **138**, 2715–2720.
Holo, H., Jeknic, Z., Daeschel, M., Stevanovic, S. & Nes, I.F. (2001) Plantaricin W from *Lactobacillus plantarum* belongs to a new family of two-peptide lantibiotics. *Microbiology*, **147**, 643–651.
Holzapfel, W.H. & Schillinger, U. (2002) Introduction to pre- and probiotics. *Food Research International*, **35**, 109–116.
Holzapfel, W.H., Haberer, P., Snel, J., Schillinger, U. & Huis in't Veld, J.H. (1998) Overview of gut flora and probiotics. *International Journal of Food Microbiology*, **41**, 85–101.
van den Hooven, H.W., Doeland, C.C., van de Kamp, M., Konings, R.N., Hilbers, C.W. & van de Ven, F.J. (1996a) Three-dimensional structure of the lantibiotic nisin in the presence of membrane-mimetic micelles of dodecylphosphocholine and of sodium dodecylsulphate. *European Journal of Biochemistry*, **235**, 382–393.
van den Hooven, H.W., Lagerwerf, F.M., Heerma, W., Haverkamp, J., Piard, J.C., Hilbers, C.W., Siezen, R.J., Kuipers, O.P. & Rollema, H.S. (1996b) The structure of the lantibiotic lacticin 481 produced by *Lactococcus lactis*: location of the thioether bridges. *FEBS Letters*, **391**, 317–322.
Horn, N., Martinez, M.I., Martinez, J.M., Hernandez, P.E., Gasson, M.J., Rodriguez, J.M. & Dodd, H.M. (1999) Enhanced production of pediocin PA-1 and coproduction of nisin and pediocin PA-1 by *Lactococcus lactis*. *Applied and Environmental Microbiology*, **65**, 4443–4450.

Horn, N., Fernandez, A., Dodd, H.M., Gasson, M.J. & Rodriguez, J.M. (2004) Nisin-controlled production of pediocin PA-1 and colicin V in nisin- and non-nisin-producing *Lactococcus lactis* strains. *Applied and Environmental Microbiology*, **70**, 5030–5032.

Hosono, A., Lee, J., Ametani, A., Natsume, M., Hirayama, M., Adachi, T. & Kaminogawa, S. (1997) Characterisation of a water soluble polysaccharide fraction with immunopotentiating activity from *Bifidobacterium adolescentis* M101-4. *Bioscience, Biotechnology and Biochemistry*, **61**, 312–316.

Hou, J.W., Yu, R.C. & Chou, C.C. (2000) Changes in some components of soymilk during fermentation with bifidobacteria. *Food Research International*, **33**, 393–397.

Hugenholtz, J., Hunik, J., Santos, H. & Smid, E.J. (2002a) Neutraceutical production by propionibacteria. *Lait*, **82**, 103–112.

Hugenholtz, J., Sybesma, W., Groot, M.N., Wisselink, W., Ladero, V., Burgess, K., van Dinderen, D., Piard, J.-C., Eggink, G., Smid, E.J., Savoy, G., Sesma, F., Jansen, T., Hols, P. & Kleerebezem, M. (2002b) Metabolic engineering of lactic acid bacteria for the production of nutraceuticals. *Antonie van Leeuwenhoek*, **82**, 217–235.

Huhne, K., Axelsson, L., Holck, A. & Krockel, L. (1996) Analysis of the sakacin P gene cluster from *Lactobacillus sake* Lb674 and its expression in sakacin-negative *Lb. sake* strains. *Microbiology*, **142**, 1437–1448.

Hunik, J. (2002) *Process for the production of vitamin $B_{12}$*. United States Patent No. 6492141B1.

Jack, R.W. & Sahl, H.G. (1995) Unique peptide modifications involved in the biosynthesis of lantibiotics. *Trends in Biotechnology*, **13**, 269–278.

Jiang, J., Bjorck, L. & Fonden, R. (1998) Production of conjugated linoleic acid by dairy starter cultures. *Journal of Applied Microbiology*, **85**, 95–102.

Jimenez-Diaz, R., Ruiz-Barba, J.L., Cathcart, D.P., Holo, H., Nes, I.F., Sletten, K.H. & Warner, P.J. (1995) Purification and partial amino acid sequence of plantaricin S, a bacteriocin produced by *Lactobacillus plantarum* LPCO10, the activity of which depends on the complementary action of two peptides. *Applied and Environmental Microbiology*, **61**, 4459–4463.

Joerger, M.C. & Klaenhammer, T.R. (1986) Characterization and purification of helveticin J and evidence for a chromosomally determined bacteriocin produced by *Lactobacillus helveticus* 481. *Journal of Bacteriology*, **167**, 439–446.

Jolly, L., Vincent, S.J., Duboc, P. & Neeser, J.R. (2002) Exploiting expolysaccharides from lactic acid bacteria. *Antonie van Leeuwenhoek*, **82**, 367–374.

Jung, G. (1991) Lantibiotics, a survey. *Nisin and Novel Lantibiotics*. (eds. Jung, G. & Sahl, H.-G. ), pp. 1–34. ESCOM, Leiden, the Netherlands.

Kaiser, A. & Montville, T.J. (1996) Purification of the bacteriocin bavaricin MN and characterization of its mode of action against *Listeria monocytogenes* Scott A cells and lipid vesicles. *Applied and Environmental Microbiology*, **62**, 4529–4535.

Kaletta, C. & Entian, K.D. (1989) Nisin, a peptide antibiotic: cloning and sequencing of the *nisA* gene and post-translational processing of its peptide product. *Journal of Bacteriology*, **171**, 1597–1601.

Kanatani, K., Oshimura, M. & Sano, K. (1995) Isolation and characterization of acidocin A and cloning of the bacteriocin gene from *Lactobacillus acidophilus*. *Applied and Environmental Microbiology*, **61**, 1061–1067.

Kawai, Y., Saito, T., Toba, T., Samant, S.K. & Itoh, T. (1994) Isolation and characterization of a highly hydrophobic new bacteriocin (gassericin A) from *Lactobacillus gasseri* LA39. *Bioscience, Biotechnology and Biochemistry*, **58**, 1218–1221.

Kellner, R., Jung, G. & Sahl, H.G. (1991) Structure elucidation of the tricyclic lantibiotic Pep5 containing eight positively charged amino acids. *Nisin and Novel Lantibiotics*, (eds G. Jung and H.G. Sahl), pp. 141–158, ESCOM Scientific Publishers BV, Leiden, the Netherlands.

Kitazawa, H., Harata, T., Uemura, J., Saito, T., Kaneko, T. & Itoh, T. (1998) Phosphate group requirement for mitogenic activation of lymphocytes by an extracellular phosphopolysaccharide from *Lactobacillus delbrueckii* ssp. *bulgaricus*. *International Journal of Food Microbiology*, **40**, 169–175.

Kitazawa, H., Toba, T., Kumano, N., Adachi, S. & Yamaguchi, T. (1991) Antitumoral activity of slime forming encapsulated *Lactococcus lactis* ssp. *cremoris* isolated from Scandinavian ropy sour milk 'viili'. *Animal Science Technology*, **62**, 277–283.

Kitazawa, H., Yamaguchi, T. & Itoh, T. (1992) B-cell mitogenic activity of slime products produced from slime-forming, encapsulated *Lactococcus lactis* ssp. *cremoris*. *Journal of Dairy Science*, **75**, 2946–2951.

Kitazawa, H., Itoh, T., Tomioka, Y., Mizugaki, M. & Yamaguchi, T. (1996) Induction of IFN-gamma and IL-1 alpha production in macrophages stimulated with phosphopolysaccharide produced by *Lactococcus lactis* ssp. *cremoris*. *International Journal of Food Microbiology*, **31**, 99–106.

Kitazawa, H., Ishii, Y., Uemura, J., Kawai, Y., Saito, T., Kaneko, T., Noda, K. & Itoh, T. (2000) Augmentation of macrophage functions by an extracellular phosphopolysaccharide from *Lactobacillus delbrueckii* ssp. *bulgaricus*. *Food Microbiology*, **17**, 109–118.
Korakli, M., Ganzle, M.G. & Vogel, R.F. (2002) Metabolism by bifidobacteria and lactic acid bacteria of polysaccharides from wheat and rye, and exopolysaccharides produced by *Lactobacillus sanfranciscensis*. *Journal of Applied Microbiology*, **92**, 958–965.
van Kranenburg, R., van Swam, I., Marugg, J.D., Kleerebezem, M. & de Vos, W. (1999a) Exopolysaccharide biosynthesis in *Lactococcus lactis* NIZO B40: Functional analysis of the glycosyltransferase genes involved in synthesis of the polysaccharide backbone. *Journal of Bacteriology*, **181**, 338–340.
van Kranenburg, R., Vos, H.R., van Swam, I.I., Kleerebezem, M. & de Vos, W.M. (1999b) Functional analysis of glycosyltransferase genes from *Lactococcus lactis* and other Gram-positive cocci: complementation, expression, and diversity. *Journal of Bacteriology*, **181**, 6347–6353.
Krause, L.J., Forsberg, C.W, & O'Connor, D.L. (1996) Feeding human milk to rats increases *Bifidobacterium* in the cecum and colon which correlates with enhanced folate status. *Journal of Nutrition*, **126**, 1505–1511.
Kuipers, O.P., Beerthuyzen, M.M., Siezen, R.J. & de Vos W.M. (1993) Characterization of the nisin gene cluster nisABTCIPR of *Lactococcus lactis*. Requirement of expression of the *nisA* and *nisI* genes for development of immunity. *European Journal of Biochemistry*, **216**, 281–291.
Lakshini, A.V. (1998) Riboflavin metabolism-relevance to human nutrition. *Industrial Journal of Medical Research*, **108**, 182–190.
Larsen, A.G., Vogensen, F.K. & Josephsen, J. (1993) Antimicrobial activity of lactic acid bacteria isolated from sour doughs: purification and characterization of bavaricin A, a bacteriocin produced by *Lactobacillus bavaricus* MI401. *Journal of Applied Bacteriology*, **75**, 113–122.
Laws, A., Gu, Y. & Marshall, V. (2001) Biosynthesis, characterisation, and design of bacterial exopolysaccharides from lactic acid bacteria. *Biotechnology Advances*, **19**, 1–28.
Leer, R.J., van der Vossen, J.M., van Giezen, M., van Noort, J.M., & Pouwels, P.H. (1995) Genetic analysis of acidocin B, a novel bacteriocin produced by *Lactobacillus acidophilus*. *Microbiology*, **141**, 1629–1635.
Levander, F., Svensson, M. & Radstrom, P. (2002) Enhanced exopolysaccharide production by metabolic engineering of *Streptococcus thermophilus*. *Applied and Environmental Microbiology*, **68**, 784–790.
Lievin, V., Peiffer, I., Hudault, S., Rochat, F., Brassart, D., Neeser, J.R. & Servin, A.L. (2000) *Bifidobacterium* strains from resident infant human gastrointestinal microflora exert antimicrobial activity. *Gut*, **47**, 646–652.
Lin, M.Y. & Young, C.M. (2000) Folate levels in cultures of lactic acid bacteria. *International Dairy Journal*, **10**, 409–413.
Lin, T.Y., Lin, C.W. & Lee, C.H. (1999) Conjugated linoleic acid concentration as effected by lactic cultures and added linoleic acid. *Food Chemistry*, **67**, 1–5.
Looijesteijn, P.J., Trapet, L., de Vries, E., Abee, T. & Hugenholtz, J. (2001) Physiological function of exopolysaccharides produced by *Lactococcus lactis*. *International Journal of Food Microbiology*, **64**, 71–80.
Maldonado, A., Ruiz-Barba, J.L., & Jimenez-Diaz, R. (2003) Purification and genetic characterisation of plantaricin NC8 a novel coculture-inducible two-peptide from *Lactobacillus plantarum* NC8. *Applied and Environmental Microbiology*, **69**, 383–389.
Maldonado, A., Jimenez-Diaz, R. & Ruiz-Barba, J.L. (2004a) Induction of plantaricin production in *Lactobacillus plantarum* NC8 after coculture with specific Gram-positive bacteria is mediated by an autoinduction mechanism. *Journal of Bacteriology*, **186**, 1556–1564.
Maldonado, A., Ruiz-Barba, J.L. & Jimenez-Diaz, R. (2004b) Production of plantaricin NC8 by *Lactobacillus plantarum* NC8 is induced in the presence of different types of Gram-positive bacteria. *Archives of Microbiology*, **181**, 8–16.
Majhenic, A.C., Matijasic, B.B. & Rogelj, I. (2003) Chromosomal location of the genetic determinanta for bacteriocins produced by *Lactobacillus gasseri* K&. *Journal of Dairy Research*, **70**, 199–203.
Majhenic, A.C., Venema, K., Allison, G.E., Matijasic, B.B., Rogelj, I., & Klaenhammer, T.R. (2004) DNA analysis of the genes encoding acidocin LF221 A and acidocin LF221 B, two bacteriocins produced by *Lactobacillus gasseri* LF221. *Applied Microbiology and Biotechnology*, **63**, 705–714.
Marin, M.L., Tejada-Simon, M.V., Lee, J.H., Murtha, J., Ustunol, Z. & Pestka, J.J. (1998) Stimulation of cytokine production in clonal macrophage and T-cell models by *Streptococcus thermophilus*: comparison with *Bifidobacterium* sp. and *Lactobacillus bulgaricus*. *Journal of Food Protection*, **61**, 859–864.
Marki, F., Hanni, E., Fredenhagen, A. & van Oostrum, J. (1991) Mode of action of the lanthionine-containing peptide antibiotics duramycin, duramycin B and C, and cinnamycin as indirect inhibitors of phospholipase A2. *Biochemical Pharmacology*, **42**, 2027–2035.

Martin, N.I., Sprules, T., Carpenter, M.R., Cotter, P.D., Hill, C., Ross, R.P. & Vederas, J.C. (2004) Structural characterization of lacticin 3147, a two-peptide lantibiotic with synergistic activity. *Biochemistry*, **43**, 3049–3056.

Martinez, J.M., Kok, J., Sanders, J.W. & Hernandez, P.E. (2000) Heterologous coproduction of enterocin A and pediocin PA-1 by *Lactococcus lactis*: detection by specific peptide-directed antibodies. *Applied and Environmental Microbiology*, **66**, 3543–3549.

Marugg, J.D., Gonzalez, C.F., Kunka, B.S., Ledeboer, A.M., Pucci, M.J., Toonen, M.Y., Walker, S.A., Zoetmulder, L.C. & Vandenbergh, P.A. (1992) Cloning, expression, and nucleotide sequence of genes involved in production of pediocin PA-1, and bacteriocin from *Pediococcus acidilactici* PAC10. *Applied and Environmental Microbiology*, **58**, 2360–2367.

Mastromarino, P., Brigidi, P., Macchia, S., Maggi, L., Pirovano, F., Trinchieri, V., Conte, U. & Matteuzzi, D. (2002) Characterization and selection of vaginal *Lactobacillus* strains for the preparation of vaginal tablets. *Journal of Applied Microbiology*, **93**, 884–893.

Mathews, C.K. & van Holde, K.E. (1990) *Biochemistry*, The Benjamin/Cummings Publishing Company, Redwood City.

McAuliffe, O., Hill, C. & Ross, R.P. (2000) Identification and overexpression of *ltnl*, a novel gene which confers immunity to the two-component lantibiotic lacticin 3147. *Microbiology*, **146**, 129–138.

McGroarty, J.A. & Reid, G. (1988) Detection of a *Lactobacillus* substance that inhibits *Escherichia coli*. *Canadian Journal of Microbiology*, **34**, 974–978.

Michetti, P., Dorta, G., Wiesel, P.H., Brassart, D., Verdu, E., Herranz, M., Felley, C., Porta, N., Rouvet, M., Blum, A.L. & Corthesy-Theulaz, I. (1999) Effect of whey-based culture supernatant of *Lactobacillus acidophilus* (*johnsonii*) La1 on *Helicobacter pylori* infection in humans. *Digestion*, **60**, 203–209.

Moll, G., Ubbink-Kok, T., Hildeng-Hauge, H., Nissen-Meyer, J., Nes, I.F., Konings, W.N. & Driessen, A.J. (1996) Lactococcin G is a potassium ion-conducting, two-component bacteriocin. *Journal of Bacteriology*, **178**, 600–605.

Monsan, P., Bozonnet, S., Albenne, C., Joucla, G., Willemot, R.M. & Remaud-Simeon, M. (2001) Homopolysaccharides from lactic acid bacteria. *International Dairy Journal*, **11**, 675–685.

Montville, T.J. & Bruno, M.E. (1994) Evidence that dissipation of proton motive force is a common mechanism of action for bacteriocins and other antimicrobial proteins. *International Journal of Food Microbiology*, **24**, 53–74.

Morishita, T., Tamura, N., Makino, T. & Kudo, S. (1999) Production of menaquinones by lactic acid bacteria. *Journal of Dairy Science*, **82**, 1897–1903.

Morrison, H.I., Schaubel, D., Desmeules, M. & Wigle, D.T. (1996) Serum folate and risk of fatal coronary heart disease. *Journal of American Medical Association*, **275**, 1893–1896.

Mortvedt, C.I., Nissen-Meyer, J., Sletten, K. & Nes, I.F. (1991) Purification and amino acid sequence of lactocin S, a bacteriocin produced by *Lactobacillus sake* L45. *Applied and Environmental Microbiology*, **57**, 1829–1834.

Nakajima, H., Suzuki, Y., Kaizu, H. & Hirota, T. (1992) Cholesterol lowering activity in ropy fermented milk. *Journal of Food Science*, **57**, 1327–1329.

Nakajima, H., Toba, T. & Toyoda, S. (1995) Enhancement of antigen-specific antibody production by extracellular slime products from slime-forming *Lactococcus lactis* subspecies *cremoris* SBT 0495 in mice. *International Journal of Food Microbiology*, **25**, 153–158.

Navaratna, M.A., Sahl, H.G. & Tagg, J.R. (1998) Two-component anti-*Staphylococcus aureus* lantibiotic activity produced by *Staphylococcus aureus* C55. *Applied and Environmental Microbiology*, **64**, 4803–4808.

Nishimura-Uemura, J., Kitazawa, H., Kawai, Y., Itoh, T., Oda, M. & Saito, T. (2003) Functional alteration of murine macrophages stimulated with extracellular polysaccharides from *Lactobacillus delbrueckii* ssp. *bulgaricus* OLL1073R-1. *Food Microbiology*, **20**, 267–273.

O'Riordan, K. & Fitzgerald, G.F. (1998) Evaluation of bifidobacteria for the production of antimicrobial compounds and assessment of performance in cottage cheese at refrigeration temperature. *Journal of Applied Microbiology*, **85**, 103–114.

Ocana, V.S., Pesce, de Ruiz Holgado, A.A. & Nader-Macias, M.E. (1999) Characterization of a bacteriocin-like substance produced by a vaginal *Lactobacillus salivarius* strain. *Applied and Environmental Microbiology*, **65**, 5631–5635.

Olson, R.E. (1984) The function and metabolism of vitamin K. *Annual Reviews of Nutrition*, **4**, 281–337.

Olson, J.A. (1994) Vitamins: The tortuous path from needs to fantasies. *Journal of Nutrition*, **124**, 1771s–1776s.

Palacios, J., Vignolo, G., Farias, M.E., de Ruiz Holgado, A.P., Oliver, G. & Sesma, F. (1999) Purification and amino acid sequence of lactocin 705, a bacteriocin produced by *Lactobacillus casei* CRL 705. *Microbiological Research*, **154**, 199–204.

Perdigon, G., Valdez, G.F. & Rachid, M. (1998) Antitumour activity of yogurt: study of possible immune mechanisms. *Journal of Dairy Research*, **65**, 129–138.

Perez, P.F., Minnaard, J., Rouvet, M., Knabenhans, C., Brassart, D., De Antoni, G.L. & Schiffrin, E.J. (2001) Inhibition of *Giardia intestinalis* by extracellular factors from lactobacilli: an *in vitro* study. *Applied and Environmental Microbiology*, **67**, 5037–5042.

Prasch, T., Naumann, T., Markert, R.L., Sattler, M., Schubert, W., Schaal, S., Bauch, M., Kogler, H. & Griesinger, C. (1997) Constitution and solution conformation of the antibiotic mersacidin determined by NMR and molecular dynamics. *European Journal of Biochemistry*, **244**, 501–512.

Price, P.A., Otsuka, A.A., Poser, J.W., Kristaponis, J. & Raman, N. (1976) Characterization of a gamma-carboxyglutamic acid-containing protein from bone. *Proceedings of the National Academy of Science, USA*, **73**, 1447–1451.

Quesada-Chanto, A., Afschar, A.S. & Wagner, F. (1994) Microbial production of propionic acid and vitamin $B_{12}$ using molasses or sugar. *Applied Microbiology and Biotechnology*, **41**, 378–383.

Rao, D.R., Reddy, A.V., Pulusani, S.R. & Cornwell, P.E. (1984) Biosynthesis and utilization of folic acid and vitamin $B_{12}$ by lactic cultures in skim milk. *Journal of Dairy Science*, **67**, 1169–1174.

Ricciardi, A. & Clementi, F. (2000) Exopolysaccharides from lactic acid bacteria: structure production and technological applications. *Italian Journal of Food Science*, **12**, 23–45.

Rogers, L.A. & Whittier, E.D. (1928) Limiting factors in lactic fermentation. *Journal of Bacteriology*, **16**, 211–229.

Ruas-Madiedo, P., Hugenholtz, J. & Zoon, P. (2002) An overview of the functionality of exopolysaccharides produced by lactic acid bacteria. *International Dairy Journal*, **12**, 163–171.

Ruijssenaars, H.J., Stingele, F. & Hartmans, S. (2000) Biodegradability of food-associated extracellular polysaccharides. *Current Microbiology*, **40**, 194–199.

Ruiz-Barba, J.L. & Jimenez-Diaz, R. (1994) Vitamin and amino acid requirements of *Lactobacillus plantarum* strains isolated from green olive fermentations. *Journal of Applied Bacteriology*, **76**, 350–355.

Ruiz-Barba, J.L., Cathcart, D.P., Warner P.J. & Jimenez-Diaz, R. (1994) Use of *Lactobacillus plantarum* LPCO10, a bacteriocin producer, as a starter culture in Spanish-style green olive fermentations. *Applied and Environmental Microbiology*, **60**, 2059–2064.

Ryan, M.P., Rea, M.C., Hill, C. & Ross, R.P. (1996) An application in cheddar cheese manufacture for a strain of *Lactococcus lactis* producing a novel broad-spectrum bacteriocin, lacticin 3147. *Applied and Environmental Microbiology*, **62**, 612–619.

Ryan, M.P., Jack, R.W., Josten, M., Sahl, H.G., Jung, G., Ross, R.P. & Hill, C. (1999) Extensive post-translational modification, including serine to D-alanine conversion, in the two-component lantibiotic, lacticin 3147. *Journal of Biological Chemistry*, **274**, 37544–37550.

Sahl, H.G., Jack, R.W. & Bierbaum, G. (1995) Biosynthesis and biological activities of lantibiotics with unique post-translational modifications. *European Journal of Biochemistry*, **230**, 827–853.

Saito, T. (2004) Selection of useful probiotic lactic acid bacteria from the *Lactobacillus acidophilus* group and their applications to functional foods. *Animal Science Journal*, **75**, 1–13.

Saleh, F.A. & El-Sayed, E.M. (2004) Isolation and characterization of bacteriocins produced by *Bifidobacterium lactis* BB-12 and *Bifidobacterium longum* BB-46. *9th Egyptian Conference for Dairy Science and Technology*, Cairo, 9–11 October 2004, Research Papers I, 323–337.

Sarma, J.D. & Duttagupta, C. (1995) Improved microbiological assay for folic acid based on microtiter plating with *Streptococcus faecalis*. *Food Biological Contaminants*, **78**, 1173–1176.

Sato, T., Nishimura-Uemura, J., Shimosato, T., Kawai, Y., Kitazawa, H. & Saito, T. (2004) Dextran from *Leuconostoc mesenteroides* augments immunostimulatory effects by the introduction of phosphate groups. *Journal of Food Protection*, **67**, 1719–1724.

Schnell, N., Entian, K.D., Schneider, U., Gotz, F., Zahner, H., Kellner, R. & Jung, G. (1988) Prepeptide sequence of epidermin, a ribosomally synthesized antibiotic with four sulphide-rings. *Nature*, **333**, 276–278.

Shu, Q., Lin, H., Rutherfurd, K.J., Fenwick, S.G., Prasad, J., Gopal, P.K. & Gill, H.S. (2000) Dietary *Bifidobacterium lactis* (HN019) enhances resistance to oral *Salmonella typhimurium* infection in mice. *Microbiology and Immunology*, **44**, 213–222.

Silva, A.M., Barbosa, F.H., Duarte, R., Vieira, L.Q., Arantes, R.M. & Nicoli, J.R. (2004) Effect of *Bifidobacterium longum* ingestion on experimental salmonellosis in mice. *Journal of Applied Microbiology*, **97**, 29–37.

Simon, L., Fremaux, C., Cenatiempo, Y. & Berjeaud, M. (2002) Sakacin G, a new type of anti-listerial bacteriocin. *Applied and Environmental Microbiology*, **68**, 6416–6420.

Skaugen, M., Nissen-Meyer, J., Jung, G., Stevanovic, S., Sletten, K., Inger, C., Abildgaard, M. & Nes, I.F. (1994) In vivo conversion of L-serine to D-alanine in a ribosomally synthesized polypeptide. *Journal of Biological Chemistry*, **269**, 27183–27185.

Skaugen, M., Christie, V.H. & Nes, I.F. (1999) Biosynthesis of the lantibiotic lactocin S: heterologous expression of a two-operon gene cluster encoding regulatory, biosynthetic and immunity functions. *Genetics Metabolism and Applications. FEMS, Abstracts of the Sixth Symposium on Lactic Acid Bacteria*, pp. C86, Veldhoven.

Smid, E.J., Starrenburg, M., Mireau, I., Sybesma, W. & Hugenholtz, J. (2001) Increase of folate levels in fermented foods. *Innovations in Food Technology*, **February/March**, 13–15.

Sreekumar, O. & Hosono, A. (1998) The antimutagenic properties of a polysaccharide produced by *Bifidobacterium longum* and its cultured milk against some heterocyclic amines. *Canadian Journal of Microbiology*, **44**, 1029–1036.

Stingele, F., Vincent, S.J., Faber, E.J., Newell, J.W., Kamerling, J.P. & Neeser, J.R. (1999) Introduction of the exopolysaccharide gene cluster from *Streptococcus thermophilus* Sfi6 into *Lactococcus lactis* MG1363: production and characterization of an altered polysaccharide. *Molecular Microbiology*, **32**, 1287–1295.

Suttie, J. (1985) Vitamin K, *The Fat-Soluble Vitamins*, (ed. A. Diplock), pp. 225–311, William Heinemann, London.

Sybesma, W., Starrenburg, M., Kleerebezem, M., Mierau, I., de Vos, W.M. & Hugenholtz, J. (2003a) Increased production of folate by metabolic engineering of *Lactococcus lactis*. *Applied and Environmental Microbiology*, **69**, 3069–3076.

Sybesma, W., Starrenburg, M., Tijsseling, L., Hoefnagel, M.H. & Hugenholtz, J. (2003b) Effects of cultivation conditions on folate production by lactic acid bacteria. *Applied and Environmental Microbiology*, **69**, 4542–4548.

Sybesma, W., van Den Born, E., Starrenburg, M., Mierau, I., Kleerebezem, M., de Vos, W.M. & Hugenholtz, J. (2003c) Controlled modulation of folate polyglutamyl tail length by metabolic engineering of *Lactococcus lactis*. *Applied and Environmental Microbiology*, **69**, 7101–7107.

Sybesma, W., Burgess, C., Starrenburg, M., van Sinderen, D. & Hugenholtz, J. (2004) Multivitamin production in *Lactococcus lactis* using metabolic engineering. *Metabolic Engineering*, **6**, 109–115.

Tahara, T., Oshimura, M., Umezawa, C & Kanatani, K. (1996) Isolation, partial characterization and mode of action of Acidocin J1132, a two-component bacteriocin produced by *Lactobacillus acidophilus* JCM 1132. *Applied and Environmental Microbiology*, **62**, 892–897.

Takada, H. & Kotani, S. (1989) Structural requirements of lipid A for endotoxicity and other biological activities. *Critical Reviews in Microbiology*, **16**, 477–523.

Takeda, K., Saito, T., Kitazawa, H. & Itoh, T. (1997) Mitogenic activity of whole cells and cell wall components of *Lactobacillus acidophilus* group lactic acid bacteria on murine spleen and Peyer's patch cells. *Milchwissenschaft*, **51**, 21–25.

Tamime, A.Y. (ed.) (2005) *Fermented Milks*, Blackwell Publishing, Oxford (in press).

Taranto, M.P., Medici, M., Perdigon, G., Ruiz Holgado, A.P. & Valdez, G.F. (2000) Effect of *Lactobacillus reuteri* on the prevention of hypercholesterolemia in mice. *Journal of Dairy Science*, **83**, 401–403.

Taranto, M.P., Vera, J.L., Hugenholtz, J., de Valdez, G.F. & Sesma, F. (2003) *Lactobacillus reuteri* CRL1098 produces cobalamin. *Journal of Bacteriology*, **185**, 5643–5647.

Tichaczek, P.S., Nissen-Meyer, J., Nes, I.F., Vogel, R. F. & Hammes, W.P. (1992) Characterization of the bacteriocins curvacin A from *Lactobacillus curvatus* LTH1174 and sakacin P from *L. sake* LTH673. *Systematic and Applied Microbiology*, **15**, 460–468.

Tichaczek, P.S., Vogel, R.F. & Hammes, W.P. (1993) Cloning and sequencing of *curA* encoding curvacin A, the bacteriocin produced by *Lactobacillus curvatus* LTH1174. *Archives in Microbiology*, **160**, 279–283.

Timmerman, H.M., Koning, C.J.M., Mulder, L., Rombouts, F.M. & Beynen, A.C. (2004) Monostrain, multistrain and multispecies probiotics – A comparison of functionality and efficacy. *International Journal of Food Microbiology*, **96**(3), 219–233.

Toure, R., Kheadr, E., Lacroix, C., Moroni, O. & Fliss, I. (2003) Production of antibacterial substances by bifidobacterial isolates from infant stool active against *Listeria monocytogenes*. *Journal of Applied Microbiology*, **95**, 1058–1069.

Turner, D.L., Brennan, L., Meyer, H.E., Lohaus, C., Siethoff, C., Costa, H.S., Gonzalez, B., Santos, H. & Suarez, J.E. (1999) Solution structure of plantaricin C, a novel lantibiotic. *European Journal of Biochemistry*, **264**, 833–839.

Twomey, D., Ross, R.P., Ryan, M., Meaney, B. & Hill, C. (2002) Lantibiotics produced by lactic acid bacteria: structure, function and applications. *Antonie van Leeuwenhoek*, **82**, 165–185.

Vaughan, E.E., Daly, C. & Fitzgerald, G.F. (1992) Identification and characterization of helveticin V-1829, a bacteriocin produced by *Lactobacillus helveticus* 1829. *Journal of Applied Bacteriology*, **73**, 299–308.

Vaughan, E.E., Heilig, H.G.H.J., Zoetendal, E.G., Satokari, R., Collins, J.K., Akkermanns, A.D.L. & de Vos, W.M. (1999) Molecular approaches to study probiotic bacteria. *Trends in Food Science and Technology*, **10**, 400–404.

Venema, K., Kok, J., Marugg, J.D., Toonen, M.Y., Ledeboer, A.M., Venema, G. & Chikindas, M.L. (1995) Functional analysis of the pediocin operon of *Pediococcus acidilactici* PAC1.0: PedB is the immunity protein and PedD is the precursor processing enzyme. *Molecular Microbiology*, **17**, 515–522.

Vignolo, G., Fadda, S., de Kairuz, M.N., de Ruiz Holgado, A.A. & Oliver, G. (1996) Control of *Listeria monocytogenes* in ground beef by 'Lactocin 705', a bacteriocin produced by *Lactobacillus casei* CRL 705. *International Journal of Food Microbiology*, **29**, 397–402.

Wegkamp, A., Starrenburg, M., de Vos, W.M., Hugenholtz, J. & Sybesma, W. (2004) Transformation of folate-consuming *Lactobacillus gasseri* into a folate producer. *Applied and Environmental Microbiology*, **70**, 3146–3148.

Winkowski, K., Crandall, A.D. & Montville, T.J. (1993) Inhibition of *Listeria monocytogenes* by *Lactobacillus bavaricus* MN in beef systems at refrigeration temperatures. *Applied and Environmental Microbiology*, **59**, 2552–2557.

Yasui, H., Kiyoshima, J. & Ushijima, H. (1995) Passive protection against rotavirus-induced diarrhea of mouse pups born to and nursed by dams fed *Bifidobacterium breve* YIT4064. *Journal of Infectious Diseases*, **172**, 403–409.

Yildirim, Z. & Johnson, M.G. (1998) Characterization and antimicrobial spectrum of bifidocin B, a bacteriocin produced by *Bifidobacterium bifidum* NCFB 1454. *Journal of Food Protection*, **61**, 47–51.

Yildirim, Z., Winters, D.K. & Johnson, M.G. (1999) Purification, amino acid sequence and mode of action of bifidocin B produced by *Bifidobacterium bifidum* NCFB 1454. *Journal of Applied Microbiology*, **86**, 45–54.

Zarate, G., Morata, D., Chaia, A.P. & Gonzalez, S.N. (2002a) Adhesion of dairy propionibacteria to intestinal epithelial tissue *in vitro* and *in vivo*. *Journal of Food Protection*, **65**, 534–539.

Zarate, G., Morata de Ambrosini, V., Perez Chaia, A. & Gonzalez, S. (2002b) Some factors affecting the adherence of probiotic *Propionibacterium acidipropionici* CRL 1198 to intestinal epithelial cells. *Canadian Journal of Microbiology*, **48**, 449–457.

Zarate, G., Gonzalez, S. & Perez Chaia, A.P. (2004) Assessing survival of dairy propionibacteria in gastrointestinal conditions and adherence to intestinal epithelia. *Methods in Molecular Biology*, **268**, 423–432.

# 9 Future Development of Probiotic Dairy Products

T. Mattila-Sandholm, M. Saarela and W.M. de Vos

## 9.1 Background

Probiotics have become an important and rapidly expanding segment of the food market as processed food manufacturers seek to improve market share by promoting the health benefits provided by functional ingredients in their products. Scientific progress has been expanding the knowledge on how foods influence consumers in relation to specific health parameters. Probiotic foods targeted towards improving the balance and activity of the intestinal microbiota, currently, provide the largest segment of the functional food market in Europe, Japan and Australia. The leading position of the European Union (EU) industries is evident from their predominant share in the functional food market worldwide that amounts to sales in 2002 of well over 30 billion Euros, and growing penetration in the American markets (Anonymous, 2004). The market pull for functional foods has been so rapid that it has resulted in certain knowledge gaps in the scientific understanding of the mechanisms by which these foods impact health. This has a bearing on the consumer who cannot make an informed choice, the producer who is limited in the reliable development of new and improved products, and authorities that implement new legislation on food claims (Anonymous, 2003). This specifically relates to the growing number of probiotic and other functional foods that modulate the human gut.

The human gastrointestinal (GI) tract is the metabolically most active body organ, and it is the prime target of present and future functional foods. It is not only the primary site of food conversion and uptake, but also the location where the mucosal immune system is primed and various local and systemic disorders become manifest. While sterile at birth, the GI tract is colonised by a myriad of microbes – collectively called microbiota – that contribute to food conversion, digestion, communication with the host, and induction of specific biological responses important in many critical physiological functions. Hence, the GI tract represents the site where the host's well-being is mostly affected by food intake in both a direct way, but also through modulating the interplay between intestinal microbes and the host (Gibson & Macfarlane, 1995; Roberts *et al.*, 2001; Freitas *et al.*, 2003; de Vos *et al.*, 2004).

Various sequential steps are essential in the development of efficacious functional foods. A prerequisite for mechanistic studies of action is an understanding of the composition and activity of the intestinal microbiota as well as interactions with the host in both healthy and diseased individuals. High-throughput molecular methods are required to examine the intestinal microbiota, and to track the location and activity of probiotic strains in the intestinal tract (Vaughan *et al.*, 2000; Blaut *et al.*, 2002). An understanding of the mechanisms by which probiotics exert beneficial effects allows the selection of strains of lactic acid bacteria (LAB) that can be tested in human clinical studies following demonstration of their safety. An important area of research includes the technologies to maximise the stability of functional traits of probiotics during manufacture, formulation, storage, and in the GI tract. Finally, an understanding of the most appropriate approaches to communicate the benefits of probiotic foods to consumers, and the influence of health messages on consumer choice is essential to ensure that products are appropriately applied and targeted to benefit specific populations (Lähteenmäki, 2004). In addition, consumers are now increasingly exposed to probiotic foods, and they can rightfully demand and expect mechanistic explanations of effects. Furthermore, the expected changes in legislation with regard to substantiated health claims will require a better insight in the mechanisms behind the functionality of these bacteria.

## 9.2 Recent EU research activities in the gut health area: the PROEUHEALTH cluster

The Food, GI tract Functionality and Human Health Cluster (PROEUHEALTH) is a consortium of eight complementary multicentre EU research projects, which started in February 2001 and will run over 5 years, ending in 2005. The PROEUHEALTH cluster is subsidised by the European Commission's 5th Framework, Quality of Life and Management of Living Resources Programme. The cluster has brought together 64 research partners from 16 European countries in the quest to obtain greater knowledge of the role of the intestinal microbiota in human health and disease, and to develop new functional foods and therapies. The research covers all aspects in the development of new probiotic foods, such as from designing molecular tools to study the ecology of the intestinal microbiota, to understanding the mechanisms of bacterium–host interactions, providing solutions to food technology issues, and finally to conducting human clinical trials to assess efficacy in preventing or treating disease (Mattila-Sandholm & Mercenier, 2001).

The following research areas demonstrate the impact of the PROEUHEALTH cluster:

- New molecular research tools for studying the composition and activity of the intestinal microbiota have been developed. These tools consist of high-throughput microarrays, use of cell sorting, monitoring *in situ* functional gene expression, and development of functional probes and probes for clinical trials.
- A molecular understanding of immune modulation by probiotic bacteria and applicability of probiotics as vaccine delivery vehicles are under investigation. Results have shown protective effects of LAB in a colitis model, improved delivery of antigens, and delivery of anti-inflammatory molecules [interleukin (IL)-10] in humans.
- Biosafety evaluation of probiotic and other LAB, including enterococci, for human consumption (900 strains) will result in the creation of a database on safety parameters, genetic stability, antibiotic resistance genes and possible virulence properties of LAB.
- New therapeutic and prophylactic treatments for intestinal infections and chronic intestinal diseases such as inflammatory bowel disease (IBD), *Helicobacter* spp. and rotavirus infection are being investigated.
- Several clinical trials are ongoing within the European Union, which gives a comparison between North-Mid-South citizens. The trials are dealing with IBD and *Helicobacter* spp. infections, and using either probiotics, prebiotics or their combinations to alleviate the symptoms of the disease/infection. Furthermore, different consumer groups, such as the elderly, have been targeted. In the elderly population, the possible linkages between ages, intestinal microbiota, functionality and diet have been studied; specific attention

has been given to the establishment of probiotic-prebiotic products targeted for the elderly.

- Probiotic and prebiotic technology is creating a database on stability, viability, and stress tolerance of probiotic strains. In addition, new prebiotic modifications have been produced.
- Host–microbe balance combines the effects of probiotics and prebiotics, and utilises novel model systems to study the functionality and cross-talk between microbes and host.
- A clearer understanding of the relationships between food, intestinal bacteria, and human health and disease has been obtained by providing an extensive list of references and scientific publications.

The research innovations produced by the PROEUHEALTH cluster have been disseminated to target audiences through three platforms (see Fig. 9.1), which can be summarised as follows:

- The Science Platform has provided the foundation for internal dissemination and networking, including exchange of data, biomaterials and personnel, and fostered external and international exposure.
- The Industry Platform has enabled the cluster to disseminate its research innovations to probiotic industries throughout Europe and the world while maximising the potential for commercial exploitation of results from the cluster's research.

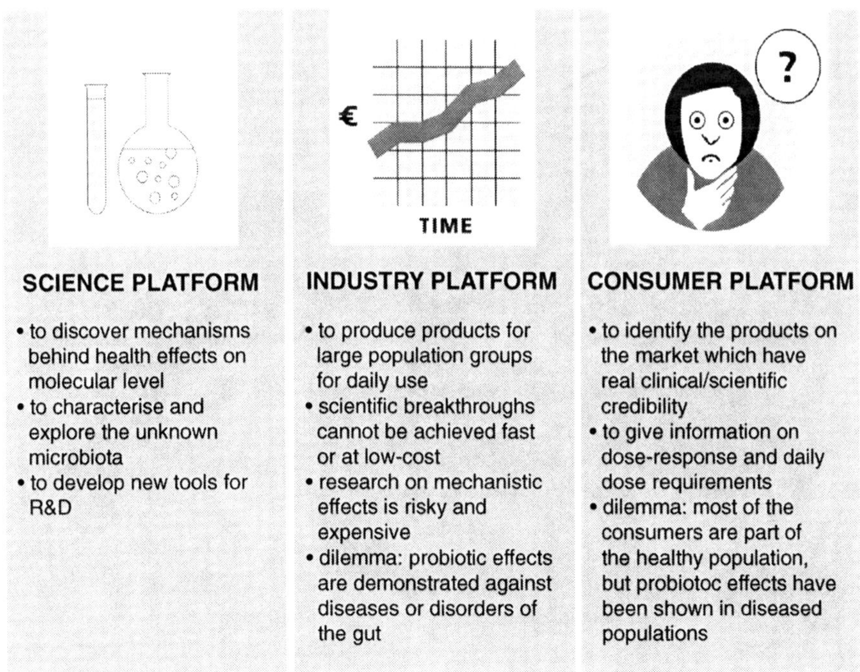


**Fig. 9.1** PROEUHEALTH cluster platform features and activities.

- The Consumer Platform has provided information to consumers about the cluster and its innovations in an appropriately tailored format, which ensures that the general public are kept informed and benefit from the research by their own language.

## 9.3 Beyond PROEUHEALTH: from products to mechanisms and back

There is an increasing demand for developing foods with specific functionalities that improve the health and well-being, stimulate mental functions, and help to prevent disease in consumers. This requirement originates from the fact that consumers at critical ages – for example young and old – require specific foods that attain these values. Moreover, there is an increased incidence of food-related disease and gastrointestinal disorders, many of which can be prevented or alleviated through foods with gut functionality. Probiotic research activities have resulted in considerable progress in demonstrating the effectiveness of probiotic and other functional microbes that avert intestinal syndromes, provide therapy for intestinal infections, and alleviate allergy symptoms (Alvarez-Olmos & Oberhelman, 2001; Kalliomäki *et al.*, 2003; Reid *et al.*, 2003a; Shanahan, 2004). In addition, the future research should capitalise on the accumulating knowledge of specific nutritional interactions between intestinal microbes and particular ingredients, such as prebiotics (Reid *et al.*, 2003b). However, in contrast to solely determining effects, one needs to provide molecular biomarkers for such effects and to establish mechanistic inter-relations between microbes and ingredients and gut health and functionality, thus providing an essential contribution to the food innovation cycle (see Fig. 9.2). This can be achieved through incorporating an array of genomic and post-genomic approaches in humans, model animals, and microbes where functional and comparative genomics have been well developed (de Vos, 2001). The application of holistic genomic approaches results in an insight into how established signalling pathways and their molecular components contribute towards functionality of consumed microbes or ingredients, as well as leading to the discovery of completely new pathways and mechanisms. High-throughput and functional technologies should be developed and should

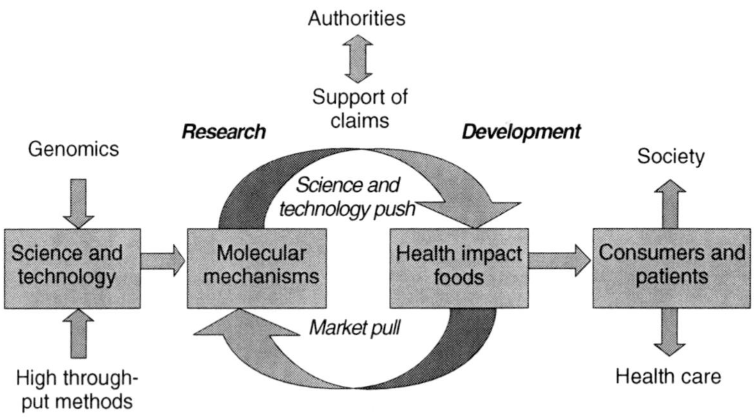


**Fig. 9.2** The position of probiotic dairy foods in the functional food innovation cycle consisting of market pull, science and technology push, its relation to the input of new technologies, and the position of a wide audience of stakeholders that are targets for knowledge sharing.

be instrumental in mechanism-based strain improvement, functionality screening, rational product development, and control of quality and safety (de Vos *et al.*, 2004).

In the future, the acceptance of probiotic, prebiotic and synbiotic products will depend on the solid proof of the health benefits they promise at the moment. Therefore, research that provides scientifically sound evidence to back-up the health claims is needed. This requires a comprehensive insight into the mechanisms by which these foods are functional, and exploitation of this knowledge base by developing new generations of more effective health impact foods. Building on the wealth of information provided by previous and still ongoing EU projects in the gut health area, this could be realised by capitalising on genomics-based and high-throughput technologies. However, it is imperative that the knowledge produced by science is comprehensible to the consumer (Lähteenmäki, 2004). This is the major challenge, as the laymen thinking is based on approximation and black-and-white views – whereas scientific thinking deals with probabilities and degrees of uncertainty. Consumers' trust on the information depends on the source and content of message. The critical point is that producers of probiotic foods can gradually build and ensure consumer trust.

## 9.4 The strategic leap towards mechanistic studies: future target areas for research

The human oro-gastrointestinal tract (especially the oral cavity and colon) is heavily colonised by the resident microbiota. The application of 16S rRNA-based technologies has generated considerable insights into the composition of the human intestinal microbiota during its development from newborn to old age (Blaut *et al.*, 2002). A large inter-individual variation, with functional redundancy, has been found to characterise the dominant microbiota. However, current understanding of the activities of the intestinal microbiota is very limited, and metagenomics approaches at the sequence and the post-genomic level (transcription, protein and metabolite) are needed to definitely determine the metabolites that impact health and provide reliable biomarkers for the selection of functional diets or ingredients (de Vos *et al.*, 2004).

Epithelial cells of the intestinal mucosa form a physical barrier against the microbiota, and can discriminate between harmless bacteria and invasive pathogens, thereby impacting upon disease outcome (Haller *et al.*, 2000). While non-pathogenic bacteria are known to greatly influence regulation of the intestinal barrier, their mechanisms of action remain unknown. Global studies of host–bacterium relationships in the intestine of mice have started (Hooper *et al.*, 2002), but comparative analysis and systematic post-genomic technologies remain to be fully exploited. This includes characterisation of the common molecular signatures of gene expression in host epithelial cells and tissue in response to bacterial strains, and their loss- or gain-of-function mutants. This will permit the definition of gene functions, signalling cascades, pathways, and other plausible targets relevant to health. In the long term, mechanisms can be defined by functional characterisation of the genes encoding host targets. Cross-talk imposes changes in expression of the bacterial strains and these need to be analysed at the (post)-genomic level.

The intestinal microbiota interfaces with epithelial cells, and the underlying mucosal immune system including regulatory T cells and dendritic cells (DCs). This exerts a sentinel function against invading pathogens as well as serving regulatory functions in mucosal immunity (Nagler-Anderson, 2001). While it is unresolved as to how intestinal DCs contact microbial antigens, it is likely that DCs sample translocated bacteria as a mechanism to sustain a primitive T-cell-independent IgA response, and induce tolerance to commensal bacteria (Macpherson *et al.*, 2000). This needs to be further studied by analysing the role of secretory IgA on the host–microbe interaction, and determining the effect of bacterial strains on the genome-wide expression profiles of DCs and other isolated cells involved in innate and adaptive immunity, such as B and T lymphocytes (Ricciardi-Castagnoli & Granucci, 2002). Cytokine production is known to influence neuroendocrine development and the functioning of neuronal networks, but this is poorly understood. Hence, there is an increasing need to study the interaction between microbes and the neuroendocrine system, largely because of the long-term potential to develop functional foods for mood.

It has been well established that notably pathogenic microbial cells and their components are recognised by specific cellular receptors that interact with multiple signalling pathways involving canonical anti-inflammatory nuclear transcription factors (Takeda *et al.*, 2003). New receptor-signalling pathways are emerging (Girardin *et al.*, 2003; Radutoiu *et al.*, 2003; Kelly *et al.*, 2004). Although it is presently not known how beneficial bacteria are perceived in mammalian systems, it is likely that these and new, not yet discovered, pathways are involved. Hence, the development of high-throughput screening approaches where such pathways, as well as other genotypic screening systems, are exploited with culture collections and metagenomic expression libraries will provide new opportunities for the identification of functional strains applied to industrial use.

In order for consumers to gain health-benefits from probiotics, the quality and safety of probiotic products offered to them must be guaranteed. Although the research on probiotic health effects is progressing rapidly, several surveys on the probiotic products on the market have indicated that there are still technological problems related both to qualitative (which probiotic strain is in the product) and quantitative (how many living probiotic cells are in the product) issues (Canganella *et al.*, 1997; Hamilton-Miller *et al.*, 1999; Fasoli *et al.*, 2003; Temmerman *et al.*, 2003a, 2003b). To maintain and increase the consumers' interest and trust in probiotic products, these problems must be resolved. The first requirement for a feasible probiotic strain is that it has to be amenable to large-scale industrial production and down-stream processing. Second, the probiotic strain has to retain its viability during formulation, storage and consumption. The industrial production of probiotic strains inevitably causes stress in bacterial cells, and this stress in even more evident in strains that are not well adapted to live outside their unique living environment (e.g. probiotic strains are usually adapted to the conditions in the GI tract). For this reason, stress responses of lactobacilli and bifidobacteria are increasingly studied with the focus on improving the stress tolerance of the strains by subjecting them to suitable sublethal stress treatments prior to, for example, harsh processing steps (Desmond *et al.*, 2002; Derzelle *et al.*, 2003; Saarela *et al.*, 2004). Another way of protecting probiotic cells against harsh conditions is encapsulation, which has been – and remains – a keenly studied area of probiotic technology (Kailasapathy, 2002; Krasaekoopt *et al.*, 2003). In spite of the progress in the probiotic technological research, current technologies do not yet provide sufficient protection for probiotic cells against environmental stresses. Therefore, a set of new functional technologies is needed to deliver existing and new strains of lactobacilli and bifidobacteria in the appropriate physiological state to the consumer. This also requires analysis of the safety of these strains and their derivatives under a set of defined conditions including processing events. The availability of post-genomics tools and identification of genes that code for health-promoting effects can be used to aid in the optimisation of probiotic production technologies.

There is a rapid increase in the number of known genomes of lactobacilli and bifidobacteria (Klaenhammer *et al.*, 2002; Schell *et al.*, 2002; Kleerebezem *et al.*, 2003; Pridmore *et al.*, 2004). However, insight in the expression and function of genes in the gut is only just emerging (Barrangou *et al.*, 2003; de Vos *et al.*, 2004). There is an imminent need to develop new bioinformatics tools and experimental approaches to detect functional sequences in beneficial microbes, notably those involved in the establishment, persistence and adaptation to the host, as well as those involved in its signalling. To enable a functional and comparative genomics approach, studying strains with developed genetic tools and sequence information is vital. This approach allows finally the screening and development of novel strains that can be instrumental in further mechanistic, and proof-of-principle studies, as well as new industrial strains, with improved delivery of health benefits.

## 9.5 The future functional food industry will approach the consumer in a different way

Functional foods, which improve the gut health, are receiving major interest from consumers and the food industry alike. Many of these foods contain functional microbes or ingredients that provide a health benefit in the human gut. However, the way by which such health improvements are manifested is not understood. This markedly hampers further industrial development for present and future gut health foods, as well as the required scientific support for functional food claims. Hence, the rational mechanisms, by which functional microbes and ingredients affect human gut health, are important aspects for the future, and provide avenues for their further development and safe production.

At present, probiotics and prebiotics are marketed successfully because they appeal to the consumer, are considered to have positive attributes and, therefore, can be included in the so-called Hi-Touch product category. However, there is a lack of scientific basis for their rational development, and for a mechanistic explanation of their effects. This prevents further innovation of marketable products that have a clearly demonstrated benefit to the consumer. Gut-related diseases and disorders are extremely prevalent, and the prophylactic management of these problems has a huge market potential. Hence, the current Lo-Tech products need to be complemented and underpinned by the understanding of the scientific basis of their positive attributes to become competitive in the long run and, thus, meet the demands of European Authorities such as DG Sanco and EFSA, since claims will need mechanistic based evidence of proof in future marketing strategies (Regulation 178/2002/EC, Directives 84/450/EEC, 90/496/EEC, 2000/13/EC and COM (2003) 424, final).

The probiotic dairy foods provide an ideal basis for combining Hi-Tech and Hi-Touch approaches in the development of future foods for improved gut health and well-being. Thus, the food industry will be further strengthened by the development of new added-value functional foods. Fundamental knowledge on intestinal microbiota functionality provides realistic targets for the development and application of robust health foods that will have identifiable impacts. This is essential for further innovations in this important area, where significant scientific progress on health impact has hitherto been hampered by the lack of reliable mechanistic explanations of effect. Reliable research tools are now available to overcome this drawback, but need to be used and combined in an effective manner (critical mass approach). Predominant scientific developments include genomics-related and high-throughput technologies that generate new approaches aimed towards providing detailed insights into the molecular mechanisms of food functionality, gut health and microbial function. These can be instrumental in developing new generations of health foods and other functional foods, through an effective harnessing of multidisciplinary research skills.

## 9.6 Conclusions

The rapidly moving post-genomic revolution is markedly influencing the nutritional sciences. Key to this is a more thorough understanding of intestinal microbiota and its interactions with the host (Fig. 9.3). The time has arrived to harness the molecular- and gut-driven research into a usable output for the consumers. The scientific understanding of the role of the intestinal microbiota will underpin new strategies to improve the health status of consumers and will contribute to a reduction in healthcare costs, particularly in ageing populations. In addition, the results will open up new horizons for the industry of the next generation of functional food products that will fit with the commercial objectives of both large-scale industries and niche-market companies.

Probiotic dairy foods have enormous potential to benefit a very wide spectrum of consumers. They can build on consumers' very strong interest in 'health' foods and the increased awareness of the role of diet in health. The major burden of gut diseases in Europe was highlighted in a recent issue of the *Alimentary Pharmacology and Therapeutics Journal*, which was published in November 2003. While many gastrointestinal diseases affect the young and old, Europe's and North America's ageing populations are faced with an explosion of chronic diseases, and an accompanying increase in healthcare costs. It is now generally believed that individuals can be empowered to take a responsible role in their own health management, and that functional foods may be a key component in strategies targeted to maintaining health and preventing overwhelming healthcare demands. The development and social acceptance of functional foods represents an exciting opportunity to help move towards a more 'preventative' healthcare model. This approach is expected to reduce costs for healthcare and costs associated with work absenteeism, especially for diet-related diseases, and will contribute to an improved quality of life.

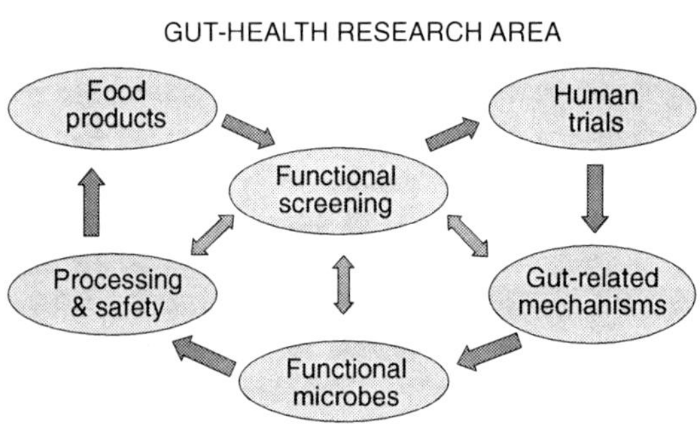


**Fig. 9.3** Different components of the gut-health research area related to the development of novel probiotic food products.

To realise this, the scientific community should address the following aspects:

- Provide biomarkers of intrinsic functionalities of the human intestinal microbiota in relation to diet and health.
- Unravel the molecular mechanisms underlying host–microbe interactions including cross-talk phenomena.
- Establish mechanisms of action of beneficial, commensal and pathogenic microbes on the immune function and neuroendocrine responses of the enteric nervous system.
- Select or generate lactobacilli and bifidobacteria with improved functionality by enhanced production and delivery of health-promoting compounds.
- Produce innovative instruments to develop functional technologies and guarantee the safety of existing and new lactobacilli and bifidobacteria during fermentation and down-stream processing.
- Develop and exploit high-throughput screening systems for the discovery of novel functional microbes and their components.
- Ensure knowledge sharing to a wide audience of stakeholders and end-users, including consumers, patient groups and authorities.
- Create scientific impact and ensure transfer of results to practical applications, thus contributing towards the product innovation and cost-effectiveness of the European food and biotechnology industries, including small- and medium-sized enterprises.

## References


Alvarez-Olmos, M.I. & Oberhelman, R.A. (2001) Probiotic agents and infectious diseases: A modern perspective on a traditional therapy. *Clinical Infectious Diseases*, **32**, 1567–1576.

Anonymous (2003) *Nutrition, Functional and Health Claims Made on Foods*. Draft EU Directive.

Anonymous (2004) The world market on functional foods and beverages. *Euromonitor International*.

Barrangou, R., Altermann, E., Hutkins, R., Cano, R. & Klaenhammer, T.R. (2003) Functional and comparative genomic analyses of an operon involved in fructo-oligosaccharide utilization by *Lactobacillus acidophilus*. *Proceedings of the National Academy of Sciences USA*, **100**, 8957–8962.

Blaut, M., Collins, M.D., Welling, G.W., Dore, J., van Loo, J. & de Vos, W. (2002) Molecular biological methods for studying the gut microbiota: the EU human gut flora project. *British Journal of Nutrition*, **87**, S203–S211.

Canganella, F., Paganini, S., Ovidi, M., Vettraino, A.M., Bevilacqua, L., Massa, S. & Trovatelli, L.D. (1997) A microbiological investigation on probiotic pharmaceutical products used for human health. *Microbiological Research*, **152**, 171–179.

Derzelle, S., Hallet, B., Ferain, T., Delcour, J. & Hols, P. (2003) Improved adaptation to cold-shock, stationary-phase, and freezing stresses in *Lactobacillus plantarum* overproducing cold-shock proteins. *Applied and Environmental Microbiology*, **69**, 4285–4290.

Desmond, C., Stanton, C., Fitzgerald, G.F., Collins, K. & Ross, R.P. (2002) Environmental adaptation of probiotic lactobacilli towards improvement of performance during spray drying. *International Dairy Journal*, **12**, 183–190.

Fasoli, S., Marzotto, M., Rizzotti, L., Rossi, F., Dellaglio, F. & Torriani, S. (2003) Bacterial composition of commercial probiotic products as evaluated by PCR-DGGE analysis. *International Journal of Food Microbiology*, **82**, 59–70.

Freitas, M., Tavan, E., Cayuela, C., Diop, L., Sapin, C. & Trugnan, G. (2003) Host-pathogens cross-talk. Indigenous bacteria and probiotics also play the game. *Biology of the Cell*, **95**, 503–506.

Gibson, G.R. & Macfarlane, G.T., Eds. (1995) *Human colonic bacteria – role in nutrition, physiology, and pathology*, CRC Press, Boca Raton.

Girardin, S.E., Hugot, J.P. & Sansonetti, P.J. (2003) Lessons from Nod2 studies: towards a link between Crohn's disease and bacterial sensing. *Trends in Immunology*, **24**, 652–658.

Hamilton-Miller, J.M.T., Shah, S. & Winkler, J.T. (1999) Public health issues arising from microbiological and labelling quality of foods and supplements containing probiotic micro-organisms. *Public Health Nutrition*, **2**, 223–229.

Haller, D., Bode, C., Hammes, W.P., Pfeifer, A.M., Schiffrin, E.J. & Blum, S. (2000) Non-pathogenic bacteria elicit differential cytokine response by intestinal epithelial cell/leukocyte co-cultures. *Gut*, **47**, 79–87.

Hooper, L.V., Midtvedt, T. & Gordon, J.I. (2002) How host-microbial interactions shape the nutrient environment of the mammalian intestine. *Annual Reviews in Nutrition*, **22**, 283–307.

Kailasapathy, K. (2002) Microencapsulation of probiotic bacteria: technology and potential applications. *Current Issues in Intestinal Microbiology*, **3**, 39–48.

Kalliomäki, M., Salminen, S., Poussa T., Arvilommi, H. & Isolauri, E. (2003) Probiotics and prevention of atopic disease: 4 year follow up of a randomised placebo-controlled trial. *Lancet*, **361**, 1869–1871.

Krasaekoopt, W., Bhandari, B. & Deeth, H. (2003) Evaluation of encapsulation techniques of probiotics for yoghurt. *International Dairy Journal*, **13**, 3–13.

Kelly, D., Campbell, J.I., King, T.P., Grant, G., Jansson, E.A., Coutts, A.G., Petterson, S. & Conway, S. (2004) Commensal anaerobic gut bacteria attenuate inflammation by regulating nuclear-cytoplasmic shuttling of PPAR-gamma and RelA. *Nature Immunology*, **5**, 104–112.

Klaenhammer, T., Altermann, E., Arigoni, F., Bolotin, A., Breidt, F., Broadbent, J., Cano, R., Chaillou, S., Deutscher, J., Gasson, M., van de Guchte, M., Guzzo, J., Hartke, A., Hawkins, T., Hols, P., Hutkins, R., Kleerebezem, M., Kok, J., Kuipers, O., Lubbers, M., Maguin, E., McKay, L., Mills, D., Nauta, A., Overbeek, R., Pel, H., Pridmore, D., Saier, D., Van Sinderen, D., Sorokin, A., Steele, J., O'Sullivan, D., de Vos, W., Weimer, B., Zagorec, M. & Siezen, R. (2002) Discovering lactic acid bacteria by genomics. *Antonie van Leeuwenhoek*, **82**, 29–58.

Kleerebezem, M., Boekhorst, J., van Kranenburg, R., Molenaar, D., Kuipers, O.P., Leer, R., Peters, S.A., Sandbrink, H.M., Fiers, M.W., Stiekema, W., Lankhorst, R.M., Bron, P.A., Hoffer, S.M., Groot, M.N., Kerkhoven, R., de Vries, M., Ursing, B., de Vos, W.M. & Siezen, R.J. (2003) Complete genome sequence of *Lactobacillus plantarum* WCFS1. *Proceedings of the National Academy of Sciences USA*, **100**, 1990–1995.

Lähteenmäki, L. (2004) Consumers and health: getting the probiotic message across. *Microbial Ecology in Health and Disease*, **16**(2-3), 145–149.

Mattila-Sandholm, T. & Mercenier, A. (2001) The food, GI tract functionality and human health European research cluster, PROEUHEALTH. *Nutritional Metabolic Cardiovascular Disease*, **11**(4S), 109–113.

Macpherson, A.J., Gatto, D., Sainsbury, E., Harriman, G.R., Hengartner, H. & Zinkernagel, R.M. (2000) A primitive T cell-independent mechanism of intestinal mucosal IgA responses to commensal bacteria. *Science*, **288**, 2222–2226.

Nagler-Anderson, C. (2001) Man the Barrier! Strategic defenses in intestinal mucosa. *Nature Reviews Immunology*, **1**, 59–67.

Pridmore, R.D., Berger, B., Desiere, F., Vilanova, D., Baretto, C., Pittet, A.-C., Zwahlen, M.-C., Rouvet, M., Altermann, E., Barrangou, R., Mollet, B., Mercenier, A., Klaenhammer, T., Arigoni, F. & Schell, M.A. (2004) The genome sequence of the probiotic intestinal bacterium *Lactobacillus johnsonii* NCC 533. *Proceedings of the National Academy of Sciences USA*, **101**, 2512–2517.

Radutoiu, S., Madsen, L.H., Madsen, E.B., Felle, H.H., Umehara, Y., Gronlund, M., Sato, S., Nakamura, Y., Tabata, S., Sandal, N. & Stougaard, S. (2003) Plant recognition of symbiotic bacteria requires two LysM receptor-like kinases. *Nature*, **425**, 585–592.

Reid, G., Jass, J., Sebulsky, M.T. & McCormick, J.K. (2003a) Potential uses of probiotics in clinical practice, *Clinical Microbiology Reviews*, **16**, 658–672.

Reid, G., Sanders, M.E., Gaskins, H.R., Gibson, G.R., Mercenier, A., Rastall, R., Robertfroid, M., Rowland, I., Cherbut, C. & Klaenhammer, T.R. (2003b) New scientific paradigms for probiotics and prebiotics. *Journal of Clinical Gastroenterology*, **37**, 105–118.

Ricciardi-Castagnoli, P. & Granucci, F. (2002) Opinion: Interpretation of the complexity of innate immune responses by functional genomics. *Nature Reviews Immunology*, **2**, 881–889.

Roberts, M.-A., Mutch, D.M. & German, J.B. (2001) Genomics: food and nutrition. *Current Opinion in Biotechnology*, **12**, 516–522.

Saarela, M., Rantala, M., Hallamaa, K., Nohynek, L., Virkajärvi, I. & Mättö, J. (2004) Stationary-phase acid and heat treatments for improvement of the viability of probiotic lactobacilli and bifidobacteria. *Journal of Applied Microbiology*, **96**, 1205–1214.

Shanahan, F. (2004) Probiotics in inflammatory bowel disease – therapeutic rationale and role. *Advanced Drug Delivery Reviews*, **56**, 809–818.

Schell, M.A., Karmirantzou, M., Snel, B., Vilanova, D., Berger, B., Pessi, G., Zwahlen, M.C., Bork, P., Delley, M., Pridmore, R.D. & Arigoni F (2002) The genome sequence of *Bifidobacterium longum* reflects its adaptation to the human gastrointestinal tract. *Proceedings of the National Academy of Sciences USA*, **99**, 14422–14427.

Takeda, K., Kaisho, T. & Akira, S. (2003) Toll-like receptors. *Annual Reviews in Immunology*, **21**, 335–376.

Temmerman, R., Pot, B., Huys, G. & Swings, J. (2003a) Identification and antibiotic susceptibility of bacterial isolates from probiotic products. *International Journal of Food Microbiology*, **81**, 1–9.

Temmerman, R., Scheirlinck, I., Huys, G. & Swings, J. (2003b) Culture-independent analysis of probiotic products by denaturating gradient gel electrophoresis. *Applied and Environmental Microbiology*, **69**, 220–226.

Vaughan, E.E., Schut, F., Heilig, H.G.H.J., Zoedental, E.G., de Vos, W.M. & Akkermans A.D.L. (2000) A molecular view of the intestinal ecosystem. *Current Issues in Intestinal Microbiology*, **1**, 1–12.

de Vos, W.M. (2001) Advances in genomics for microbial food fermentations and safety. *Current Opinion in Biotechnology*, **12**, 493–498.

de Vos, W.M., Bron, P.A. & Kleerebezem, M. (2004) Post-genomics of lactic acid bacteria and other food-grade bacteria to discover gut functionality. *Current Opinion in Biotechnology*, **15**, 86–93.

# Index

Note: page numbers in *italics* refer to figures and those in **bold** refer to tables.

Lightning Source UK Ltd.
Milton Keynes UK
26 September 2009

144206UK00001B/12/P

9 781405 121248